T0074575

Metaheuristic Algorithms in Industry 4.0

Advances in Metaheuristics

Anand J. Kulkarni

Symbiosis Center for Research and Innovation, Pune, India

Patrick Siarry

Universite Paris-Est Creteil, France

Handbook of AI-based Metaheuristics

Edited by Anand J. Kulkarni and Patrick Siarry

Metaheuristic Algorithms in Industry 4.0

Edited by Pritesh Shah, Ravi Sekhar, Anand J. Kulkarni, and Patrick Siarry

For more information about this series please visit: https://www.routledge.com/Advances-in-Metaheuristics/book-series/AIM

Metaheuristic Algorithms in Industry 4.0

Edited by
Pritesh Shah, Ravi Sekhar, Anand J. Kulkarni,
and Patrick Siarry

CRC Press
Taylor & Francis Group
Boca Raton London New York

CRC Press is an imprint of the
Taylor & Francis Group, an **informa** business

First edition published 2022
by CRC Press
6000 Broken Sound Parkway NW, Suite 300, Boca Raton, FL 33487-2742

and by CRC Press
2 Park Square, Milton Park, Abingdon, Oxon, OX14 4RN

© 2022 selection and editorial matter, Pritesh Shah, Ravi Sekhar, Anand J. Kulkarni, Patrick Siarry; individual chapters, the contributors

CRC Press is an imprint of Taylor & Francis Group, LLC

Library of Congress Cataloging-in-Publication Data
Names: Shah, Pritesh, editor. | Sekhar, Ravi, editor. | Kulkarni, Anand Jayant, editor. | Siarry, Patrick, editor.
Title: Metaheuristic algorithms in industry 4.0 / edited by Pritesh Shah, Ravi Sekhar, Anand J. Kulkarni, Patrick Siarry.
Description: First edition. | Boca Raton : CRC Press, 2022. | Includes bibliographical references and index.
Identifiers: LCCN 2021013813 | ISBN 9780367698393 (hardback) | ISBN 9780367698409 (paperback) | ISBN 9781003143505 (ebook)
Subjects: LCSH: Metaheuristics—Industrial applications. | Mathematical optimization. | Industry 4.0.
Classification: LCC QA76.9.A43 M47 2022 | DDC 519.6—dc23
LC record available at https://lccn.loc.gov/2021013813

ISBN: 978-0-367-69839-3 (hbk)
ISBN: 978-0-367-69840-9 (pbk)
ISBN: 978-1-003-14350-5 (ebk)

DOI: 10.1201/9781003143505

Typeset in Palatino
by codeMantra

Contents

Preface

Optimization is widely used in all disciplines for various industrial applications, including product design, static/transient analyses, plant layout, supply chain, inventory planning, production planning/scheduling, automation, robotics, building architecture, material handling, tooling, assembly line balancing, system modeling/identification, tuning of various controllers, machine learning and many more. Due to increasing Industry 4.0 practices, massive industrial process data is now available for researchers for modeling and optimization. Artificial intelligence methods can be applied on the ever-increasing process data to achieve robust control against foreseen/unforeseen system fluctuations. Smart computing techniques, machine learning, deep learning, computer vision etc. will be inseparable from the highly automated factories of tomorrow. Effective cybersecurity will be a must for all Internet of Things (IoT)-enabled work and office spaces.

Furthermore, the optimization promises to play a pivotal role in the interplay of physical machines, control systems and artificial intelligence. However, the traditional optimization methodologies are inept at handling complexities of big data analytics and new age dynamic systems. Metaheuristics are nature-inspired answers to such 21st-century-world problems. Metaheuristics are typically socio-inspired, bio-inspired or physics-derived algorithms based on simple principles that have proved effective in varied domains. With respect to advanced control systems, metaheuristics can be applied for the selection of controllers as per different applications and for auto tuning of systems. Metaheuristic algorithms are also suitable for optimizing controller performance indices, model structure selection, model structure parameter estimation, reliability analysis, fault diagnosis and closed-loop robustness analysis in different control systems.

This book aims to address metaheuristics in all aspects of Industry 4.0. Metaheuristic applications in IoT, cyber physical systems, control systems, smart computing, artificial intelligence, sensor networks, robotics, cybersecurity, smart factory, predictive analytics and more will be covered. This book will prove to be a guiding light to numerous engineers, scientists, students, faculty and professionals engaged in exploring and implementing Industry 4.0 solutions in various systems/processes. In addition, this book is intended primarily for advanced undergraduate, graduate and PhD students in Mechatronics Engineering, Data Science, Automation Engineering, Artificial Intelligence and Robotics, Computer Science and Engineering, Information Technology and interdisciplinary areas like IoT, Cyber Physical Systems and Data Analytics. The proposed book can be used as a reference book for a graduate course, for an advanced undergraduate course or for a summer school. It will serve as a self-contained handbook of metaheuristics in all aspects of the fourth industrial revolution. In total, twelve chapters have been accepted in this book. The contribution summary of every chapter is discussed below.

In **Chapter 1,** Sharma et al. provided a detailed and comprehensive bibliometric analysis of the cyber physical systems (CPS) and the smart computing (SC). This chapter concentrates on the physical computing units of the CPS along with the advanced connectivity ensuring real-time data acquisition and information feedback from the cyber space. Furthermore, the capabilities for data management, analytics and computations are also discussed in very details. The study of this chapter concludes that with the inception of Industry 4.0 and embedding of the sensors in almost all the real-world processes, data collection is no longer a challenge but the efficient and intelligent processing could be the one to be dealt with. Moreover, the study infers that it has enabled the use of SC within the support of the CPS.

Chapter 2 by Khachane and Jatti implements the important principle of noise reduction by enclosing its source. For the study, domestic mixer grinder assembly is considered as a single unit and the enclosure width is controlled using an established mathematical model for the prediction of insertion loss. Importantly, the simulations and experimental solutions are compared for the validation of the solution. In order to establish the choice of the Jaya algorithm, a thorough literature review of the need of optimization and existing algorithms is carried out. This chapter provides illustrations associated with several parameters affecting the loss function reducing the noise. Finally, this chapter concludes that the theoretical and experimental solutions are in harmony with each other with acceptable deviation. Most significantly, this chapter provides a limitation of the study and suggests further improvements in which the potential readers could be interested.

Chapter 3 by Tamba T. A. emphasizes on one of the important aspects in the design and implementation of a cyber physical system (CPS) referred to as the security assurance of its cyber elements in performing its functions in the face of potential cyber-attacks from external adversaries. Accordingly, this chapter discusses a moving target defense control approach for detecting and mitigating possible cyber-attacks from malicious adversaries on a CPS. The presented approach essentially utilizes a specialized strategy referred to as switching and stabilizing controllers under event-triggered control scheduling scheme to complicate external attempts in delivering the attacks. This chapter demonstrates the ability of particle swarm optimization (PSO) algorithm to ensure the optimality of each controller that is used in the proposed moving target defense control scheme. This chapter in detail provides the theoretical and mathematical formulation of the setup of the system and secure control problem.

The issue of performance degeneracy of any nature-inspired optimization algorithm when applied to constrained problems is underscored by Kulkarni et al. in **Chapter 4**. It discusses the salp swarm algorithm (SSA) incorporated with the dynamic penalty function approach and its effective use with the algorithm. Importantly, a real-world problem associated with the thinning of the blank during the drawing process of a sheet metal-forming process is considered. This chapter illustrates formal SSA procedure, the problem description along with the dynamic penalty function approach. Moreover, the performance of the SSA is validated by comparing the solutions with the experimental solutions. This chapter in details discusses the importance of the forming process as well as the significance of the problem solution along with the nature-inspired algorithms employed so far.

The inefficiency of the conventional optimization techniques for solving the robot path planning problems is highlighted in **Chapter 5** by Rao R. V. This chapter provides the optimum solution to path planning of a robot to a specific location from the origin to the destination in certain constrained area. The solution approximation-based optimization approaches such as Jaya, Rao-1, Rao-2 and Rao-3 are employed for solving the robotic path planning problems. The contribution reveals simplicity of the algorithms and the key characteristic of not requiring any parameter to be tuned. Four case studies of robot path planning in a static workspace with static obstacles are presented in this chapter. In addition, the results are compared with the probability and fuzzy logic (PFL), cuckoo search algorithm (CS), genetic algorithm (GA), firefly algorithm (FA), fuzzy neural network algorithm, bacteria foraging optimization (BFO), ant colony optimization (ACO) and particle swarm optimization (PSO).

Chapter 6 by Gaikwad et al. presents the investigation of surface integrity parameters in electrical discharge machining (EDM) process referred to as white layer thickness (WLT). It is worth mentioning here that the experimental evaluation of WLT is carried out using Taguchi design of an experiment along with an empirical modeling to determine the WLT by using Buckingham's pie theorem. The results between the experimental and empirical modeling methods are in close agreement with each other. In addition, the Jaya algorithm is also employed for the optimum parameter setting. This implementation underscored the necessity of the elite optimization algorithms for solving the real-world problems from the manufacturing domain in the view point of minimum utilization of the resources.

The importance of the convolution neural networks (CNN) in various fields, including computer vision and image processing, is highlighted in **Chapter 7** by Bansod et al. Accordingly, this chapter presents a thorough and critical literature survey of various CNN architectures and may act as a convenient and comprehensive solution for CNN advancements over time and the major architectural improvements in each architecture as well as associated significant improvements. More specifically, major CNN architectural innovations, a brief description about their constituent blocks and structural description are provided as use cases that pertain to applications in the Industry 4.0 are covered in this chapter. It is worth mentioning that the described limitations of every architecture reveal the directions of possible research for the readers.

Chapter 8 by Chavan et al. emphasizes on assessing the lung health of patients suffering from a variety of pulmonary diseases, including COVID-19, tuberculosis and pneumonia, by applying Earth mover's distance (EMD) algorithm to the X-ray images of the patients. In this effort, the lungs X-ray images of patients suffering from pneumonia, TB, COVID-19 and healthy persons are pooled together from various existing datasets. The preliminary data used is based on certain random images depicting each type of lung diseases such as COVID-19, tuberculosis and pneumonia. The study infers and concludes that the patients suffering from pneumonia have the highest severity as per values obtained from the EMD scale.

As the electric discharge machining (EDM) is a complex process, its material removal rate (MRR) and tool wear rate (TWR) need to be modeled mathematically, which are highlighted in **Chapter 9** by Gaikwad et al. This chapter presents a model developed to predict the MRR during cryo-treated EDM process using adaptive neuro fuzzy inference system (ANFIS). It further explains the elite experiments carried out using electrolytic copper as tool material and workpiece of NiTi alloy, both cryo-treated, varying the pulse on time, gap current, pulse off time, electrical conductivity of tool and workpiece materials. The choice of the triangular and trapezoidal membership functions is also justified by developing the prediction model.

Chapter 10 by Vanchinathan et al. presents a metaheuristic optimization algorithm-based variable speed control (VSC) of brushless DC (BLDC) motor with excellent time domain performance index as an industrial case study. The study deals with the optimization techniques based on optimal tuning of controllers, such as proportional integral derivative (PID) and fractional order PID (FOPID) controllers for VSC drives. This chapter highlights the complexity of tuning the FOPID parameters and describes the necessity of the metaheuristic optimization algorithms to achieve the robust performance of the system by minimizing the rise time, peak time, settling time, steady-state error, control efforts and performance indices. In association with this, this chapter reviews various metaheuristic optimization algorithms for tuning PID/FOPID controllers for sensorless speed control of BLDC motor drives.

Chapter 11 by Rajule et al. presents a wide survey of predictive analysis of cellular networks. It includes various analysis techniques used for the analysis of cellular traffic data. Moreover, this chapter enlightens the application areas of predictive analysis and discusses the role and importance of deep learning in the analysis of cellular networks. Furthermore, it also represents a bridge to fill the gap between emerging intelligent networks and their relevance with predictive analysis in cellular networks. This chapter covers essential cellular traffic characteristics and aspect analysis and predictive analysis as well as network analysis for special parameters such as hotspot detection, holiday traffic prediction, customer churn prediction, fault prediction, anomaly detection, etc.

The importance of the role of the optimization methods for the dental implant design is highlighted by Karnik and Dhatrak in **Chapter 12**. It aims to review and explore various optimization techniques and algorithms being employed in the field of dental implants. This chapter emphasizes that with the increase in computing power, optimization algorithms played a key role to achieve maximum performance from features like thread parameters, material properties, surface morphology and shape optimization. These features are explored in this chapter along with the various algorithms like genetic algorithm, topology optimization technique, particle swarm optimization, multiobjective optimization algorithms, uncertainty optimization and a few more, which are commonly used in dental implants. Furthermore, this chapter also reviews various complementary techniques, which assist optimization like the surrogate models which make use of artificial neural networks, Kriging interpolation and support vector regression.

MATLAB® is a registered trademark of The MathWorks, Inc. For product information, please contact:

    The MathWorks, Inc.
    3 Apple Hill Drive
    Natick, MA 01760-2098 USA
    Tel: 508-647-7000
    Fax: 508-647-7001
    E-mail: info@mathworks.com
    Web: www.mathworks.com

Editors

Pritesh Shah is an Associate Professor at the Symbiosis Institute of Technology, Symbiosis International (Deemed University), India.

Ravi Sekhar is an Assistant Professor at the Symbiosis Institute of Technology, Symbiosis International (Deemed University), India.

Anand J. Kulkarni is an Associate Professor at the Symbiosis Center for Research and Innovation, Symbiosis International (Deemed University), India.

Patrick Siarry is a Professor of Automatics and Informatics at the University of Paris-Est Créteil, where he leads the Image and Signal Processing team in the Laboratoire Images, Signaux et Systèmes Intelligents (LiSSi).

Contributors

Javier Andreu-Perez
Institute for Advancing Artificial
    Intelligence (I4AAI), UK

Gaurav Bansod
Department of Electronics and
    Telecommunication
Pune Institute of Computer Technology
Pune, India

Niranjan Chavan
Institute for Thermal Energy Technology
    and Safety (ITES) Karlsruher Institute
    for Technology
Karlsruhe, Germany.

Satish S. Chinchanikar
Department of Mechanical Engineering
Vishwakarma Institute of Information
    Technology
Pune, India

Pankaj Dhatrak
School of Mechanical Engineering
Dr. Vishwanath Karad MIT-WPU
Pune, India

Mahendra Uttam Gaikwad
Department of Mechanical Engineering
Sathyabama Institute of Science and
    Technology
Chennai, India

Vaibhav S. Gaikwad
Department of Production Engineering
K. K. Wagh Institute of Engineering
    Education and Research
Nashik, India

Prashant K. Gupta
Institute for Advancing Artificial
    Intelligence (I4AAI), UK

Rajiv Janardhanan
Laboratory of Disease Dynamics &
    Molecular Epidemiology
Amity Institute of Public Health
Amity University Uttar Pradesh
Noida, India

and

Health Data Analytics & Visualization
    Environment
Amity Institute of Public Health
Amity University Uttar Pradesh
Noida, India

Vijaykumar S. Jatti
Department of Mechanical Engineering
D.Y. Patil College of Engineering
Pune, India

G. M. Kakandikar
School of Mechanical Engineering
Dr. Vishwanath Karad MIT World
    Peace University
Pune, India

Niharika Karnik
School of Mechanical Engineering
Dr. Vishwanath Karad MIT-WPU
Pune, India

Ashish Khachane
Department of Mechanical Engineering
D.Y. Patil College of Engineering
Pune, India

Shardul Khandekar
Department of Electronics and
    Telecommunication
Pune Institute of Computer Technology
Pune, India

Soumya Khurana
Department of Electronics and
    Telecommunication
Pune Institute of Computer Technology
Pune, India

A. Krishnamoorthy
Department of Mechanical Engineering
Sathyabama Institute of Science and
    Technology
Chennai, India

Anju Kulkarni
Dr. D. Y. Patil Institute of Technology
Pune, India

Omkar Kulkarni
School of Mechanical Engineering
Dr. Vishwanath Karad MIT World Peace
    University
Pune, India

Uday Kumar
Delhi State Cancer Institute
Delhi, India

Radhika Menon
Dr. D. Y. Patil Institute of Technology
Pune, India

V. M. Nandedkar
Production Engineering
Shri Guru Gobind Singhji Institute of
    Engineering & Technology
Nanded, India

Keshav N. Nandurkar
Department of Production Engineering
K. K. Wagh Institute of Engineering
    Education and Research
Nashik, India

S. Patel
Department of Mechanical Engineering
S.V. National Institute of Technology
Ichchanath, Surat, Gujarat, India

Nilakshee Rajule
Dr. D. Y. Patil Institute of Technology
Pune, India

Priya Ranjan
SRM University
Chennai, India

R. V. Rao
Department of Mechanical Engineering
S.V. National Institute of Technology
Ichchanath, Surat, Gujarat, India

P. Sathiskumar
Applied Materials
Bangalore, India

N. Selvaganesan
Department of Avionics
Indian Institute of Space Science and
    Technology
Thiruvananthapuram, India

Deepak Sharma
Institute for Advancing Artificial
    Intelligence (I4AAI), UK

Kumar Dron Shrivastav
Health Data Analytics & Visualization
    Environment
Amity Institute of Public Health
Amity University Uttar Pradesh
Noida, India

Tua A. Tamba
Department of Electrical Engineering
Parahyangan Catholic University
Bandung, Indonesia

K. Vanchinathan
Department of Electrical and Electronics
    Engineering
Velalar College of Engineering and
    Technology
Erode, India

1

A Review on Cyber Physical Systems and Smart Computing: Bibliometric Analysis

Deepak Sharma, Prashant K. Gupta, and Javier Andreu-Perez

Institute for Advancing Artificial Intelligence (I4AAI)

CONTENTS

1.1 Introduction

Cyber physical systems (CPSs) are the class of systems that find applications in various day-to-day domains like healthcare, academia, industry, etc. Their popularity has been growing. This is evident from the numerous survey articles, where CPS is the main theme. In Ref. [1], the authors discussed history of CPS, its relation with various research fields, different types of domain-specific applications and challenges in CPS. In Ref. [2],

DOI: 10.1201/9781003143505-1

the author discussed about the state-of-the-art CPS in Industry 4.0 through 595 articles extracted from the Web of Science (WoS) database. This work focused on the industrial applications of CPS, uncovering the research trends and its challenges in Industry 4.0. Also, in another survey of CPS [3], the author presented a comprehensive overview of CPS based on the 77 articles collected from Scopus database. This survey article illustrated the role of CPS and its challenges in ten research categories, viz., agriculture, education, energy management, environmental monitoring, intelligent transportation, medical devices and systems, process control, security, smart city and smart home, and smart manufacturing. In Ref. [4], the authors shed a light on the survey of the Industrial Internet of Things (IIOT) with a perspective of CPS. It also discussed the architecture, applications, characteristics and challenges of IIOT.

CPSs consist of mainly (i) the physical computing units along with advanced connectivity to ensure the real-time data acquisition and information feedback from cyber space, and (ii) capabilities for data management, analytics and computations [5]. As the sensors are being increasingly embedded in almost all the industrial processes owing to advent of Industry 4.0, data collection is no longer a challenge. The challenge is rather a smart and intelligent processing of this data, which can be achieved by employing the smart devices, technology, etc. Smart devices, technology, objects and any software, regardless of its application domain, that epitomize artificial intelligence (AI) can be clubbed under the umbrella of the "smart computing (SC)". Although the areas of the CPS and SC have been going on hand-to-hand with each other for numerous years, little has been said about their amalgamation.

So, with this aim in mind, we present in this book chapter a bibliometric analysis of the CPSs and the SC. We have used the WoS as our primary source for data/research publication collection. We found that 4,538 publications surfaced with search criteria set as "cyber physical system" and "smart computing". We have also analysed the research data along with various other dimensions like annual publications, types of publications, possible future research directions, countries and authors. Then, we have performed analysis along with the dimension of cooperation networks such as authors, countries, organizations, etc. Finally, we have also performed the analysis about the citation structure and the burst detection.

Therefore, the scope of the current chapter/work in the nutshell is discussed in the following.

Using the WoS database, we have analysed

- Number of annual publications in the field of CPS and SC for the past 10 years (2010–2020).
- Types of publications such as book chapters, journal papers, etc.
- Top journals/sources of publications.
- Top research directions for CPS and SC. These include the computer science, engineering, telecommunications, etc. Further, an analysis on the trend in these research directions has also been put forth.
- Top countries and places, as well as productive researchers and organizations.

We have also analysed various types of collaborative relationships in the 4,538 research publications, obtained from WoS on CPS and SC, using the VOSviewer [6]. These relationships are as follows:

- Collaborative strengths amongst researchers
- Collaboration strengths of organizations
- Collaborative strengths of places.

We have also provided a citation structure analysis from different aspects, like:

- Citation landscape for research papers
- Citation landscape for researchers
- Citation landscape for organizations
- Citation landscape for places.

Finally, we have used the CiteSpace for analysing the timeline view and the burst detection, which enables the identification of the time periods when some publications, technology, etc. are popular in the field of CPS and SC. CiteSpace is a free Java-based application that facilitates trends and patterns visualization in the scientific literature. It brings forth critical time instants in the research development, in particular the intellectual turning points and pivotal points [7].

Our key findings from the bibliometric analysis of the CPS and SC in the years 2010–2020 are as follows:

- A total of 4,538 publication appeared in the WoS when searched with keywords such as *Cyber Physical Systems* and *Smart Computing*.
- The number of publications was quite small in 2010 and 2011. They more than doubled in 2012 of what was their number in 2011, kept increasing constantly to cross 1,000 in 2019 and close at 935 in October 2020. Maximum number of publications (1,018) on CPS and SC were published in the year 2019.
- The published papers were of various types like articles, proceeding papers, reviews etc. and were published in various journals. Most published types of research works were the articles (3,968 numbers or 86.78% of the total publication types) in comparison with other publication types.
- Maximum number of publications appeared in *IEEE Access* (268 numbers or 5.91% of the total publications).
- The most popular research direction was the computer science, bagging 2,404 (52.98% or more than half of total) number of publications. It was closely followed by the research direction: engineering (2,398 or 52.84% number of publications).
- The most contributing places to the publication landscape were China (1,320 publications) followed by USA. Various other countries also made their respective contributions, though comparatively quite less in number.
- Various organizations and authors contributing to these publication numbers were from different places across the globe; however, the maximum number was coming from China.
- On a global level, a total of 9,740 researchers had worked collaboratively to produce the research publications, of which 1,860 had the largest collaboration network, and many of these 1,860 researchers also had dense collaboration networks. The author LI, Y had the most dense collaborative network.

- Out of a total of 3,363 organizations, 453 had the largest connected network, of which numerous organizations also had dense collaboration networks. The SHANGHAI JIAO TONG UNIVERSITY from China secured the first position on the list of top 13 organizations with dense collaboration network. Also, the CHINESE ACADEMY OF SCIENCE and UNIVERSITY CHINESE ACADEMY OF SCIENCE were the largest collaborators (with twelve joint works).

- Globally, ninety-four countries/regions contributed at least one paper into the research area. USA and China jointly contributed the maximum number of research outputs.

- With a citation threshold of a research paper set at five, less than half of 4,538 papers qualified the criteria. In addition, ninety-nine research papers attracted more than 100 citations, thus forming the biggest network. Also, a large number of research papers were also co-cited.

- Of 9,740 researchers, 387 had the citation count crossing 100 and 384 secured the largest co-citation network. The author Wan, J had the highest citation count.

- Of 3,363 organizations, 243 had been successful in getting over 100 citations, and many had strong co-citation network also. South China University of Technology topped the list of most popular seventeen co-citation organizations.

- Out of ninety-four places (countries/regions), each one had at least one research paper with 1,511 couples and secured a total of 27,221 co-citations. Also, eighty-seven places had the largest co-citation network. USA secured the first rank with the maximum number of citations.

- The timeline analysis produced various popular keywords grouped into different clusters, with one or more keywords found to have popularity at different time instants.

- When analysed the keywords for the burst detection, it was learnt that the burst time period was almost 4–5 years for the majority of the returned keywords.

- For the reference burst analysis, it was learnt that the majority of these publications were published outside the period 2010–2020; however, they attained their bursts within this period. The highest burst strength was attained by the research paper of the year 2008, entitled as *Cyber physical systems: Design challenges* by Edward A. Lee [8].

1.2 Data Selection and Extraction

We have used WoS as our primary research data source for the purpose of this bibliometric study. It has an extraordinary research content collection in the form of scientific manuscripts and their impacts, starting 1990 up to the present day. This research data repository is used throughout the world due to numerous features. The WoS Core Collection contains numerous popular journals and publications as well as sufficient details about them. This enables easy export (and import) to (and from) any bibliometric analysis platform. It contains eight Citation Indexes, viz., Science Citation

Index Expanded (SCI-EXPANDED), Emerging Sources Citation Index (ESCI), Social Sciences Citation Index (SSCI), Conference Proceedings Citation Index-Science (CPCI-S), Conference Proceedings Citation Index-Social Science & Humanities (CPCI-SSH), Book Citation Index-Social Sciences and Humanities (BCI-SSH), Book Citation Index–Science (BCI-S), and Arts and Humanities Citation Index (AHCI)). We searched WoS Core Collection with the following criteria: keywords: "Cyber Physical System" and "Smart Computing", timespan: 2010–2020, and their relative information (the record content set as "full record and cited references"). We obtained 4,538 publications and exported them in the form of plain text on October 13, 2020. We found the most precise results from SCI-EXPANDED, ESCI, SSCI, CPCI-S and CPCI-SSH.

1.3 Distribution of Publications along Different Verticals

Here, we present the details on distribution of publications along different verticals: year-on-year basis, popular places, as well as famous organizations and authors.

1.3.1 Publications Analysed: Year-on-Year Basis

We used the WoS as our primary data source for extracting the number of publications, with the keywords CPS and SC. The trend for the number of publications from 2010 to 2020, year-on-year basis, is illustrated in Figure 1.1. As shown in Figure 1.1, during the starting 2 years (2010–2011), the total number of publications was 63. However, from 2012 onwards, the number of publications started increasing, with a figure of 111 in the same year up to almost a double number of 262 in 2015. The successive year saw approximately a doubling of the numbers within a span of a year. After that, the trend has been witnessing a constant increase, crossing a number of 1,000 in penultimate year

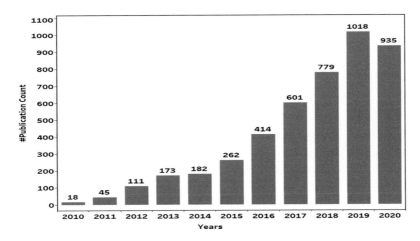

FIGURE 1.1
Annual publication from 2010 to 2020.

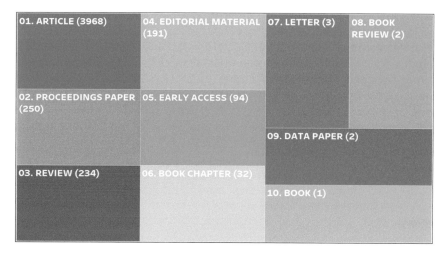

FIGURE 1.2
The detail of publications in top ten types.

of the period (2010–2020), however showing a slight dip by 83 numbers in the year 2020 (up to October 2020).

WoS has also enabled us to analyse the publications, in the years 2010–2020, along the vertical publication types such as journal articles, book chapters, book reviews etc. This analysis is pictorially depicted in Figure 1.2. From Figure 1.2, it is evident that the first position is achieved by the article type of publications, and their count is 3,968. This is ~86.78% of the total number of publications. The second position is held by the proceeding paper, with a count of 250 and 5.51% contribution. The other types of publications (and their numbers) are as follows: review (234, 5.16%), editorial material (191, 4.21%), early access (94, 2.07%), book chapter (32, 0.71%), letter (3, 0.07%), book review (2, 0.04%), data paper (2, 0.04%) and book (1, 0.02%). It can be seen that the articles are the most published research works in comparison with other publication types.

We have also analysed the research articles according to their repository of publication. The details of topmost ten repositories (source title), as returned by WoS, are given in Figure 1.3. It can be seen that the maximum numbers (5.91%) were published in *IEEE Access*, whereas 3.28% made their way into the SENSORS (the second type). The other source titles were *IEEE Transactions on Industrial Informatics* (116, 2.56%), Future Generation Computer Systems *the International Journal of Escience* (94, 2.07%), *IEEE Transactions on Automatic Control* (77, 1.70%), *Proceedings of the IEEE* (75, 1.65%), *IEEE Internet of Things Journal* (67, 1.48%), *IEEE Transactions on Smart Grid* (67, 1.48%), *ACM Transactions on Cyber Physical Systems* (64, 1.41%) and *International Journal of Distributed Sensor Networks* (59, 1.30%). Thus, majority of the source titles are from IEEE, which account for 20.12% amongst the top ten source titles in the CPSs and SC.

1.3.2 Research Directions of Publications

WoS helped us in bringing forth an analysis on the current (and possible future) research fields in which the research publications appeared. These numbers are shown in Figure 1.4. As it appears from the figure, the first and the second popular research

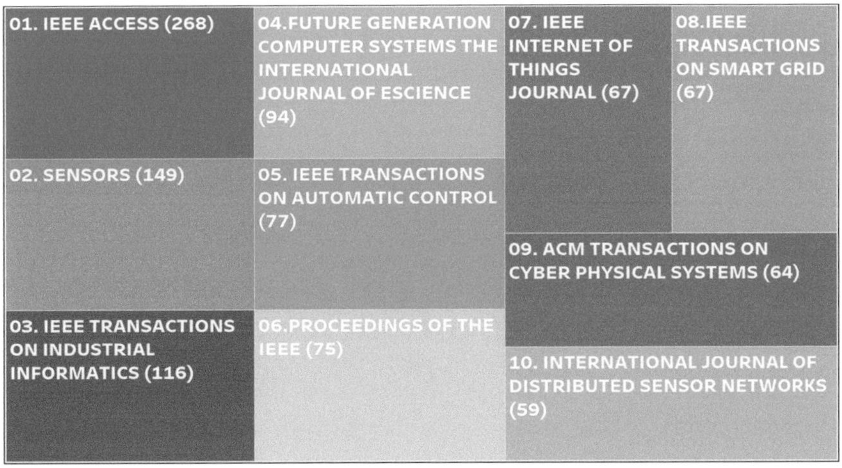

FIGURE 1.3
The top ten source titles of the publications.

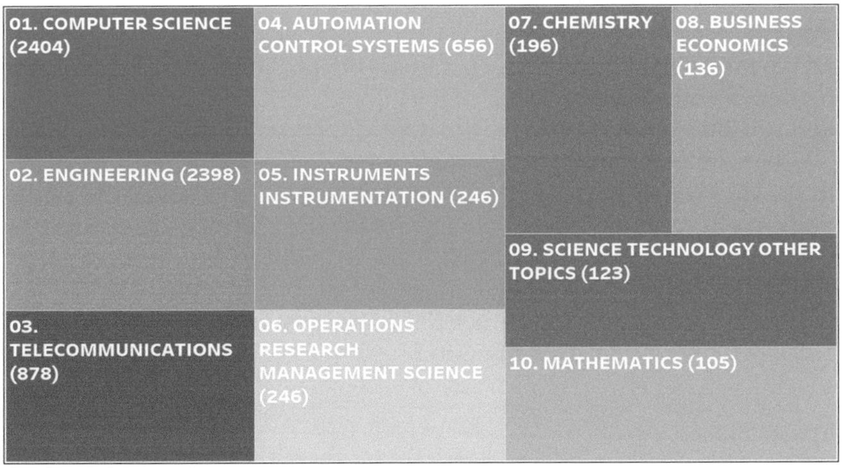

FIGURE 1.4
The popular publications' research directions.

directions are computer science and engineering, respectively, both bagging more than half of the total publications, 2,404 numbers (52.98%) and 2,398 numbers (52.84%), respectively. Further, the most popular research directions (numbers, percentage of the total proportion) were found to be the telecommunications (878, 19.35%), automation control systems (656, 14.46%), instruments instrumentation (246, 5.42%), operation research management science (246, 5.42%), chemistry (196, 4.32%), business economics (136, 3.00%), science technology and other topics (123, 2.71%) and mathematics (105, 2.31%).

Similarly, as analysed from WoS, the trends of research directions are shown in Figure 1.5. The research direction in the area of computer science and engineering demonstrates the steady increasing trend. Also, the research direction of telecommunication

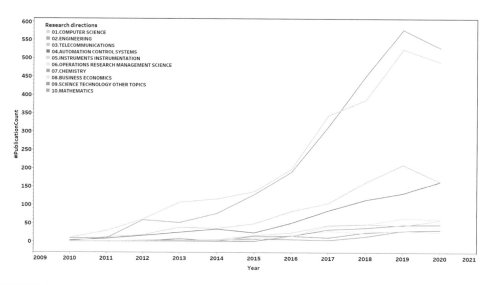

FIGURE 1.5
The trend of popular publications' research directions.

and automation control system shows a steady increase but less number of publications than the previous two research directions. Finally, the remaining areas too show increasing trends with very few publications.

This trend of increased interest in the area of computer science and engineering (in comparison with other areas) can possibly be attributed to the resurgence in AI coupled with the CPS [9]. AI came to be understood as a dissemination and proliferation into Industry 4.0, Internet of Things (IoT), etc. during the years that show a large spike in the field (as evident in Figure 1.5). Thus, CPS coupled with SC (in a form of AI) provided the researchers a new pathway for bringing their ideas into reality. Hence, the field of computer science and engineering attracted the largest number of publications (an indicator of increased interest) in the said period.

1.3.3 Popular Places

WoS has a remarkable feature that enables an analysis along the vertical of popular (most contributed) places along with their yearly publications within a specific time period. We adjusted the time period as 2010–2020, to analyse the number of yearly publications as contributed by various places, for CPS and SC. The obtained results are shown in Figure 1.6.

According to the results obtained, we found that the top position was grabbed by China, with 1,320 publications. This was followed by USA (1,300 publications), Germany (392 publications), England (248 publications), Italy (243 publications), India (223 publications), South Korea (223 publications), Canada (193 publications), Spain (192 publications) and finally Australia (183 publications). The Chinese and the American authors have been published research articles since 2010. As far as other popular places are concerned, Canada and Australia first published an article in 2010, Germany, South Korea and Spain in 2011, and nonetheless, England and Italy made a start in 2012. However, opposed to this trend, the other countries began publishing from 2013 onwards.

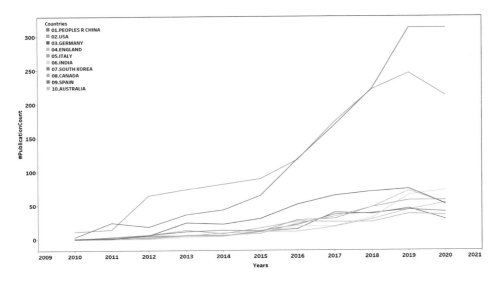

FIGURE 1.6
The top ten productive places.

From Figure 1.6, it is evident that trend is always on an increase. The top two countries, viz., China and USA, crossed the 100 publication mark in 2016 and held the top position for the next 4 years. Further, USA published more than China, almost always except in 2018–2020.

1.3.4 Productive Organizations and Researchers

Along this dimension of analysis, WoS presented us with the names of the most contributing organizations as well as researchers. The obtained results have been comprehensively provided in Figure 1.7 and Table 1.1, the former detailing the popular organizations and the latter the researchers.

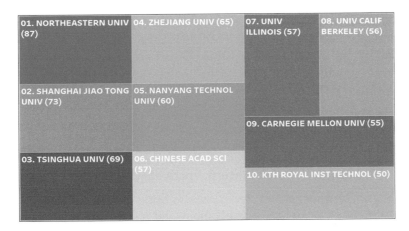

FIGURE 1.7
The top ten most research article-contributing organizations.

TABLE 1.1

The 15 Highly Contributing Authors

Position	Author	Places	P
1	Wan JF	China	35
2	Yang GH	China	35
3	Wang Y	China	31
4	Li Y	China	23
5	Li L	China	22
6	Zhang Y	Norway	22
7	Li D	China	21
8	Liu X	USA	21
9	Sun DH	China	21
10	Kim J	South Korea	20
11	Li X	China	20
12	Shi L	China	20
13	Wang C	China	20
14	Yu W	USA	20
15	Zhang H	China	20

From Figure 1.7, it can be seen that the most contributing organizations (in decreasing order of popularity) are Northeastern University, Shanghai Jiao Tong University, Tsinghua University, Zhejiang University (all four being located in the China), Nanyang Technology University (Singapore), Chinese Academy of Science (China), University of Illinois (USA), University California Berkeley (USA), Carnegie Mellon University (USA) and KTH Royal Institute Technology (Sweden). The contribution of the Chinese organizations, viz., 351 (7.73%), is almost twice of that of USA, viz., 168 (3.70%), of all the 4,538 publications, amongst the top ten productive organizations. It is pertinent to mention that half of the organizations are based in China, while the number of organizations in the countries such as USA, Singapore and Sweden are 3, 1 and 1, respectively.

Visiting Table 1.1, here the most contributing authors are listed along with their number of publications (denoted as *P*). It is evident that the maximum number of publications attained by any author is 35 and only 18% of authors published more than 20 publications. These authors come from the following countries (their publication count): China (11), USA (2), Norway (1) and South Korea (1).

1.4 Analysis along the Collaboration Vertical

The aim of the analysis along this vertical was to bring out the collaboration strength amongst the different researchers, organizations and places. This analysis was fuelled by the use of VOSviewer. Let us discuss each one in detail.

1.4.1 Collaboration Strength amongst the Researchers

The VOSviewer enabled to uncover the strength and extent of collaboration amongst the researchers, worldwide, contributing to the CPS and SC. This tool provides an easy-to-interpret output in terms of numbers. Further, 2,539 (out of total of 9,740) researchers have produced a minimum of two research outputs and 1,860 researchers out of a total of 9,740 (who published on these topics) have the largest collaboration (cooperation) network. It is evident pictorially in Figure 1.8.

In Figure 1.8, the node size corresponds to the total link strength (TLS) of 2,539 researchers. TLS is a measure of collaborative affinity (cooperation frequency) amongst the researchers. A special case of the TLS is established via analysis of cooperation frequency between two researchers, denoted as two respective nodes with a direct link. As seen from Figure 1.8, 8,894 links and 11,963 TLS are established amongst community of the researchers. Table 1.2 shows the top researchers with the strongest TLS values. Further, the cooperation network for researcher, Li, Y, is also displayed in Figure 1.9.

1.4.2 Collaboration Strength of Organizations

The collaborative strength of the organizations is depicted in Figure 1.10 (as obtained from the VOSviewer). The figure consists of the nodes (also referred as the linked items or cooperators) and the links between them. It is mentioned here that the size of a node and thickness of a link denote the cooperation numbers and the link strength, respectively, of the node.

In the nutshell, 466 (out of total of 3,363) organizations have produced a minimum of five research outputs, and 453 out of these have the largest connected collaborative networks. Table 1.3 gives the details (Position, Organization, Place, Publication numbers (P), Link and TLS) about the top 13 organizations with largest TLS values, as put forth by the VOSviewer. As seen from table, all the organizations exceed the TLS value of 50.

For interpreting the data from table, it must be followed that the TLS numbers correspond to the number of cooperative organizations. As seen from table, Shanghai Jiao Tong University has the highest P value (73), with a link number of 59 and also maximum TLS value of 105. Thus, it is evident that this university with a link number of 59 and TLS number of 105 has collaborated with one or more organizations on multiple occasions. An interesting observation from Table 1.3 is that the relative number of TLS, link number and P value for some organizations is as TLS>link >P. This is a mark of the fact that some research works are contributed jointly by more than two organizations.

Also, in Table 1.4, we also present six organizations with the strongest collaborative strength (attained working jointly) and their relative data.

From Table 1.4, it is evident that the strongest collaboration exists between the Chinese Academy of Science and University of the Chinese Academy of Sciences, with twelve joint works. Also, the Chinese Academy of Science has a collaboration strength of 21.05%. Further, the Simula Research Laboratory and the University of Oslo emerged as the strongest collaborators, with the highest collaboration strength of 78.57%.

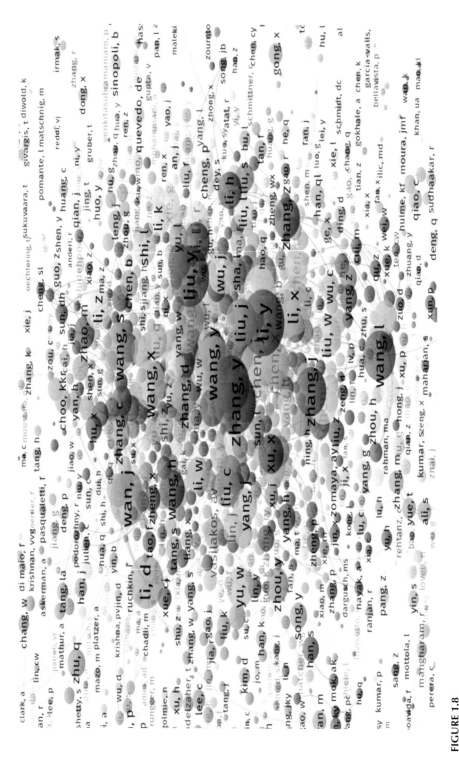

FIGURE 1.8
The collaboration network of researchers.

TABLE 1.2

The Top Ten Researchers with the Strongest TLS Values

Position	Researcher	P	Link	TLS
1	Li, Y	59	140	202
2	Zhang, Y	61	147	186
3	Liu, Y	51	121	201
4	Wang, J	50	113	153
5	Wan, J	41	67	147
6	Li, X	39	111	145
7	Wang, Y	53	117	143
8	Li, D	35	69	138
9	Chen, J	42	104	130
10	Wang, L	40	85	120

1.4.3 Collaborative Strength of Places

The analysis results about the collaborative strength of places using VOSviewer provided us ninety-four places (countries/regions) with each of this place publishing at least one research output. These results are depicted in Figure 1.11. In the figure, the link joining the two nodes and the link width correspond to the collaboration amongst the places and the strength of collaboration, respectively. The node size or its TLS is calculated by summing up all its link strengths.

As observed from figure, six places attained a TLS of over 250. In figure, ninety-one countries are grouped together into eight clusters; each cluster is labelled by a different colour. Further, the places with highest TLS are USA, China and England. Also, the strongest collaboration occurs between USA and China.

A further study into the collaborative strengths of the places provided us with a list of top six places with a respective partner in the collaboration. These results are presented in Table 1.5. From the table, it is observed that USA sits at the first position with 948 TLS numbers. Also, China is its partner in the collaboration with both having a collaboration strength of 29.54%, owing to 280 joint works. A minimum collaboration strength of 4.60% is observed between England and Germany, with England having a TLS of 326 and 15 joint works.

1.5 Analysis along the Citation Landscape

Here, we have presented the details of citation landscape pertaining to the field of CPS and SC along with the most received citations in four dimensions, viz., the research papers, researchers, organizations and places. The detailed resulting analytical findings for each dimension are provided in respective sections. The terminologies used in the sections are P (publication numbers), C (citation numbers), AC (number of average citation), Link (number of co-cited items) and TLS.

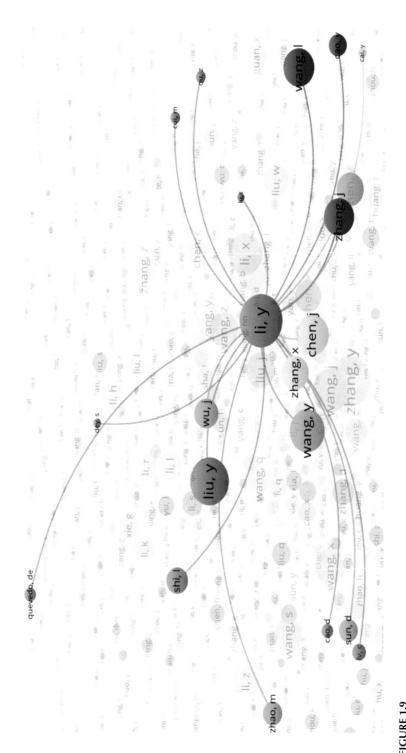

FIGURE 1.9
The strongest cooperation network of researchers.

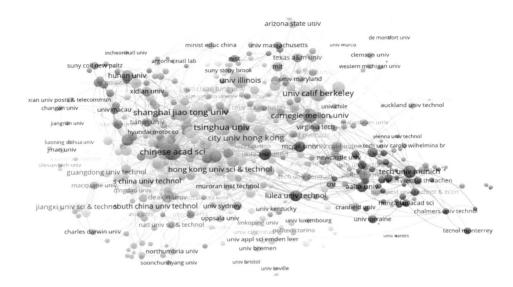

FIGURE 1.10
The cooperation network of organizations.

TABLE 1.3

The Top 13 Organizations with the Maximum TLS

Position	Organization	Place	P	Link	TLS
1	Shanghai Jiao Tong University	C	73	59	105
2	Chinese Academy of Sciences	China	57	65	100
3	Nanyang Technological University	Singapore	60	53	98
4	Tsinghua University	China	69	61	95
5	City University of Hong Kong	Hong Kong	45	58	93
6	University of California, Berkeley	USA	56	53	92
7	Zhejiang University	China	65	58	88
8	Hong Kong Polytechnic University	Hong Kong	43	47	79
9	University of Illinois	USA	57	47	77
10	Carnegie Mellon University	USA	55	57	65
11	KTH Royal Institute of Technology	Sweden	50	52	65
12	Technical University of Munich	Germany	40	39	54
13	Hong Kong University of Science and Technology	Hong Kong	28	33	52

1.5.1 The Citation Landscape for Research Papers

The VOSviewer enabled us to analyse the citation landscape for research papers published in 2010–2020 for the topics CPS and SC. The obtained findings from the citation and co-citation point of view are shown in Figure 1.12, and further details about the top ten highly cited papers are given in Table 1.6.

In Figure 1.12, the minimum number of citations of a research paper is set to 5, and 1,892 out of 4,538 research papers meet this threshold. From the figure, it is seen that

TABLE 1.4

The Top Six Pairs of Organizations with the Strongest Collaboration

Position	Organization	P	Link	TLS	Total Collaboration Strength	Organization	Link Strength	Collaboration Strength
1	Chinese Academy of Sciences	57	65	100	175.44%	University of the Chinese Academy of Sciences	12	12.00%
2	Simula Research Laboratory	14	8	18	128.57%	University of Oslo	11	61.11%
3	Towson University	36	9	24	66.67%	XI an Jiao Tong University	11	45.83%
4	Shanghai Jiao Tong University	73	59	105	143.84%	Yanshan University	10	9.52%
5	Aalto University	22	18	33	150.00%	Lulea University of Technology	8	24.24%
6	Hunan University	29	25	45	155.17%	Suny College New Paltz	8	17.77%

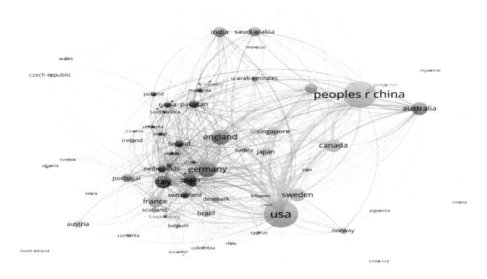

FIGURE 1.11

The collaborative network of places.

TABLE 1.5

The Top Six Places with the Strongest Collaboration

Position	Place	P	Link	TLS	Total Collaboration Strength	Place	Link Strength	Collaboration Strength
1	USA	1,299	60	948	72.98%	China	280	29.54%
						South Korea	46	4.85%
						Germany	45	4.75%
2	China	1,320	56	917	69.47%	USA	280	30.53%
						Canada	81	8.83%
						England	71	7.74%
3	Germany	391	41	334	85.42%	USA	45	13.47%
						China	26	7.78%
						Italy	23	6.89%
4	England	248	51	326	131.45%	China	74	22.70%
						USA	41	12.58%
						Germany	15	4.60%
5	Australia	182	43	258	141.76%	China	63	24.42%
						USA	36	13.95%
						Japan	14	5.43%
6	Italy	243	44	255	104.94	USA	32	12.55%
						France	23	9.02%
						Germany	23	9.02%

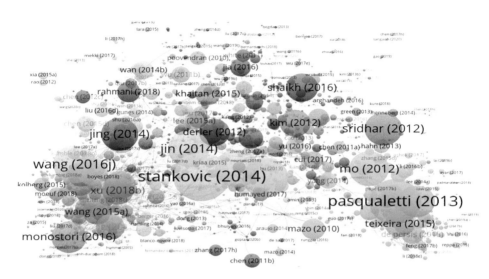

FIGURE 1.12
The co-citation graph of research papers.

ninety-nine out of 1,528 research papers were cited more than 100 times and form the biggest connected network. Further, the link between the research papers has surfaced due to their co-citation, and a size of the node is a measure of a research paper's citation count. The VOSviewer clubs the results into 23 groups; each is annotated by a different colour.

Returning to Table 1.6, which projects the top ten highly cited research papers, it is seen that nine research papers result from the collaborative works. The only exception is the single author work, by Stankovic, JA, entitled *Research Directions for the Internet of Things* [10]. Another interesting finding from the table is that all are journal articles and contribute to the following research directions: engineering, electrical and electronics, computer science, information systems, telecommunications, and automation and control systems. Further, it is mentioned that all the listed research papers have been cited more than 400 times and were published mainly in 2014.

1.5.2 The Citation Landscape for Researchers

The analysis along with the citation pathway of the researchers resulted in numerous interesting findings, which are shown in Figure 1.13. In the figure, the node size is a measure of the citation count received by the researcher. The link joining the two nodes corresponds to the two researchers that begin co-cited: the thicker the link, the greater has been their co-citations.

It is evident from the figure that all 9,740 researchers produced research papers that added value to the area. Of these, 387 bagged more than a hundred citation and 384 had the largest co-citation network. These 384 researchers have been shown as differently coloured clusters.

Further analysis into the citation landscape of the researchers provided the details about top 13 of them, whose details are given in Table 1.7. From table, it can be seen that

TABLE 1.6

The Top Ten Highly Cited Research Papers

Position	Research Papers	Researchers	Publication Year	Research Direction	C	Link
1	Research directions for the Internet of things [10]	Stankovic, JA	2014	Computer science, information systems Engineering, electrical & electronics; telecommunications	775	23
2	Attack detection and identification in cyber-physical systems [11]	Pasqualetti, F; Doerfler, F; Bullo, F	2013	Automation & control systems; engineering, electrical & electronic	706	95
3	For the grid and through the grid: The role of power line communications in the smart grid [13]	Galli, S; Scaglione, A; Wang, ZF	2011	Engineering, electrical & electronic	586	5
4	A survey on internet of things: Architecture, enabling technologies, security and privacy, and applications [13]	Lin, J; Yu, W; Zhang, N; Yang, XY; Zhang, HL; Zhao, W	2017	Computer science, information systems; engineering, electrical & electronic; telecommunications	556	22
5	Implementing smart factory of Industry 4.0: An outlook [14]	Wang, SY; Wan, JF; Li, D; Zhang, CH	2016	Computer science, information systems; telecommunications	549	42
6	An information framework for creating a smart city through internet of things [15]	Jin, J; Gubbi, J; Marusic, S; Palaniswami, M	2014	Computer science, information systems; engineering, electrical & electronic; telecommunications	520	9
7	Cyber-physical security of a smart grid infrastructure [16]	Mo, YL; Kim, THJ; Brancik, K; Dickinson, D; Lee, H; Perrig, A; Sinopoli, B	2012	Engineering, electrical & electronic	487	68
8	Cyber-physical system security for the electric power grid [17]	Sridhar, S; Hahn, A; Govindarasu, M	2012	Engineering, electrical & electronic	450	65
9	Security of the internet of things: Perspectives and challenges [18]	Jing, Q; Vasilakos, AV; Wan, JF; Lu, JW; Qiu, DC	2014	Computer science, information systems; engineering, electrical & electronic; telecommunications	449	18
10	Secure estimation and control for cyber-physical systems under adversarial attacks [19]	Fawzi, H; Tabuada, P; Diggavi, S	2014	Automation & control systems; engineering, electrical & electronic	433	74

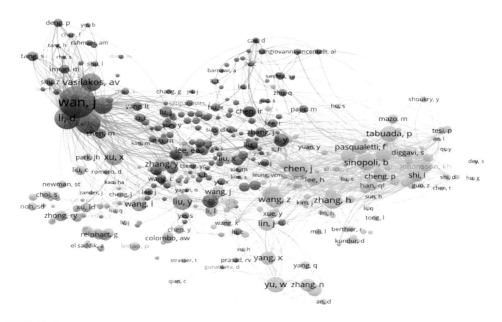

FIGURE 1.13
The co-citation network of authors.

TABLE 1.7

The Metrics of Highly Cited Researchers

Position	Researcher	Places	P	C	AC	TLS
1	Wan, J	China	41	3,382	82.48	2,499
2	Li, D	USA	35	1,874	53.54	1,739
3	Wang, S	China	31	1,797	57.96	1,233
4	Vasilakos, AV	Sweden	14	1,315	93.92	701
5	Chen, J	China	42	1,200	28.57	1,256
6	Zhang, D	China	19	1,168	61.47	1,233
7	Zhang, H	China	33	1,116	33.81	1,139
8	Wang, Z	China	34	1,052	30.94	1,083
9	Yu, W	USA	22	1,046	47.54	777
10	Zhang, C	China	20	1,039	51.95	832
11	Tabuada, P	USA	16	1,039	64.93	954
12	Xu, X	China	32	1,037	32.40	1,066
13	Liu, Y	China	51	1,036	20.31	1,182

all of the researchers had a citation count exceeding one thousand. Also, from the table, we can see that the researchers published 20 or more research papers in the research area (except Vasilakos, AV; Tabuada, P; and Zhang, D), thereby gaining the largest number of citations. The TLS of one author is 2,499; of eight authors, it is above 1,000; and for the remaining four authors, it is below 1,000. Also, the majority of the most influential thirteen authors are from China.

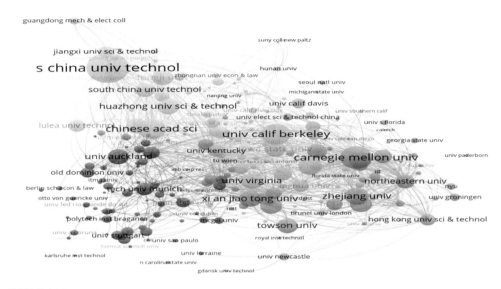

FIGURE 1.14
The citation network of organizations.

1.5.3 The Citation Landscape for Organizations

The citation landscape along with the organizations is shown in Figure 1.14. In the figure, it is shown that 3,363 organizations have published research papers relevant to the area, with 466 amongst them publishing a minimum of five papers. Further, 243 organizations attracted more than 100 citations. Also, it can be seen that 243 organizations were marked as differently coloured eight clusters on account of being co-cited.

A further detailed analysis on the co-citation of top 17 organizations is provided in Table 1.8. From the table, it can be seen that South China University of Technology collaborated 989 times with 129 organizations, thus making it to the top of the list. The last rank in the order is obtained by Towson University with a link number of 90 and TLS of 310. Also, from the table, it can be seen that six organizations are located in China and seven in USA. The others are in Australia, Sweden, Saudi Arabia and New Zealand.

1.5.4 The Citation Landscape for Places

The details about the analysis along with this dimension, as obtained from VOSviewer, are shown in Figure 1.15. From the figure, it can be seen that ninety-four places (countries/regions) have at least one research paper with 1,511 couples and secured a total of 27,221 co-citations. Also, from the figure, it can be seen that eighty-seven places have the largest co-citation network. Of these, twenty-one most cited ones have over thousand citations, and their detailed metrics are provided in Table 1.9.

From the table, it can be seen that USA has secured the first rank with a maximum number of citations and an average citation count of ≈20. China has managed to publish 1,320 research papers in the area and received 17,900 citations on them, thus grabbing the second position on both counts. Saudi Arabia, New Zealand and Australia are the

TABLE 1.8

The Metrics of Highly Cited Organizations

Position	Researcher	Places	P	C	AC	Link	TLS
1	South China University of Technology	China	22	2,562	116.45	129	989
2	Carnegie Mellon University	USA	55	1,734	31.53	156	788
3	University of California, Berkeley	USA	56	1,621	28.95	162	840
4	Chinese Academy of Sciences	China	57	1,531	26.86	156	630
5	Zhejiang University	China	65	1,247	19.18	154	781
6	University of Illinois	USA	57	1,244	21.82	128	348
7	XI'an Jiao Tong University	China	36	1,136	31.56	113	429
8	University of California, Los Angeles	USA	25	1,111	44.44	121	616
9	Swinburne University of Technology	Australia	14	1,105	78.93	118	442
10	Huazhong University of Science and Technology	China	35	1,100	31.43	153	578
11	The University of Virginia	USA	22	1,097	49.86	105	248
12	KTH Royal Institute of Technology	Sweden	50	1,094	21.88	160	844
13	Tongji University	China	24	1,070	44.58	142	758
14	King Saud University	Saudi Arabia	32	1,069	33.41	149	542
15	The University of Auckland	New Zealand	36	1,059	29.42	124	424
16	Iowa State University	USA	15	1,037	69.13	138	419
17	Towson University	USA	20	1,029	51.45	90	310

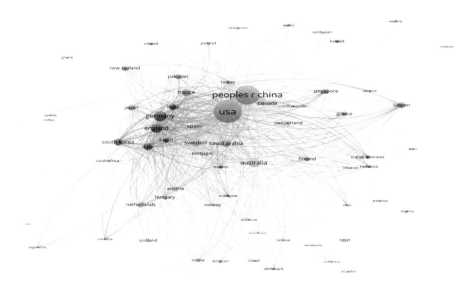

FIGURE 1.15
The citation graph of places.

top three countries with the respective highest average citations of 27.19, 23.85 and 21.47. They have been, respectively, cited 2,148, 1,097 and 3,908 times, while the numbers of their publications are 79, 46 and 182, respectively. From Table 1.9, it infers that USA and China are the two countries having the highest co-citation with the other countries.

TABLE 1.9

The Top 21 Most Influential Places

Position	Places	P	C	AC	Link	TLS
1	USA	1,299	25,967	19.99	82	10,322
2	China	1,320	17,900	13.56	80	9,783
3	Germany	391	4,747	12.14	75	2,518
4	Australia	182	3,908	21.47	70	2,153
5	England	248	3,695	14.9	75	2,931
6	Canada	193	3,666	18.99	64	1,849
7	Sweden	155	2,924	18.86	62	1,916
8	France	150	2,154	14.36	72	1,371
9	Saudi Arabia	79	2,148	27.19	60	1,097
10	South Korea	223	2,035	9.13	63	1,367
11	Italy	243	1,937	7.97	64	1,831
12	Spain	192	1,869	9.73	66	1,320
13	Taiwan	91	1,448	15.91	62	857
14	India	223	1,443	6.47	66	1,535
15	Japan	111	1,398	12.59	55	836
16	Singapore	100	1,277	12.77	48	1,091
17	Netherlands	85	1,276	15.01	47	798
18	Austria	105	1,272	12.11	60	636
19	New Zealand	46	1,097	23.85	48	523
20	Pakistan	64	1,045	16.33	50	517
21	Brazil	86	1,008	11.72	61	783

1.6 Timeline Analysis and Burst Detection

In this section, we present the analytical details along with the verticals of timeline and burst detection, as obtained from the CiteSpace tool. The timeline analysis provides information on the change of research trend with time, whereas the burst detection details the sudden spurts in the research area. We provide the details about each in the respective section.

1.6.1 Timeline Review Analysis

The results pertaining to the timeline analysis as obtained from the CiteSpace are shown in Figure 1.16. As seen from the figure, the timeline is divided into ten clusters, and each one is marked in a different colour. The topics of the clusters are the ones that remained popular throughout the years 2010–2020. They are sustainable manufacturing, research issue, smart city, of-service attack, smart grid, special section, future cyber physical energy system, traffic lattice and unmanned aerial vehicle. The less popular one in the period was electric-vehicle frequency regulation. Also, from the figure, the dotted circles on the timeline correspond to the prominent research topics (keywords as picked from the research papers) during that period.

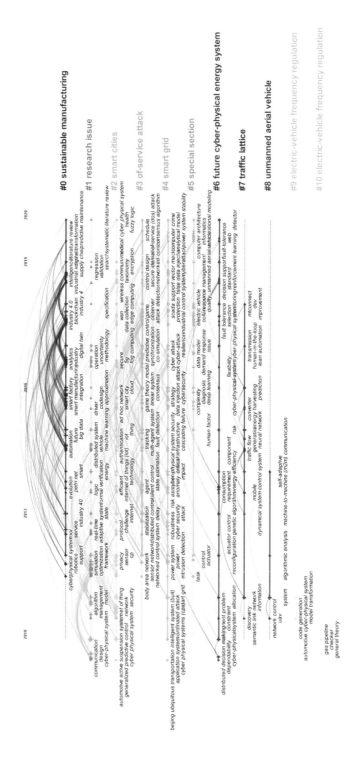

FIGURE 1.16
Timeline view of keywords.

TABLE 1.10

The Prominent Keywords with the Highest Citation Burst Strength

Position	Keyword	Strength	Begin	End	2010–2020
1	Algorithm	10.94	2011	2015	Image
2	Wireless sensor network	10.07	2011	2015	Image
3	Smart grid	9.70	2011	2014	Image
4	Verification	8.98	2012	2015	Image
5	Hybrid system	7.91	2012	2016	Image
6	Performance	7.42	2012	2015	Image
7	Sensor network	6.18	2012	2017	Image
8	AD HOC	5.83	2011	2015	Image
9	Embedded system	5.11	2010	2016	Image
10	Middleware	4.74	2011	2018	Image
11	Modelling	4.30	2012	2015	Image
12	Wireless	4.04	2013	2016	Image
13	Localization	3.85	2012	2016	Image
14	Pervasive computing	3.75	2011	2014	Image
15	Experimentation	3.48	2010	2015	Image
16	Access	2.90	2011	2016	Image
17	Behaviour	2.79	2014	2017	Image
18	Real-time system	2.69	2013	2016	Image
19	Real time	2.44	2012	2017	Image
20	Aggregation	2.41	2011	2017	Image
21	Control	2.29	2012	2016	Image
22	Logic	2.16	2014	2018	Image
23	Model checking	2.12	2013	2017	Image
24	Computation	2.08	2012	2016	Image

1.6.2 Keyword Burst Detection

An analysis (using the CiteSpace) about the keywords with the strongest citation burst during the period 2010–2020 provided twenty-four keywords as well as their metrics. All these details are presented in Table 1.10. From the table, it can be seen that the maximum burst occurred for the keyword Algorithm in the period 2011–2015 with a burst strength of 10.94. Following the suite, the keyword Wireless Sensor Network also secured almost the same burst strength in the same period of 5 years. Further, the lowest burst strength of 2.08 was secured by Computation in the period 2012–2016.

1.6.3 References Burst Detection

Now, we present the prominent references and their burst strength as returned by the CiteSpace in Table 1.11. From the table, it can be seen that the highest burst strength was attained by the research paper of the year 2008, entitled as *Cyber physical systems: Design challenges* by Edward A. Lee [8], in the period of 6 years starting 2011 with a burst strength of 26.78. Further, the lowest burst strength of 2.20 was secured by 2009 publication entitled as *Cloud computing and emerging it platforms: Vision, hype, and reality for delivering computing as the 5th utility* by R. Buyya et al. [20], in 2014–2018. It is mentioned here

TABLE 1.11

The Top 21 References with the Strongest Citation Bursts

Position	Title of Articles	Year	Strength	Begin	End	2010–2020
1	Cyber physical systems: Design challenges [8]	2008	26.78	2011	2016	Image
2	Introduction to embedded systems: A cyber-physical system approach [21]	2011	10.80	2013	2017	Image
3	Advances in cyber-physical systems research [22]	2011	10.79	2012	2017	Image
4	A survey of recent results in networked control systems [23]	2007	10.30	2011	2015	Image
5	Cyber-physical systems: Close encounters between two parallel worlds [point of view] [24]	2010	10.23	2011	2016	Image
6	Vulnerability assessment of cybersecurity for SCADA systems [25]	2008	9.13	2012	2017	Image
7	A survey of cyber-physical systems [26]	2011	8.24	2013	2018	Image
8	Cyber-physical systems: A new frontier [27]	2009	8.20	2013	2017	Image
9	A theory of timed automata [28]	1994	7.98	2012	2016	Image
10	Scheduling algorithms for multi-programming in a hard-real-time environment [29]	1973	7.78	2012	2016	Image
11	Cyber-physical system security for the electric power grid [17]	2012	7.43	2013	2017	Image
12	Cyber-physical systems – are computing foundations adequate [30]	2006	7.06	2010	2015	Image
13	An online optimization approach for control and communication codesign in networked cyber-physical systems [31]	2013	6.02	2013	2017	Image
14	Ensuring safety, security, and sustainability of mission-critical cyber-physical systems [32]	2012	5.734	2013	2017	Image
15	Wireless sensor networks: A survey [33]	2002	4.27	2012	2016	Image
16	Power system stability and control [34]	1994	4.13	2014	2018	Image
17	Rdds: A real-time data distribution service for cyber-physical systems [35]	2012	4.13	2014	2018	Image
18	Spatio-temporal event model for cyber-physical systems [36]	2009	3.65	2011	2017	Image
19	Metropolis: An integrated electronic system design environment [37]	2003	3.20	2012	2016	Image
20	Cyber security analysis of state estimators in electric power systems [38]	2010	2.99	2014	2018	Image
21	Cloud computing and emerging it platforms: Vision, hype, and reality for delivering computing as the 5th utility [20]	2009	2.20	2014	2018	Image

that the majority of these publications were published outside the period 2010–2020, but attained their bursts within this period.

1.7 Conclusion

In this chapter, we have put forward a detailed and comprehensive bibliometric analysis of the CPS and the SC. CPSs consist of the physical computing units along with the advanced connectivity to ensure the real-time data acquisition and information feedback from the cyber space and the capabilities for data management, analytics and computations [5]. With the onset of Industry 4.0 and embedding of the sensors in almost all the real-life processes, data collection is no longer a challenge but the intelligent processing of this data is. This has enabled the use of SC within the aegis of the CPS. The areas of the CPS and SC, though, have been going on hand-to-hand with each other for numerous years; little has been said about what their amalgamation can accomplish. This motivated us to bring forth this bibliometric analysis.

We have performed the bibliometric analysis using the respective tools whose details are as follows:

- Using the WoS database, we have analysed
  - The number of annual publications in the field of CPS and SC for the past 10 years (2010–2020).
  - The types of publications such as book chapters, journal papers, etc.
  - The top journals/sources of publications.
  - The top research directions for CPS and SC. These include the computer science, engineering, telecommunications, etc. Further, an analysis of the trend in these research directions has also been put forth.
  - The top countries and the regions, as well as the productive authors and the organizations.
- Using the VOSviewer, we have analysed various types of cooperation relationships. These include cooperation relationships, highlighting
  - The collaborative strength amongst the researchers
  - The collaboration strength of organizations
  - The collaborative strength of places.
- Using the VOSviewer, we have also provided a citation structure analysis from different aspects, which include:
  - The citation landscape for research papers
  - The citation landscape for researchers
  - The citation landscape for organizations
  - The citation landscape for places.
- Also, using the CiteSpace, we have also put forth the analytical details from the verticals of
  - The timeline view
  - The keywords burst detection
  - The references burst detection.

The key findings from the analysis presented in this chapter for the period 2010–2020 (up to October 2020) are as follows:

- A total of 4,538 publication appeared in the WoS when searched with keywords such as *Cyber Physical Systems* and *Smart Computing*.
- The number of publications was quite small in 2010 and 2011. They more than doubled in 2012 of what was their number in 2011, kept increasing constantly to

cross 1,000 in 2019 and close at 935 in October 2020. Maximum number of publications (1,018) on CPS and SC were published in the year 2019.

- The published papers were of various types like articles, proceeding papers, reviews etc. and were published in various journals. Most published type of research works was the articles (3,968 numbers, or 86.78% of the total publication types) in comparison with other publication types.

- Maximum number of publications appeared in *IEEE Access* (268 numbers, or 5.91% of the total publications).

- The most popular research direction was the computer science, bagging 2,404 (52.98%, or more than half of total) number of publications. It was closely followed by the research direction: engineering (2,398, or 52.84% number of publications).

- The most contributing places to the publication landscape were China (1,320 publications) followed by USA. Various other countries also made their respective contributions, though comparatively quite less in number.

- Various organizations and authors, contributing to these publication numbers, were from different places across the globe; however, the maximum number was coming from China.

- On a global level, a total of 9,740 researchers had worked collaboratively to produce the research publication, of which 1,860 had the largest collaboration network, and many of these 1,860 researchers also had dense collaboration network. The author Li, Y had the most dense collaborative network.

- Out of a total 3,363 organizations, 453 had the largest connected network, of which numerous organizations also had dense collaboration network. The Shanghai Jiao Tong University from China secured the first position on the list of top 13 organizations with dense collaboration network. Also, the Chinese Academy of Sciences and University of the Chinese Academy of Sciences were the largest collaborators (with twelve joint works).

- Globally, ninety-four countries/regions contributed at least one paper into the research area. USA and China jointly contributed to the maximum number of research outputs.

- With a citation threshold of a research paper set at five, less than half of 4,538 papers qualified the criteria. In addition, ninety-nine research papers attracted more than 100 citations, thus forming the biggest network. Also, a large number of research papers were co-cited.

- Of 9,740 researchers, 387 had the citation count crossing 100 and 384 secured the largest co-citation network. The author Wan, J had the highest citation count.

- Of 3,363 organizations, 243 had been successful in getting over 100 citations, and many had strong co-citation network also. South China University of Technology topped the list of most popular seventeen co-citation organizations.

- Out of ninety-four places (countries/regions), each one had at least one research paper with 1,511 couples and secured a total of 27,221 co-citations. Also, eighty-seven places had the largest co-citation networks. USA secured the first rank with the maximum number of citations.

- The timeline analysis produced various popular keywords grouped into different clusters, with one or more keywords found to have popularity at different time instants.

- When the keywords were analysed for the burst detection, it was learnt that the burst time period was almost 4–5 years for the majority of the returned keywords.

- From the reference burst analysis, it was learnt that the majority of these publications were published outside the period 2010–2020, but attained their bursts within this period. The highest burst strength was attained by the research paper of the year 2008, entitled as *Cyber physical systems: Design challenges* by Edward A. Lee [8].

Therefore, based on these facts stated above, it is evident that the trend is towards the area of sustainable manufacturing, and the key contributors to it are twin concepts of algorithms and the wireless sensor networks (the most popular keywords). China and USA have driven the research on CPS and SC in the previous years. It seems that they will continue to do so in the coming years, also. Hence, CPS and SC hold a great potential for future research, with the number of contributions from researchers, organizations, and places poised to increase in the coming years. This will also open new doors of opportunities for collaborative works. Finally, we strongly feel that this bibliometric review gives a holistic view to any researcher beginning his/her journey into the area of CPS and SC.

References

1. Volkan Gunes, Steffen Peter, Tony Givargis, and Frank Vahid, A survey on concepts, applications, and challenges in cyber-physical systems. *KSII Transactions on Internet & Information Systems*, 8(12):4242–4268, 2014.
2. Yang Lu. Cyber physical system (CPS)-based industry 4.0: A survey. *Journal of Industrial Integration and Management*, 2(03):1750014, 2017.
3. Hong Chen. Applications of cyber-physical system: a literature review. *Journal of Industrial Integration and Management*, 2(03):1750012, 2017.
4. Hansong Xu, Wei Yu, David Griffith, and Nada Golmie. A survey on industrial internet of things: A cyber-physical systems perspective. *IEEE Access*, 6:78238–78259, 2018.
5. Jay Lee, Behrad Bagheri, and Hung-An Kao. A cyber-physical systems architecture for industry 4.0-based manufacturing systems. *Manufacturing Letters*, 3:18–23, 2015.
6. Nees Jan Van Eck, and Ludo Waltman. Software survey: VOSviewer, a computer program for bibliometric mapping. *Scientometrics*, 84(2):523–538, 2010.
7. Chaomei Chen. CiteSpace: Visualizing patterns and trends in scientific literature. http://cluster.cis.drexel.edu/cchen/citespace/ (Accessed on 01/23/2021).
8. Edward A Lee. Cyber physical systems: Design challenges. *In 2008 11th IEEE International Symposium on Object and Component-Oriented Real-Time Distributed Computing (ISORC)*, pp. 363–369, IEEE, Orlando, FL, USA, 2008.
9. Petar Radanliev, David De Roure, Max Van Kleek, Omar Santos, and Uchenna Ani. Artificial intelligence in cyber physical systems. *AI & Society*, 1–14, 2020.

10. John A Stankovic. Research directions for the internet of things. *IEEE Internet of Things Journal*, 1(1):3–9, 2014.

11. Fabio Pasqualetti, Florian Dörfler, and Francesco Bullo. Attack detection and identification in cyber-physical systems. *IEEE Transactions on Automatic Control*, 58(11): 2715–2729, 2013.

12. Stefano Galli, Anna Scaglione, and Zhifang Wang. For the grid and through the grid: The role of power line communications in the smart grid. *Proceedings of the IEEE*, 99(6):998–1027, 2011.

13. Jie Lin, Wei Yu, Nan Zhang, Xinyu Yang, Hanlin Zhang, and Wei Zhao. A survey on internet of things: Architecture, enabling technologies, security and privacy, and applications. *IEEE Internet of Things Journal*, 4(5): 1125–1142, 2017.

14. Shiyong Wang, Jiafu Wan, Di Li, and Chunhua Zhang. Implementing smart factory of Industry 4.0: An outlook. *International Journal of Distributed Sensor Networks*, 12(1):3159805, 2016.

15. Jiong Jin, Jayavardhana Gubbi, Slaven Marusic, and Marimuthu Palaniswami. An information framework for creating a smart city through internet of things. *IEEE Internet of Things Journal*, 1(2):112–121, 2014.

16. Yilin Mo, Tiffany Hyun-Jin Kim, Kenneth Brancik, Dona Dickinson, Heejo Lee, Adrian Perrig, and Bruno Sinopoli. Cyber–physical security of a smart grid infrastructure. *Proceedings of the IEEE*, 100(1):195–209, 2011.

17. Siddharth Sridhar, Adam Hahn, and Manimaran Govindarasu. Cyber–physical system security for the electric power grid. *Proceedings of the IEEE*, 100(1):210–224, 2011.

18. Qi Jing, Athanasios V Vasilakos, Jiafu Wan, Jingwei Lu, and Dechao Qiu. Security of the internet of things: Perspectives and challenges. *Wireless Networks*, 20(8):2481–2501, 2014.

19. Hamza Fawzi, Paulo Tabuada, and Suhas Diggavi. Secure estimation and control for cyber-physical systems under adversarial attacks. *IEEE Transactions on Automatic Control*, 59(6):1454–1467, 2014.

20. Rajkumar Buyya, Chee Shin Yeo, Srikumar Venugopal, James Broberg, and Ivona Brandic. Cloud computing and emerging it platforms: Vision, hype, and reality for delivering computing as the 5th utility. *Future Generation Computer Systems*, 25(6):599–616, 2009.

21. Edward A Lee and Sanjit A Seshia. Introduction to embedded systems-a cyber-physical systems approach, 2011. http://LeeSeshia. org.

22. Jiafu Wan, Hehua Yan, Hui Suo, and Fang Li. Advances in cyber-physical systems research. *KSII Transactions on Internet & Information Systems*, 5(11):1891–1908, 2011.

23. Joo P Hespanha, Payam Naghshtabrizi, and Yonggang Xu. A survey of recent results in networked control systems. *Proceedings of the IEEE*, 95(1):138–162, 2007.

24. RADHA Poovendran. Cyber–physical systems: Close encounters between two parallel worlds [point of view]. *Proceedings of the IEEE*, 98(8):1363–1366, 2010.

25. Chee-Wooi Ten, Chen-Ching Liu, and Govindarasu Manimaran. Vulnerability assessment of cybersecurity for SCADA systems. *IEEE Transactions on Power Systems*, 23(4):1836–1846, 2008.

26. Jianhua Shi, Jiafu Wan, Hehua Yan, and Hui Suo. A survey of cyber-physical systems. *In 2011 International Conference on Wireless Communications and Signal Processing (WCSP)*, pp. 1–6, IEEE, Nanjing, China, 2011.

27. Lui Sha, Sathish Gopalakrishnan, Xue Liu, and Qixin Wang. Cyber-physical systems: A new frontier. *In 2008 IEEE International Conference on Sensor Networks, Ubiquitous, and Trustworthy Computing (SUTC 2008)*, pp. 1–9, IEEE, Taichung, Taiwan, 2008.

28. Rajeev Alur and David L Dill. A theory of timed automata. *Theoretical Computer Science*, 126(2):183–235, 1994.

29. Chung Laung Liu and James W Layland. Scheduling algorithms for multiprogramming in a hard-real-time environment. *Journal of the ACM (JACM)*, 20(1):46–61, 1973.

30. Edward A Lee. Cyber-physical systems-are computing foundations adequate. *In Position Paper for NSF Workshop on Cyber-Physical Systems: Research Motivation, Techniques and Roadmap,* vol. 2, pp. 1–9, Citeseer, 2006.

31. Xianghui Cao, Peng Cheng, Jiming Chen, and Youxian Sun. An online optimization approach for control and communication codesign in networked cyber-physical systems. *IEEE Transactions on Industrial Informatics,* 9(1):439–450, 2012.

32. Ayan Banerjee, Krishna K Venkatasubramanian, Tridib Mukherjee, and Sandeep Kumar S Gupta. Ensuring safety, security, and sustainability of mission-critical cyber–physical systems. *Proceedings of the IEEE,* 100(1):283–299, 2011.

33. Ian F Akyildiz, Weilian Su, Yogesh Sankarasubramaniam, and Erdal Cayirci. Wireless sensor networks: A survey. *Computer Networks,* 38(4):393–422, 2002.

34. Prabha Kundur, Neal J Balu, and Mark G Lauby. *Power System Stability and Control,* vol. 7. McGraw-hill, New York, 1994.

35. Woochul Kang, Krasimira Kapitanova, and Sang Hyuk Son. Rdds: A real-time data distribution service for cyber-physical systems. *IEEE Transactions on Industrial Informatics,* 8(2):393–405, 2012.

36. Ying Tan, Mehmet C Vuran, and Steve Goddard. Spatio-temporal event model for cyber-physical systems. *In 2009 29th IEEE International Conference on Distributed Computing Systems Workshops,* pp. 44–50, IEEE, 2009.

37. Felice Balarin, Yosinori Watanabe, Harry Hsieh, Luciano Lavagno, Claudio Passerone, and Alberto Sangiovanni-Vincentelli. Metropolis: An integrated electronic system design environment. *Computer,* 36(4):45–52, 2003.

38. André Teixeira, Saurabh Amin, Henrik Sandberg, Karl H Johansson, and Shankar S Sastry. Cyber security analysis of state estimators in electric power systems. *In 49th IEEE Conference on Decision and Control (CDC),* pp. 5991–5998, IEEE, Atlanta, GA, USA, 2010.

2

Design Optimization of Close-Fitting Free-Standing Acoustic Enclosure Using Jaya Algorithm

Ashish Khachane and Vijaykumar S. Jatti

D.Y. Patil College of Engineering

CONTENTS

2.1 Introduction

Sound emanates from a vibrating body, propagates through a medium usually air and reaches our ears (generation-propagation-reception). Curbing noise at its source is the best and at the receivers end is the last resort [1]. In the transmission path, sound barriers, dampers and enclosures can be used for diminishing noise. This chapter deals with only free-standing close-fitting and single-layer enclosure of small size. Free standing means that there are no mechanical connections or links between the enclosure and the sound source, and close fitting means that noise source is closely surrounded by enclosure or most of the enclosure volume is occupied by the equipment or machinery from which sound is emitted. This kind of enclosure is generally used for portable generators,

DOI: 10.1201/9781003143505-2

compressors, engines, etc. where less space is available for noise reduction [2]. The cost of enclosure depends upon its size; therefore from an economic point of view, enclosure should be as compact as possible.

There are two types of enclosure resonances that are significant. The first is mechanical or structural resonance of the enclosure panels, whereas the second is acoustic resonance of the air space between an enclosed machine and the enclosure wall [3]. Free-standing close-fitting enclosure panels are excited through airborne path only. Insertion loss (IL) is the difference between sound pressure levels (SPL) measured without enclosure and with enclosure surrounding the sound source. It is expressed in decibels (dB). Many materials like mild steel and copper offer large IL but at the expense of very high enclosure weight. Enclosure material has direct effect on its weight, cost, portability, heat resistance, corrosion resistance and aesthetics. Therefore, material selection is one of the most important tasks in design of enclosure. Availability factor should also be considered while selecting a material for enclosure. Once material is selected, it becomes very important to decide its shape and dimension, as there is always a space constraint for enclosure system. Jackson [4,5] had introduced IL as a performance measure for enclosures for the first time and developed an empirical model by assuming the source and the enclosure to be as infinite parallel plates. The first plate vibrates with constant velocity and acts as a source, and the second plate acts as an enclosure wall. Jackson considered an IL as a function of vibration level only, but later Oldham and Hillarby [6] found that IL is not only a function of vibration level of the panel; rather, it is dependent on its material and geometrical properties. Oldham and Hillarby had a detailed study of close-fitting free-standing enclosure. The performance of small-size close-fitting acoustic enclosure to reduce the noise from machines was considered. One mathematical model related to the performance of the small enclosures at lower frequencies was considered for simply supported boundary condition. The second mathematical model related to the performance of the small enclosures at higher frequencies was considered for clamped boundary condition. Further, it was reported that there is no practical model, which completely satisfies either of the two boundary conditions, but it actually lies between two boundary conditions. The theoretical models were experimentally verified and discussed in Oldham and Hillarby [7]. The higher-frequency model was found to be more successful than the lower-frequency model. Blanks [8] used the mathematical models developed by Oldham and Hillarby for the optimization of enclosure design by the pattern search method. The effects of variation in the material properties and source to panel distance of the noise source on IL were discussed. Chiu [9] discussed simulated annealing as an optimization method for IL and SPL. For optimization, again theoretical model developed by Oldham and Hillarby was used. A case study was carried out on aluminium acoustic hood to study the sensitivity of IL with respect to thickness of hood (h), source to panel distance (d) and internal damping coefficient (η) of enclosure material. IL was maximized for pure tone noise, and total SPL was minimized for broadband noise. Rigid rectangular-shaped enclosure having no mechanical connection with sound source was considered.

In this study, analysis is carried out on mixer grinder as the source of noise with regular rectangular-shaped enclosure. The geometrical parameters are optimized using Jaya algorithm. Jaya, a newly developed optimization technique, is a search optimization algorithm. It does not require any algorithm-specific parameters [10]. Khachane and Jatti [11] applied Jaya algorithm to optimize electrical discharge machining (EDM) process parameters. Their study shows that Jaya algorithm is a promising method of optimization that requires very few iterations and hence less function evaluations for convergence to the optimum solution.

2.2 Insertion Loss

This section mainly describes mathematical model for IL given by Oldham and Hillarby [7]. This is followed by the selection of material for enclosure. Need of optimizing various geometrical parameters and how the optimization using Jaya algorithm can be performed is also discussed.

2.2.1 Mathematical Model for Prediction of Insertion Loss

Oldham and Hillarby expressed IL in terms of Young's modulus (E), Poisson's ratio (μ), flexural rigidity of the enclosure wall (D), density of enclosure material (ρ), thickness (h), source to panel distance (d) and frequency of sound (f). To take into account the effect of internal damping, complex flexural rigidity (D^1) is introduced. The expression for IL given in [6] for the low-frequency region is as follows:

$$\text{IL} = 10 * \log_{10} \left\{ \cos(kd) + \left(\frac{\pi^2}{K\omega\rho_0 c} \right) * \sin(kd) \right\}^2 \tag{2.1}$$

where

$$K = \frac{16}{\pi^2 \left\{ D^1 * \pi^4 \left(1/a^4 + 2/a^2b^2 + 1/b^4 \right) - \omega^2 \rho h \right\}} \tag{2.2}$$

Bulk modulus for simply supported boundary condition

$$K = \frac{1.35}{3.86D^1 * \left(129.6/a^4 + 78.4/a^2b^2 + 129.6/b^4 \right) - \omega^2 \rho h} \tag{2.3}$$

Bulk modulus for clamped boundary condition

$$D^1 = D * (1 + i\eta) \tag{2.4}$$

$$D = \frac{Eh^3}{12(1 - v^2)} \tag{2.5}$$

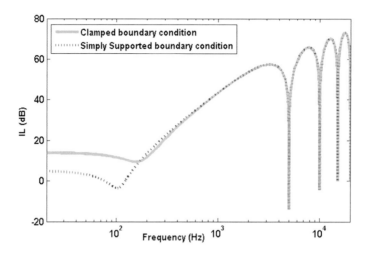

FIGURE 2.1
IL for clamped and simply supported boundary conditions [6].

a=height of the enclosure (m),
b=length of the enclosure (m),
ρ_o (specific acoustic impedance for air)=408 (Rayl),
k (wave number)=$2\pi/\lambda$ (m^{-1}),
ω (circular frequency)=$2\pi f$ (rad/s),
c (speed of sound in air)=340 (m/s),
η=internal damping coefficient constant,
μ=Poisson's ratio.

The validity of the lower-frequency model extends to higher frequencies for small spacing (source to panel), and the validity of the higher-frequency model extends to lower frequencies for the larger spacing [7]. The case discussed in this paper is of the former type. Figure 2.1 shows the comparison of IL obtained for simply supported and clamped boundary conditions. It is seen that the difference lies only in the position of first null and IL curves take the same path after 300 Hz. As there no such practical case exists which satisfies either of the two conditions completely [6], so clamped boundary condition is considered in this study.

2.2.2 Need for Optimization

IL mainly depends upon source to panel distance (d), thickness of enclosure wall (h) and internal damping coefficient (η). Therefore, the effect of change in these parameters on IL is assessed considering polypropylene as a material for acoustic hood having fixed length and height (Chiu [9]).

From Table 2.1, it can be inferred that IL though is large for materials such as aluminium, steel, glass (soda), copper alloys, fibreglass and brass, but their weight is very high. For materials like acrylic, polyvinyl chloride, polyethylene (HDPE) and polypropylene, IL is almost the same but the cost of acrylic, polyvinyl chloride and polyethylene (HDPE) is high. Hence in this study, polypropylene is considered as the material for

TABLE 2.1

Values of Insertion Loss and Weight of Enclosure Panel for Different Materials

Sr. No	Material	Density (kg/m³)	Young's Modulus (GPa)	Poisson's Ratio	Insertion Loss (in dB) ($f=315\,$Hz)	Weight (kg)
1	Aluminium	2,700	69	0.33	37.6092	19.656
2	Steel, mild 1020	7,800	210	0.29	47.117	56.784
3	Acrylic	1,180	3.2	0.35	24.6593	8.5904
4	Glass (soda)	2,500	50	0.23	31.7889	18.2
5	Polyvinyl chloride	1,400	2.8	0.42	26.8517	10.192
6	Polyethylene (HDPE)	950	0.7	0.42	24.6326	6.916
7	Copper alloys	8,300	135	0.35	39.2133	60.424
8	Fibreglass	2,160	51.7	0.22	33.6648	15.7248
9	Brass (70Cu30Zn)	8,400	130	0.33	37.6537	61.152
10	Polypropylene	905	1.75	0.4	22.9677	6.5884

enclosure. Geometrical dimensions are optimized for polypropylene enclosure panel to maximize the overall noise reduction.

2.2.2.1 Effect of Variation in Panel Thickness (h) on IL

Figure 2.2 shows the effect on IL due to a variation in panel thickness. As the panel thickness (*h*, in metres) increases, the value of IL increases, which is desirable. The nulls (minima) occurring for different '*h*' coincide, i.e., the occurrences of nulls are independent of '*h*'. Preferably, higher value of '*h*' should be selected but this leads to an increase in weight of the enclosure. Thus, this parameter should be optimized taking into consideration the weight of enclosure along with IL.

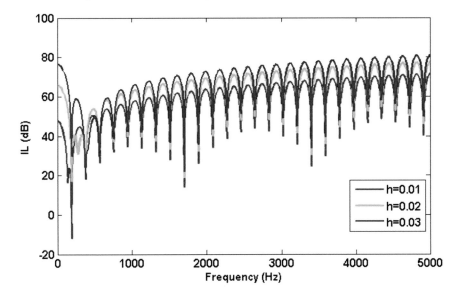

FIGURE 2.2

Effect of variation in panel thickness (*h*) on IL.

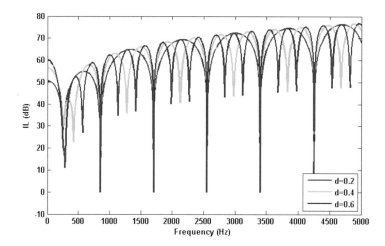

FIGURE 2.3
Effect of variation in source to panel distance (*d*) on IL.

2.2.2.2 Effect of Variation in Source to Panel Distance (d) on IL

Figure 2.3 shows the effect on IL due to a variation in source to panel distance (*d*, in metres). The minima occurring are the 'null' points wherein the IL is very small. This is to be avoided. It can be seen that as '*d*' increases, it causes more occurrences of nulls. This hints lower value of '*d*'. But if '*d*' is decreased, then the value of IL at nulls is further dropped, which is not desirable. This calls for the optimization of source to panel distance for the maximum IL over the frequency band.

2.2.2.3 Effect of Variation in Internal Damping Coefficient (η) on IL

Figure 2.4 shows that there is no appreciable effect of change in '*η*' either on the IL value or on the frequencies at which nulls occur.

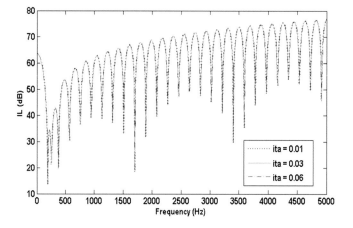

FIGURE 2.4
Effect of variation in internal damping coefficient (*η*) on IL.

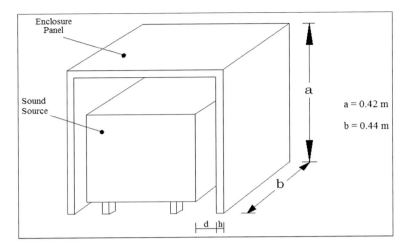

FIGURE 2.5
Sound source enclosed with closely fitted enclosure.

2.2.3 Optimization

The values of '*h*', '*d*' and '*η*' should be such that the IL is maximum. Internal damping coefficient '*η*' is the material property, so it is fixed for particular type of material. Also, variation in '*η*' doesn't have much effect on IL value. Therefore, the value of '*η*' obtained from the optimization program can be disregarded. Jaya algorithm is used to find the optimum solution within the search range of design variables.

2.2.3.1 Formulation of Optimization Problem

The sound source is a household mixer grinder of 500 W, 230 V (AC), 50 Hz. An enclosure is to be made such that two dimensions '*a*' and '*b*' are fixed so that it closely fits the grinder as shown in Figure 2.5. IL is maximized at a particular frequency as the function of design variables: panel thickness (*h*) and source to panel distance (*d*). The length and height of the enclosure are considered 0.42 and 0.44 m, respectively. The width of the enclosure is controlled by optimizing panel thickness and source to panel distance. The search range for the design variables is given in Table 2.2.

Therefore, the optimization problem can be written in the mathematical form as:

$$\text{Maximize, IL} = f(h, d)$$

Subjected to $0.001 \leq h \leq 0.01$, $0.015 \leq d \leq 0.04$.

TABLE 2.2

Search Range for Design Variables

No.	Design Variables	Lower Bound	Upper Bound
1	Panel thickness (*h*)	0.001	0.01
2	Source to panel distance (*d*)	0.015	0.04

TABLE 2.3

Optimized Sets of Design Variables

Sr No.	Frequency (Hz)	'h' (m)	'd' (m)
1	250	0.01	0.04
2	315	0.01	0.04
3	400	0.01	0.04
4	500	0.01	0.04
5	630	0.01	0.04
6	800	0.01	0.04
7	1,000	0.01	0.04
8	1,250	0.01	0.04
9	1,600	0.01	0.04
10	2,000	0.01	0.04
11	2,500	0.01	0.034
12	3,150	0.01	0.027
13	4,000	0.01	0.021

2.2.3.2 Optimization by Jaya Algorithm

Jaya algorithm is nonconventional search optimization method proposed by Rao [10]. This algorithm does not stuck at the local minima, and eventually solution gets better. The IL is considered as the fitness function, while panel thickness and source to panel distance are design variables.

As IL is the function of frequency, dimensions can be optimized for a particular frequency. If optimization is done only at a particular frequency, then it is not necessary that it will prove to be a good case for all other frequencies over the entire frequency band. Therefore, optimization is done at thirteen different one-third octave band frequencies (viz. 250, 315, 400, 500, 630, 800, 1,000, 1,250, 1,600, 2,000, 2,500, 3,150 and 4,000 Hz) where sound is relatively more noticeable to human ear. This leads to 13 runs of algorithm for each frequency. Every run will provide a set of optimized parameters such as thickness of panel (h) and source to panel distance (d). It is observed in few cases that sets of h, d are having the same values and the total numbers of distinct sets obtained are four. The optimum values of h and d for each frequency are given in Table 2.3.

2.2.3.3 Final Dimensions of the Enclosure

Figure 2.6 represents the graph which shows the value of IL at one-third octave band frequencies up to 4,000 Hz for each distinct set of h and d obtained. Also, as mentioned in Section 2.2, this study is dealing with the low-frequency model.

It can be inferred that for lower frequencies, curve shown in blue gives the highest value of IL amongst all. However, it dives sharply after 3,000 Hz. Curve shown in cyan gives good result above 2,500 Hz but it performs poorly at low frequencies. Thus, it is realized that the curve shown in green is performing better over the entire frequency range considered. Hence, the enclosure is fabricated with values 'h' and 'd' to be 0.01 and 0.034 m, respectively.

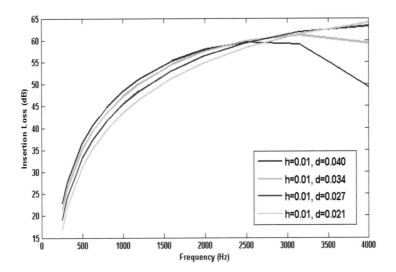

FIGURE 2.6
IL vs frequency graph for different sets of design variables.

2.3 Experimentation

With the values of '*h*' and '*d*' obtained, IL at one-third octave band frequencies is computed. This theoretical data generated is compared with the experimental data. The SPL is measured through microphone, and data is acquired by DAQ hardware (DEWE 43A, Maker: DEWESoft, Slovenia). The data obtained is further processed (fast Fourier transformation (FFT) analysis) in the software associated: DEWESoft X2.

2.3.1 Experimental Set-up

The experiment was carried out in a free-field environment over a hard-reflecting surface, and the measurement grid is designed according to BIS standards [12]. This method, known as survey method, is in line with ISO 3746. The measurement grid is shown in Figure 2.7.

2.3.2 Experimental Procedure

For enclosure material, polypropylene is utilized. Measurement of SPL is done at five different locations as shown in Figure 2.7 with and without enclosing the source of sound, which gives IL values of enclosure.

The detailed measurement procedure is reported as under the following conditions:

I. Initially, background noise (when the equipment is not operating) was measured at test location. This is to ensure that the difference between the background noise and the total noise when equipment is in ON mode is greater than 6 dB [2], in accordance with ISO 3746.

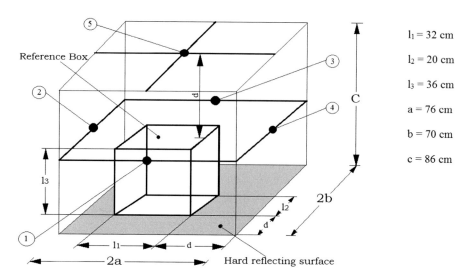

FIGURE 2.7
Measurement grid dimensions.

FIGURE 2.8
SPL at test location 3 without enclosure.

II. Instead of employing arrays of microphones, only one microphone is used to measure SPL at all test location points. The microphone is supported by stand while measuring SPL.

III. SPL was recorded at locations 1, 2, 3, 4 and 5 without enclosure. Figure 2.8 displays the SPL at test location 3 without enclosing the sound source.

IV. With the help of data acquisition software (DEWESoft X2 V 7.2.3), SPL values at various frequencies were tabulated over the entire one-third octave band.

V. SPL was recorded at locations 1, 2, 3, 4 and 5 with enclosure. Figure 2.9 displays the SPL at test point 3 with enclosing the sound source.

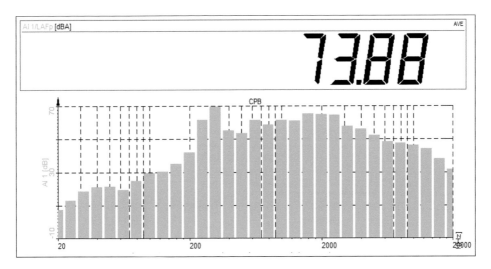

FIGURE 2.9
SPL at test location 3 with enclosure.

2.4 Results and Discussion

Table 2.4 shows the SPL with and without enclosure at all test locations. Noise reduction was calculated by taking difference between average SPL with and without enclosure, where average SPL was calculated simply by arithmetic mean of SPL values at all test locations.

Total noise reduction = Average SPL (without enclosure) – Average SPL (with enclosure)

$$= 90.71 - 73.52 = 17.19 \text{ dBA } (18.95\% \text{ noise reduction})$$

2.4.1 Theoretically Predicted Vs Experimentally Obtained Results

Figure 2.10 gives the comparison of theoretically predicted (lavender-line) and experimentally obtained (green-line) results at 23 different frequencies of one-third octave band from 125 to 20,000 Hz [13–14]. The observed SPL curve is plotted from experimental

TABLE 2.4

Comparison of Average SPL at Each Measurement Position

Position	SPL (dBA), without Enclosure	SPL (dBA), with Enclosure
1	92.4	73.43
2	91.44	74.31
3	91.55	73.88
4	89.89	73.76
5	88.28	72.24
Average	90.71	73.52

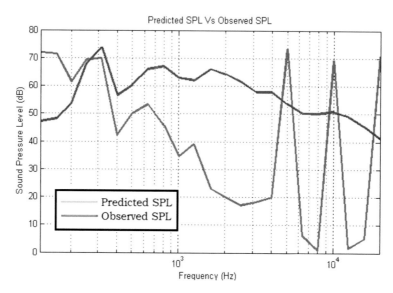

FIGURE 2.10
Theoretically predicted vs experimentally obtained SPL (dB).

readings in the presence of enclosure, and the predicted curve shows the expected SPL level with enclosure computed from the mathematical model of IL.

Following observation can be noted from Figure 2.10.

i. The overall noise reduction actually observed is less than that of the predicted case. This is because the enclosure fabricated was not perfectly sealed air-tight. Also, the hard-reflecting surface and the bottom surfaces of the enclosure panel were not perfectly flat. This has also contributed to the difference between the predicted and actual values.

ii. The trend of the predicted and observed SPL curve is fairly similar in the range 200–4,000 Hz. The deviation of the observed curve increases at higher frequency.

iii. The sudden rise in the predicted values of SPL at the frequencies 5,000, 10,000 and 20,000 Hz is because of the occurrences of nulls. This is not found in actual SPL curve.

2.5 Conclusions and Future Scope

The total noise reduction by enclosing the sound source by polypropylene enclosure is found to be 17.19 dBA. According to Oldham's theory, there is no practical model that satisfies the simply supported or clamped boundary condition perfectly, but lies between them; hence, the practical results tend to deviate from the theoretical results as clamped boundary condition is used for prediction. Since the trend of the predicted and observed (actual) curve is fairly similar in the range 200–4,000 Hz, it is concluded that Oldham's model is a fair predictor in this range. The deviation in the theoretical and observed values is more at higher frequencies. Thus, the model is a good predictor at lower- to medium-frequency range. The enclosure in the present study had manufacturing limitations. Precision in manufacturing would have yielded better results. The experimental work, if carried out in acoustic chambers, will provide accurate recordings of SPL. Further, acoustic finite element analysis and boundary element analysis can also be carried out to verify the results. This chapter deals with only one-layer enclosure, mathematical model for two-layered like foam-panel and multilayered species of enclosure can be developed. Other analysis techniques like transfer matrix method can also be employed.

References

1. R. F. Barron, *"Industrial Noise Control and Acoustics,"* Marcel Dekker, Inc., New York, 2001.
2. L. L. Beranek, I. L. Ver, *"Noise and Vibration Control Engineering,"* John Wiley and Sons, Inc., Hoboken, NJ, 2005.
3. D. Bies, C. Hansen, *"Engineering Noise Control,"* Spon Press, Taylor & Francis Group, London, 2009.
4. R. S. Jackson, "The performance of acoustic hoods at low frequencies," *Acta Acustica United with Acustica*, 12(3), 139–152, 1962.
5. R. S. Jackson, "Some aspects of the performance of acoustic hoods," *Journal of Sound and Vibration*, 3(1), 82–94, 1966.
6. D. J. Oldham, S. N. Hillarby, "The acoustical performance of small close fitting enclosure, part 1: Theoretical models," *Journal of Sound and Vibration*, 150(2), 261–281, 1991.
7. D. J. Oldham, S. N. Hillarby, "The acoustical performance of small close fitting enclosure, part 2: Experimental investigation," *Journal of Sound and Vibration*, 150(2), 283–300, 1991.
8. J. E. Blanks, "Optimal design of an enclosure for a portable generator," Thesis, Virginia Polytechnic Institute and State University, 1997.
9. M. C. Chiu, "Optimal design on one-layer close-fitting acoustical hoods using a simulated annealing method," *Journal of Marine Science and Technology*, 22(2), 211–217, 2014.
10. R. V. Rao, "Jaya: A simple and new optimization algorithm for constrained and unconstrained optimization problems," *International Journal of Industrial Engineering Computations*, 7, 19–34, 2016.
11. A. Khachane, V. Jatti, "Multi-objective optimization of EDM process parameters using Jaya algorithm combined with grey relational analysis," *AEOTIT*, Surat, August 3–5, 2018.
12. BIS. "Methods of measurement of airborne noise emitted by rotating electrical machinery, part-1 and part-2," 12998, 1991.
13. N. H. Bhingare, S. Prakash, V. S. Jatti, "A review on natural and waste material composite as acoustic material", *Polymer Testing*, 80, 106142, 2019.
14. H. Rammal, J. Lavrentjev, "Acoustic study of novel eco-friendly material for vehicle NVH applications", *Materials Today Proceedings*, 28(4), 2331–2337, 2020.

3

A Metaheuristic Scheme for Secure Control of Cyber-Physical Systems

Tua A. Tamba

Parahyangan Catholic University

CONTENTS

3.1 Introduction

Cyber-physical system (CPS) has now become an important framework, which underlies the design and synthesis of modern control systems. Through the utilization of recent advances and development in sensing, computation and communication technologies, the CPS framework combines the physical (e.g. plant, sensors, actuators) and cyber (e.g. computational engine, communication networks) elements of the control system in a seamless manner and thereby provides an integrated abstraction, modeling and design techniques for the whole interacting components of the control system. Today, the CPS framework has been utilized in various real-life applications, ranging from simple household automation process to large-scale and complex industrial control system architecture [1,26,27].

An important characteristic of the CPS is its strong reliance on the used cyber elements for performing fast and real-time control computation and data transmission tasks.

Despite their important roles, these cyber elements have, however, also been recognized as a prominent source of vulnerability in CPS implementations [5,21]. As reported in a number of literature works, different types of cyber attacks which target CPSs that are responsible for managing a nation's critical infrastructure have occurred more frequently in recent years (see e.g. [6,9,13,20,21,30,40,41] among others). Due to the strong couplings and interactions between the physical and cyber components within the CPS, any fault or problem that occurs in either component will cause direct degrading impacts on the performance of, and the services provided by, the system as a whole. With these realizations, various issues related to the security of CPS in the face of possible cyber attacks from malicious adversaries have recently became among the most active research topics, which drew significant attention from researchers in computer science and engineering fields.

From the viewpoint of computer and communication network systems, it is generally well understood that vulnerabilities of a deployed cyber security system are unavoidable and therefore, rather than trying to alleviate them once and for all, one can only attempt to minimize the risk they (may) cause. In this regard, the majority of the currently existing cyber security methods/strategies usually rely on what can be referred to as a *static defense* mechanism (SDM). In an SDM, heavily secured perimeter firewalls and intrusion detection systems are deployed to minimize the risk of being attacked/compromised by malicious adversaries. However, it is now widely known that such an SDM often fails to protect a cyber system due to the presence of the so-called information asymmetry on the defender side during the attacker-defender interaction. More specifically, during such an interaction, an intelligent attacker may deploy customized backdoors to scan and collect information about the defender's cyber structure and vulnerabilities. Since the defender uses the SDM whereby its cyber structures/configurations remain the same and do not change over time, given a sufficiently enough time duration, the attacker will eventually learn the defender's cyber vulnerabilities and then exploit them to effectively deliver the attacks or compromise the system. The SDM-based defender, on the other hand, left only with the option of using the installed firewall or intrusion detection system to process a huge number of known or unknown attacks and perform complex analysis to protect itself. It is clear the attacker has an advantage over the defender in such an interaction, and thus, better alternatives of defense schemes are needed to ensure a CPS's cyber system security [5,9].

Recently, an alternative strategy called *moving target defense* (MTD) has been proposed to overcome the limitations of the SDM [11,23,33,43]. Generally speaking, the basic idea in MTD strategy is to vary a cyber system's actual appearance (also known as *surface*) by creating a moving target that will produce a time-varying service availability under different system configurations. Such a change in appearance is intended towards blocking adversaries from tracking the system configuration to carry out the attacks, or at least force them to spend significant amount of time and resources to perform the attacks. In this regard, the MTD strategy balances the information availability constraints among actors in the attacker-defender interaction by imposing similar information asymmetry on the attacker side.

As discussed in various reports and literature works, the MTD strategies can outperform and overcome the aforementioned fundamental limitations of the currently used SDM and have the potential to be a future choice of cyber security design framework and deployment platform [3,4,12,28,39]. Over the last few years, issues related to the study and development of MTD strategies have also drawn considerable attentions from researchers, but mainly centered to and discussed extensively by cyber systems (computer/communication systems) communities. Considering the fact that the physical and cyber components in a CPS are tightly coupled through especially the communication and controls functions, it is then of interest to find ways in which the security strategy of the used communication system can be implemented at the control systems design level.

This chapter aims to examine a control design framework that supports the MTD-based secure cyber systems implementation in a CPS. More specifically, this chapter further develops a recently proposed MTD control framework in Ref. [24] through the examination of its implementation under event-triggered (ET) scheduling of control actions. We point out that the basic idea of the MTD control framework as proposed in Ref. [24] is that of using a family of switched stabilizing controllers [19] (instead of a single controller) on the considered plant to obfuscate the system configuration/surface appearance as observed by adversaries. In this paper, we further develop such an idea by (i) introducing an ET control update strategy and (ii) implementing particle swarm optimization (PSO)-based controller tuning [18,29,31,36] to the analysis and controller synthesis of the closed-loop CPS. On the one hand, the purpose of adopting the ET control update scheme is to reduce the computational and communication loads that are required by the CPS to perform its function [38]. On the other hand, the motivation for using the PSO-based control tuning is to achieve better control performance for implementing the proposed MTD scheme [37]. Under the proposed MTD scheme, we first derive a condition on the switching frequency of the switched controllers that will guarantee the asymptotic stability of the resulting closed-loop CPS. We then propose the use of PSO-based tuning method to optimize the performance of each controller. Simulation example is finally presented to illustrate the effectiveness of the proposed MTD scheme.

This chapter is structured as follows. Section 3.2 describes the setup of the system and then formulates the considered secure control problem. This section also briefly recalls the basic idea of the MTD control framework that was developed in Ref. [24]. Section 3.3 presents the stability analysis of the considered system model in the presence of actuator intrusion. An actuator intrusion detection scheme based on the boundedness characteristics of the closed-loop system trajectories is presented in Section 3.4.1. Section 3.5 then proposes a PSO-based control tuning approach for constructing the set of optimal candidate actuating modes in the proposed MTD scheme. Simulation example of the proposed MTD scheme is presented in Section 3.6. Section 3.7 concludes this chapter with remarks and discussions.

3.2 Setup and Preliminaries

3.2.1 System Description

Consider the trajectories $x(t)$ of a CPS which for time $t \geq 0$ satisfy a linear time-invariant (LTI) system model of the form:

$$\dot{x}(t) = Ax(t) + Bu(t),$$

$$= Ax(t) + b_i u_i(t), \qquad x(0) = x_0, \tag{3.1}$$

where $x \in \mathbb{R}^n$ and $u \in \mathbb{R}^m$ are the state and input vectors, respectively, $A \in \mathbb{R}^{n \times n}$ is the state matrix, and $B \in \mathbb{R}^{n \times m}$ is the input matrix whose ith column b_i corresponds to the ith control input $u_i(t)$ on the ith actuator. It is assumed that each element of the actuators of (3.1) is subject to potential attacks from adversaries. In this regard, (3.1) may be rewritten as

$$\dot{x}(t) = Ax(t) + B\tilde{u}(t), \tag{3.2}$$

where $\tilde{u}(t)$ denotes an input signal that is being emulated by a time-varying attack function $\gamma(t)$ of the form:

$$\tilde{u}(t) := \gamma(t)u(t) = \left(\operatorname{diag}\{\gamma_{ii}(t)\}_{i=1}^m\right)u(t). \tag{3.3}$$

Assumption 3.1 summarizes the assumed properties of the attack function $\gamma(t)$.

Assumption 3.1

For any closed time interval $[t_1, t_2]$ such that $0 \le t_1 \le t_2$, then

1. $\gamma(t)$ is locally integrable
2. $\operatorname{supp}(\gamma(t)) < m$.

Note that the second condition in Assumption 3.1 essentially states that the attacker cannot simultaneously compromise all actuators at the same time. It can also be observed that $\gamma_{ii} = 1$ in Equation (3.3) implies that the control input is not being compromised.

Let us consider the set $\mathcal{B} := \{b_i, i = 1, \dots, m\}$ of m actuators of (3.1), which forms a set $\mathcal{P}(\mathcal{B}) = 2^{\mathcal{B}}$ of possible actuators combinations in the system. Then, each input matrix $B_j(j = 1, \dots, 2^m)$ in (3.1) with column vector elements $b_i(i = 1, \dots, m)$ as in (3.1) is an element of the power set $\mathcal{P}(\mathcal{B})$. Our focus in this chapter will be on those matrices B_j, which belong to a subset $\mathcal{B}_c$ of $\mathcal{B}$ defined as follows:

$$\mathcal{B}_c = \left\{B_j \in 2^{\mathcal{B}} : \operatorname{rank}\left(\left[B_j \ AB_j \cdots A^{n-1}B_j\right]\right) = n\right\}. \tag{3.4}$$

Note that $\mathcal{B}_c$ in (3.4) is the set of *candidate actuating modes*, the element of which corresponds to a pair (A, B_j) in (3.1) that is fully controllable.

Now, for a particular set of actuating modes $B_i \in \mathcal{B}_c$ with $i = 1, \dots, m$, the closed-loop system (3.1) for all time $t \ge 0$ is given as

$$\dot{x}(t) = Ax(t) + B_i u_i(t), \quad x(0) = x_0. \tag{3.5}$$

This chapter considers the case where each control signal $u_i(t)$ in Equation (3.5) is computed based on the linear quadratic regulator (LQR) scheme and is implemented through an ET updating scheme. Thus, on the one hand, the control signal of each actuating mode is chosen to be that which minimizes the following infinite time cost function J_i:

$$J_i = \min_{u_i} \int_0^\infty \left(x^T(t) Q_i x(t) + \rho_i u_i^T(t) R_i u_i(t) \right) d\tau, \quad i = 1, \ldots, |\mathcal{B}_c| \tag{3.6}$$

where $Q_i \geq 0$, $R_i > 0$, and $\rho_i > 0$. On the other hand, the update sequence of each control signal is triggered based on two functions, namely, (i) an *event function* $\xi : \mathbb{R}^n \times \mathbb{R}^n \mapsto \mathbb{R}$, which determines if the control input should be updated (if $\xi \leq 0$) or not (if $\xi > 0$), and (ii) a *feedback control law* $u_i(t) : \mathbb{R}^n \mapsto \mathbb{R}^m$ to be applied on each actuating mode. Let t_k denote the time instant (event) of the kth control update and define the control signal in (3.6) to be $u_i(t)$ for all $t \in [t_k, t_{k+1})$. Then for the case where each pair (A, B_i) in (3.5) is stabilizable, the work in Ref. [31] showed that an optimal control signal $u_i(t)$ for each actuating mode of the form

$$\hat{u}_i(t) = -K_i x(t_k) = -2\rho R_i^{-1} B_i^T P_i x(t_k), \tag{3.7}$$

can be obtained under the following ET control update rule:

$$\begin{aligned} \xi\left(x(t), x(t_k)\right) &= (\nu - 1) x^T(t) \left[A^T P_i + P_i A \right] x(t) \\ &\quad - 4\rho_i x^T(t) P_i B_i R_i^{-1} B_i^T P_i \left[\nu x(t) - x(t_k) \right], \end{aligned} \tag{3.8}$$

where $\nu \in (0,1)$ is an ET parameter and each $P_i \geq 0$ is a symmetric matrix that satisfies the *algebraic Riccati equation* (ARE) of the form:

$$A^T P_i + P_i A - 4\rho P_i B_i R_i^{-1} B_i^T P_i + Q_i = 0. \tag{3.9}$$

Then $\forall t \in [t_k, t_{k+1}[$, the closed-loop system (3.5) for each actuating mode i satisfies

$$\dot{x}(t) = Ax(t) - B_i K_i x(t_k), \quad x(0) = x_0, \tag{3.10}$$

with a sequence of control update instances that is defined by an ET rule of the form:

$$t_0 := 0, \quad t_{k+1} = \left\{ t > t_k \,|\, \xi\left(x(t), x(t_k)\right) \geq 0 \right\}. \tag{3.11}$$

It can be seen that the resulting closed-loop system in (3.10) and (3.11) defines an event-triggered switched linear system (ET-SLS). In the remainder of this chapter, we use $\mathcal{K}$ with $|\mathcal{K}| = |\mathcal{B}_c|$ to denote the set of optimal feedback control gains of the form (3.7), which correspond to the set of actuating modes in (3.4).

3.2.2 A Moving Target Defense Scheme Using Switching Controllers

The ET-SLS in (3.10)–(3.11) can be used to develop an MTD strategy to detect and mitigate potential cyber attacks on the system from malicious external adversaries. More specifically, under a choice of switching rule which orchestrates a sequence of activation of each stabilizing controller in $\mathcal{K}$, the resulting closed-loop system will produce surface randomizations to obfuscate potential attackers.

To this end, consider a switching signal $\sigma(t): [0,\infty) \to \mathcal{I}$, where $\mathcal{I} = \{1,\ldots,|\mathcal{K}|\}$ that is right-continuous and piecewise constant function. Now set $\sigma(t) = i$ with $i \in \mathcal{I}$. It is clear that $\sigma(t)$ essentially defines a mapping from t to the set of candidate actuating modes of the SLS in (3.10). In this regard, the ET-SLS in (3.10) and (3.11) may also be written for all $t \in [t_k, t_{k+1}[$ as

$$\dot{x}(t) = Ax(t) - B_{\sigma(t)}K_{\sigma(t)}x(t_k). \tag{3.12}$$

The work in Ref. [24] showed that the SLS structure of the form (3.12) provides a mean to perform the MTD strategy. Such a strategy essentially boils down to the design of a switching signal $\sigma(t)$, which simultaneously ensures the minimization of the LQR cost function (3.6) and the maximization of an information entropy function, which captures the unpredictability of the closed-loop system appearance. This design requirement essentially produces a trade-off between system optimality (captured by the optimal cost function J_i^*) and unpredictability (captured by the system entropy $\mathcal{H}(p) = -p^T \log(p)$ under a probability simplex p that describes the probability that each controller gain K_i is active). Such a trade-off can be represented by a probability function $\mathcal{P}_i$ of the form Ref. [24, Theorem 1]

$$P_i = \exp\left[-\frac{J_i^*}{\varepsilon} - 1 - \varepsilon \log\left(e^{-1}\Sigma_{i=1}^{|K|} e^{\frac{J_i^*}{\varepsilon}} \right) \right], \tag{3.13}$$

which measures the likelihood of the ith controller gain K_i to be activated. The parameter $\epsilon > 0$ in (3.13) denotes a weighting factor of the unpredictability/entropy of the resulting ET-SLS.

3.2.3 Problem Formulation

Motivated by the aforementioned MTD control framework as proposed in Ref. [24], our objective in this chapter is to develop a secure control strategy for a CPS under the potential threat of cyber attacks from external adversaries. The proposed strategy will include stability analysis, intrusion detection mechanism design, and stabilizing switching controller synthesis for the resulting closed-loop ET-SLS.

To achieve such an objective, let us first consider an error signal $e(t) = x(t_k) - x(t)$ for all $t \in [t_k, t_{k+1}[$, which denotes the difference between the values of the system states at time t and at the last control update instant t_k. Under the definition of $e(t)$, the dynamics of the ET-SLS in (3.12) for all $t \in [t_k, t_{k+1}[$ may also be written as

$$\dot{x}(t) = \mathcal{A}_{\sigma(t)}x(t) - B_{\sigma(t)}K_{\sigma(t)}e(t), \tag{3.14}$$

where $\mathcal{A}_{\sigma(t)} = A - B_{\sigma(t)}K_{\sigma(t)}$. This chapter assumes that the switching signal $\sigma(t)$ in (3.14) is *regular* (in terms of switching number) in the sense of Assumption 3.2.

Assumption 3.2

Given the ET-SLS in (3.14) and a time interval (t_1, t_2) with $t_2 \geq t_1 \geq 0$, the number of switches $N_\sigma(t_1, t_2)$ of $\sigma(t)$ over (t_1, t_2) satisfies [19]

$$N_\sigma(t_1, t_2) \leq N_0 + \frac{t_2 - t_1}{\tau_D}, \tag{3.15}$$

where $N_0 > 0$ and $\tau_D > 0$ are the *chatter bound* and the *average dwell time* of $\sigma(t)$, respectively.

Under the assumed property of the switching signal $\sigma(t)$, the stability analysis of the closed-loop ET-SLS (3.14) in the presence of cyber attacks will be examined using the notion of exponential stability as stated in Definition 3.1.

Definition 3.1: Exponential Stability

System (3.14) is said to be *globally exponentially stable* (GES) under a switching signal $\sigma(t)$ if there exist constants c_1, $c_2 > 0$ such that all of the system solutions $x(t)$ for any initial condition $x(0) = x_0$ satisfy

$$\| x(t) \| \leq c_1 e^{-c_2(t-t_0)} \| x_0 \|, \quad \forall t \geq 0. \tag{3.16}$$

3.3 System Analysis in the Absence of Cyber Attack

In this section of this chapter, we analyze the stability of the ET-SLS (3.14) in the absence of cyber attack/intrusion. More specifically, this section derives the conditions for the stability of the ET-SLS (3.14) when no attack intrudes the system actuators.

To derive the conditions for stability of the ET-SLS (3.14), we first examine the boundedness property of the inter-event time of the ET-SLS (3.14). Such a boundedness property is characterized in Lemma 3.1.

Lemma 3.1

Consider the ET-SLS (3.14) with ET rule (3.11) in the absence of actuator attack. Then for any update instant t_k and all time $t \in [t_k, t_{k+1}[$, the inter-event time $\Delta t_k^{k+1} := t_{k+1} - t_k$ is bounded from below by a strictly positive constant of the form:

$$\Delta t_k^{k+1} = \frac{\ln(1 + \kappa_1 \kappa_2)}{\gamma_1} \tag{3.17}$$

where $\kappa_1 = \omega_1 / (\omega_1 + \omega_2)$, $\kappa_2 = \theta_1 / (\theta_1 + \theta_2)$, $\omega_1 = \| 4\rho P_i B_i R_i^{-1} B_i^T P_i \|$, $\omega_2 = \| (v-1) Q_i \|$, $\theta_1 = \| A \|$, a $\theta_2 = \max_{i \in \mathcal{I}} \| B_i K_i \|$.

Proof 3.1

Using the definition of the error signal $e(t) = x(t_k) - x(t)$, its derivative with respect to time $t \in [t_k, t_{k+1}[$ is given as follows:

$$\dot{e}(t) = -\dot{x}(t) = -\left(A_{\sigma(t)}x(t) - B_{\sigma(t)}K_{\sigma(t)}e(t)\right)$$

$$= -\left(A - B_{\sigma(t)}K_{\sigma(t)}\right)x(t) + B_{\sigma(t)}K_{\sigma(t)}e(t)$$

$$= -Ax(t) + B_{\sigma(t)}K_{\sigma(t)}\left[x(t) + e(t)\right]$$

$$= -Ax(t) + B_{\sigma(t)}K_{\sigma(t)}x(t_k) \tag{3.18}$$

The norm of the time derivative of $e(t)$ thus satisfies the following inequality:

$$\frac{d}{dt} \| e(t) \| \leq \| \dot{e}(t) \| = \| -Ax(t) + B_{\sigma(t)}K_{\sigma(t)}x(t_k) \|$$

$$\leq \| Ax(t) \| + \| B_{\sigma(t)}K_{\sigma(t)}x(t_k) \|$$

$$\leq \| A \| \| x(t) \| + \| B_{\sigma(t)}K_{\sigma(t)} \| \| x(t_k) \|$$

$$\leq \theta_1 \| x(t) \| + \theta_2 \| x(t_k) \| \tag{3.19}$$

Since the definition $x(t) = x(t_k) - e(t)$ implies $\| x(t) \| \leq \| e(t) \| + \| x(t_k) \|$, the above norm may then be rewritten as follows:

$$\frac{d}{dt} \| e(t) \| \leq \theta_1 \left(\| e(t) \| + \| x(t_k) \| \right) + \theta_2 \| x(t_k) \|$$

$$\leq \theta_1 \| e(t) \| + (\theta_1 + \theta_2) \| x(t_k) \| \tag{3.20}$$

As a result, the evolution of the normed error function $\| e(t) \|$ with initial condition $e(t_k) = 0$ over the time interval $[t_k, t_{k+1}[$ can be bounded as follows:

$$\| e(t) \| \leq e^{\theta_1(t-t_k)} \| e(t_k) \| + \int_{t_k}^{t} e^{\theta_1(t-\tau)}(\theta_1 + \theta_2) \| x(t_k) \| d\tau \tag{3.21}$$

On the other hand, note that the definition $e(t) = x(t_k) - x(t)$ also implies that the satisfaction of the ET rule (3.11) of the form

$$\omega_1 \| e(t) \| \leq \omega_2 \| x(t) \| \tag{3.22}$$

can be guaranteed to hold by the following sufficient condition:

$$\| e(t) \| \leq \frac{\omega_1}{\omega_1 + \omega_2} \| x(t_k) \| \tag{3.23}$$

Hence, in the case that $x(t_k) \neq 0$, then (3.21) and (3.23) suggest that the following must hold before an event is generated:

$$\frac{\omega_1}{\omega_1 + \omega_2} \| x(t_k) \| = \int_{t_k}^{t} e^{\gamma_1(t-\tau)} (\gamma_1 + \gamma_2) \| x(t_k) \| d\tau$$

$$= (\theta_1 + \theta_2) \| x(t_k) \| \int_{t_k}^{t} e^{\theta_1(t-\tau)} d\tau$$

$$= (\theta_1 + \theta_2) \| x(t_k) \| \left[\frac{1}{\theta_1} \left(e^{\theta_1(t-\tau)} \Big|_{\tau=t}^{\tau=t_k} \right) \right]$$

$$= \frac{(\theta_1 + \theta_2)}{\theta_1} \left(e^{\theta_1(t-t_k)} - 1 \right) \| x(t_k) \| \tag{3.24}$$

From the comparison of the coefficients of both sides of the above equality, we have

$$\frac{\omega_1}{\omega_1 + \omega_2} = \frac{(\theta_1 + \theta_2)}{\theta_1} \left(e^{\theta_1(t-t_k)} - 1 \right) \tag{3.25}$$

which through rearrangement and simplification can be rewritten as:

$$e^{\gamma_1(t-t_k)} = 1 + \frac{\theta_1 \omega_1}{(\theta_1 + \theta_2)(\omega_1 + \omega_2)} \tag{3.26}$$

$$= 1 + \kappa_1 \kappa_2$$

where κ_1 and κ_2 are as defined in the lemma. Taking the logarithm of (3.26) and using the definition of the inter-event time $\Delta t_k^{k+1} = t_{k+1} - t_k$, we then have that

$$\gamma_1 \Delta t_k^{k+1} = \ln\left(1 + \kappa_1 \kappa_2\right) \tag{3.27}$$

from which we conclude (3.17). The proof is complete.

Note that Lemma 3.1 basically shows that the inter-event times that result from the use of the ET rule (3.11) are bounded from below by a strictly positive constant whose value is defined in (3.17). The existence of such a lower bound essentially implies that the used ET scheme will control update instances without producing Zeno-type behaviors (i.e. for a finite time interval, the used ET rule will only generate a finite number of control update instances).

Having shown that the used ET rule excludes the Zeno behavior from the control update instances, we now derive a condition on the switching signal $\sigma(t)$ that will guarantee the ET-SLS (3.14) to be exponentially stable.

Theorem 3.1

Consider the ET-SLS in (3.14) when no attacks are present. For each $i \in \mathcal{I}$, let $\sigma(t) = i$ be the switching signal, and K_i be the corresponding feedback control gain of the form (3.7). Then for any $\sigma(t)$ that satisfies (3.15), the ET-SLS is GES if the average dwell time τ_D of the switchings satisfies the following lower bound:

$$\tau_D > \frac{\ln \alpha}{\beta_i} := \frac{\ln \left[\max_{(i,\,i') \in \mathcal{I}} \left(\bar{\lambda}(P_i) / \underline{\lambda}(P_{i'}) \right) \right]}{v \max_{i \in \mathcal{I}} \left(\bar{\lambda}(Q_i) / \underline{\lambda}(P_i) \right)} \tag{3.28}$$

Proof 3.2

For each active mode $\sigma(t) = i$ where $i \in \mathcal{I}$, let us consider the following candidate Lyapunov function:

$$V_i(x) = x^T(t) P_i x(t). \tag{3.29}$$

Note for any $i, i' \in \mathcal{I}$ that (3.29) implies

$$\underline{\lambda}(P_i) \parallel x(t) \parallel^2 \le V_i(x) = x^T P_i x \le \bar{\lambda}(P_i) \parallel x(t) \parallel^2, \tag{3.30}$$

$$\underline{\lambda}(P_{i'}) \parallel x(t) \parallel^2 \le V_{i'}(x) = x^T P_{i'} x \le \bar{\lambda}(P_{i'}) \parallel x(t) \parallel^2. \tag{3.31}$$

Combining (3.30) and (3.31), we may write that

$$V_i(x) \le \bar{\lambda}(P_i) \parallel x(t) \parallel^2 \le \bar{\lambda}(P_i) \left(\frac{V_{i'}(x)}{\underline{\lambda}(P_{i'})} \right), \tag{3.32}$$

such that the following holds for any pair $(i, i') \in \mathcal{I}$

$$V_i(x) \le \alpha V_{i'}(x) \tag{3.33}$$

where $\alpha = \max_{(i,\,i') \in \mathcal{I}} \left(\bar{\lambda}(P_i) / \underline{\lambda}(P_{i'}) \right)$. In what follows next, we examine the ET-SLS dynamics both for a particular mode and for different modes.

The directional derivative of (3.29) along the trajectories of (3.14) is given by

$$\dot{V}_i(x) = \dot{x}^T(t) P_i x(t) + x^T(t) P_i \dot{x}(t)$$

$$= \left[A_i x(t) - B_i K_i e(t) \right]^T P_i x(t) + x^T(t) P_i \left[A_i x(t) - B_i K_i e(t) \right]$$

$$= x^T(t) \left[A_i^T P_i + P_i A_i \right] x(t) - 2 x^T(t) P_i B_i K_i e(t)$$

Using the definition of $A_i = A - B_i K_i$, where the gain K_i is defined as in (3.7), then

$$\dot{V}_i = x^T \left[\left(A^T - 2\rho P_i B_i R_i^{-1} B_i^T \right) P_i + P_i \left(A - 2\rho B_i R_i^{-1} B_i^T P_i \right) \right] x(t)$$

$$- 2 x^T(t) P_i B_i \left(2\rho R_i^{-1} B_i^T P_i \right) e(t) \tag{3.34}$$

$$\dot{V}_i = x^T \left[A^T P_i + P_i A - 4\rho P_i B_i R_i^{-1} B_i^T P_i \right] x(t) - 4\rho x^T(t) P_i B_i R_i^{-1} B_i^T P_i e(t) \qquad (3.35)$$

$$\dot{V}_i = -x^T(t) Q_i x(t) - 4\rho x^T(t) P_i B_i R_i^{-1} B_i^T P_i e(t) \qquad (3.36)$$

Note that (3.36) was obtained from (3.35) using the ARE in (3.9).

We now evaluate (3.36) with regard to the occurrence of control update events. For the solution of the ET-SLS (3.14) within $[t_s, t_{s+1}[$ and the inter-event period $[t_k, t_{k+1}[$ of the ET rule in (3.11), let us assume the ET-SLS switches at time instant t_s from mode i to i' with $(i, i') \in \mathcal{I}$. Then, the following two cases may occur with regard to the dynamics of the ET-SLS:

- **Case 1**: This case corresponds to an instance where $t_k \le t_s$ and $t_{k+1} \ge t_{s+1}$ such that no control update event is triggered within the switching period $[t_s, t_{s+1}[$. According to (3.8) and (3.11), it then holds for $t \in [t_k, t_{k+1}[$ that:

$$(\nu - 1) x^T(t) \Phi_i x(t) = 4\rho x^T(t) P_i B_i R_i^{-1} B_i^T P_i \left[\nu x(t) - x(t_k) \right]$$

with $\Phi_i := A^T P_i + P_i A$. Using $e(t) = x(t_k) - x(t)$, the above can be rewritten as

$$(\nu - 1) x^T(t) \Phi_i x(t) = 4\rho x^T(t) P_i B_i R_i^{-1} B_i^T P_i \left[(\nu - 1) x(t) - e(t) \right]$$

which can be combined with (3.9) to obtain

$$4\rho_i x^T(t) P_i B_i R_i^{-1} B_i^T P_i = (\nu - 1) x^T(t) Q_i x(t). \qquad (3.37)$$

Substitution of (3.37) into (3.36) thus gives

$$\dot{V}_i(x) = -x^T(t) Q_i x(t) - (\nu - 1) x^T(t) Q_i x(t)$$

$$= -\nu x^T(t) Q_i x(t)$$

$$\le -\nu \bar{\lambda}(Q_i) \| x(t) \|^2 \qquad (3.38)$$

which when combined with (3.30) implies

$$\dot{V}_i(x) \le -\nu \bar{\lambda}(Q_i) \left(\frac{V_i(x)}{\underline{\lambda}(P_i)} \right) = -\nu \frac{\bar{\lambda}(Q_i)}{\underline{\lambda}(P_i)} V_i(x) \qquad (3.39)$$

Now for any $i \in \mathcal{I}$, the inequality in (3.39) will be guaranteed to hold if

$$\dot{V}_i(x) \le -\beta_i V_i(x) \qquad (3.40)$$

where $\beta_i = \nu \max_{i \in \mathcal{I}} \left(\bar{\lambda}(Q_i) / \underline{\lambda}(P_i) \right) > 0$.

- **Case 2**: This case corresponds to an instance where the interswitching duration $[t_s, t_{s+1}[$ contains (possibly many) occurrences of control update events (e.g.
  $t_k \leq t_s < t_{k+1} < t_{k+2} < \cdots < t_{k+q} \leq t_{s+1}$ for q numbers of update events within $[t_s, t_{s+1}[$).
  In this case, (3.37) remains valid on each inter-event subinterval and therefore
  (3.40) also valid within the period $[t_s, t_{s+1}[$.

We may now use (3.40) to evaluate V_i for the above two cases. For Case 1, we have

$$V_i\left(x(t)\right) \leq e^{-\beta_i(t-t_s)} V_{i(t_s)}\left(x(t_s)\right) \tag{3.41}$$

For Case 2, the solution of $V_i(x)$ on each subinterval satisfies

$$V_i\left(x(t)\right) \leq \begin{cases} e^{-\beta_i(t-t_s)} V_{i(t_s)}(x(t_s)), t \in [t_s, t_{k+1}[\\ e^{-\beta_i(t-t_{k+1})} V_{i(t_{k+1})}(x(t_{k+1})), t \in [t_{k+1}, t_{k+2}[\\ e^{-\beta_i(t-t_{k+q})} V_{i(t_{k+q})}(x(t_{k+q})), t \in [t_{k+q}, t_{k+1}[\end{cases} \tag{3.42}$$

Since the error signal $e(t)$ is bounded and piecewise continuous within the switching interval $[t_s, t_{s+1}[$ (possibly has a finite number of jump discontinuities) and
$\sigma(t_s) = \sigma(t_{k+1}) = \cdots = \sigma(t_{k+q})$, (3.42) is essentially equivalent to (3.41).

Next, let us examine (3.29) when switchings between different modes occur. Consider
the operational duration $[0, t]$ of the ET-SLS (3.14), and define within this duration the
sequence of mode switching instances $0 := t_0 \leq t_1 \leq t_2 \leq \ldots \leq t_s = t_{N_\sigma(0,t)} \leq t$, which satisfies
(3.15). Then based on (3.33), (3.40) can be rewritten as

$$V_i\left(x(t)\right) \leq \alpha e^{-\beta_i(t-t_s)} V_{i\left(t_s^-\right)}\left(x\left(t_s^-\right)\right)$$

$$\leq \alpha e^{-\beta_i(t-t_s)} e^{-\beta(t_s - t_{s-1})} V_{i(t_{s-1})}\left(x(t_{s-1})\right)$$

$$\leq \alpha^2 e^{-\beta_i(t-t_{s-1})} V_{i\left(t_{s-1}^-\right)}\left(x\left(t_{s-1}^-\right)\right)$$

$$\leq \cdots$$

$$\leq \alpha^{N_\sigma(t_0, t)} e^{-\beta_i(t-t_0)} V_{i(t_0)}\left(x(t_0)\right) := e^{\left(-\beta_i(t-t_0) + \ln \alpha^{N_\sigma(t_0, t)}\right)} V_{i(t_0)}\left(x(t_0)\right)$$

$$\leq e^{\left(-\beta_i(t-t_0) + \left(N_0 + \frac{t-t_0}{\tau_D}\right)\ln \alpha\right)} V_{i(t_0)}\left(x(t_0)\right)$$

$$\leq e^{(N_0 \ln \alpha)} e^{-\left(\beta - \frac{\ln \alpha}{\tau_D}\right)(t-t_0)} V_{i(t_0)}\left(x(t_0)\right)$$

Using the inequality in (3.30), we may infer the following from the above inequality:

$$\| x(t) \| \leq \sqrt{\frac{\bar{\lambda}(P_i)}{\underline{\lambda}(P_i)}} e^{\left(\frac{N_0 \ln \alpha}{2}\right)} e^{-\frac{1}{2}\left(\beta_i - \frac{\ln \alpha}{\tau_D}\right)(t-t_0)} \| x(t_0) \|. \tag{3.43}$$

Hence, if the average dwell time τ_D of (3.14) satisfies condition (3.28), then the ET-SLS is guaranteed to be GES in accordance with Definition 3.1 with constants $c_1 = \sqrt{\frac{\bar{\lambda}(P_i)}{\underline{\lambda}(P_i)}} e^{\left(\frac{N_0 \ln \alpha}{2}\right)}$ and $c_2 = \frac{1}{2}\left(\beta_i - \frac{\ln \alpha}{\tau_D}\right)$. The proof is complete.

3.4 System Analysis in the Presence of Cyber Attack

In this section, we analyze the stability of the closed-loop ET-SLS in (3.14) in the presence of actuator attack/intrusion. Prior to presenting such a stability analysis, below we first present a scheme for detecting the presence of such an intrusion. More specifically, we first characterize in this section of this chapter an upper bound of the ET-SLS (3.14) solutions when cyber intrusion is present on the system actuator. Subsequently, we also characterize a bound on the deviation of the compromised system's trajectories from that of the optimal trajectory when no intrusion is present. Using such bounds, we then propose a scheme for detecting the presence of cyber attack/intrusion on the actuators of the ET-SLS in (3.14). Finally, the stability analysis of the closed-loop ET-SLS is presented.

3.4.1 A Detection Scheme for the Presence of Actuator Intrusion

To derive the bound of the system trajectories in the presence of cyber intrusion, let us use $\hat{x}(t)$ and $\tilde{x}(t)$ to denote the system trajectories in the absence and presence of intrusion, respectively. Fix a time period $[t_0, t]$, and consider the dynamics of $\hat{x}(t)$ and $\tilde{x}(t)$ over this period. Then, the optimal trajectory $\hat{x}(t)$ is defined by the closed-loop system under the LQR control law (3.7) of the form:

$$\dot{\hat{x}}(t) = (A - B_i K_i)\hat{x}(t) - B_i K_i e(t), \quad x(t_0) = x_0 \tag{3.44}$$

The dynamics of the compromised trajectories $\tilde{x}(t)$ under the actuator attack parameter $\gamma(t)$, on the contrary, are given by

$$\dot{\tilde{x}}(t) = (A - B_i \gamma(t) K_i)\tilde{x}(t) - B_i \gamma(t) K_i e(t), \quad \tilde{x}(t_0) = x_0 \tag{3.45}$$

By methods of term completion and rearrangement, (3.45) can be rewritten as:

$$\dot{\tilde{x}}(t) = (A - B_i \gamma(t) K_i)\tilde{x}(t) + B_i K_i \tilde{x}(t) - B_i K_i \tilde{x}(t) - B_i \gamma(t) K_i e(t) + B_i K_i e(t) - B_i K_i e(t)$$

$$= (A - B_i K_i)\tilde{x}(t) - B_i K_i e(t) + B_i [I - \gamma(t)] K_i [\tilde{x}(t) + e(t)]$$

$$= A_i \tilde{x}(t) - B_i K_i e(t) + B_i [I - \gamma(t)] K_i [\tilde{x}(t) + e(t)] \tag{3.46}$$

where we have defined $\mathcal{A}_i = A - B_i K_i$. The compromised system trajectories $\tilde{x}(t)$ may then be determined based on (3.46) as follows:

$$\tilde{x}(t) = e^{\mathcal{A}_i(t-t_0)} x_0 - \int_{t_0}^{t} e^{\mathcal{A}_i(t-\tau)} B_i K_i e(\tau) d\tau$$

$$+ \int_{t_0}^{t} e^{\mathcal{A}_i(t-\tau)} B_i \left[I - \gamma(\tau) \right] K_i \left[\tilde{x}(\tau) + e(\tau) \right] d\tau \tag{3.47}$$

The norm of $\tilde{x}(t)$ may then be computed as follows:

$$\| \tilde{x}(t) \| \leq \| e^{A_i(t-t_0)} \| \| x_0 \| - \int_{t_0}^{t} \| e^{A_i(t-\tau)} \| \| B_i \| \| K_i \| \| e(\tau) \| d\tau$$

$$+ \left(\int_{t_0}^{t} \| e^{A_i(t-\tau)} \| \| B_i \| \| I - \gamma(\tau) \| \| K_i \| \right) \left(\| \tilde{x}(\tau) + e(\tau) \| \right) d\tau$$

$$\leq \| e^{A_i(t-t_0)} \| \| x_0 \| + \int_{t_0}^{t} \| e^{A_i(t-\tau)} \| \| B_i \| \| I - \gamma(\tau) \| \| K_i \| \| \tilde{x}(\tau) + e(\tau) \| d\tau \tag{3.48}$$

Now notice that $\mathcal{A}_i = A - B_i K_i$ under the optimal control law (3.7) is a stable matrix and therefore has a bounded transition matrix. Let us thus define the following:

$$\xi_i = \max_t \| e^{\mathcal{A}_i(t-t_0)} \|, \text{ and } \mu_i(\tau) = \| e^{\mathcal{A}_i(t-\tau)} \| \| B_i \| \| K_i \|$$

Using such ξ_i and $\mu_i(\tau)$, we may then rewrite the bound of $\| \tilde{x}(t) \|$ in (3.48) as follows:

$$\tilde{x}(t) \leq \xi_i \| x_0 \| + \int_{t_0}^{t} \mu_i(\tau) \| I - \gamma(\tau) \| \left(\tilde{x}(\tau) + e(\tau) \right) d\tau$$

$$\leq \xi_i \| x_0 \| + \int_{t_0}^{t} \mu_i(\tau) \| I - \gamma(\tau) \| \left(1 + \frac{\omega_2}{\omega_1} \right) \| \tilde{x}(\tau) \| d\tau$$

$$\leq \xi_i \| x_0 \| + \frac{1}{\kappa_1} \int_{t_0}^{t} \mu_i(\tau) \| I - \gamma(\tau) \| \| \tilde{x}(\tau) \| d\tau$$

$$\leq \xi_i \| x_0 \| e^{\frac{1}{\kappa_1} \int_{t_0}^{t} \mu_i(\tau) \| I - \gamma(\tau) \| d\tau} \tag{3.49}$$

Next, let us examine the deviation $\varepsilon(t) = \tilde{x}(t) - \hat{x}(t)$ of the compromised trajectories (under the influence of attacks) from the optimal trajectories (in the absence of attacks). The evolution of such a deviation $\varepsilon(t)$ satisfies the following equation:

$$\dot{\varepsilon}(t) = \dot{\tilde{x}}(t) - \dot{\hat{x}}(t)$$

$$= \left[A - B_i \gamma(t) K_i \right] \tilde{x}(t) - B_i \gamma(t) K_i e(t) - \left[(A - B_i K_i) x(t) - B_i K_i e(t) \right]$$

$$= \mathcal{A}_i \varepsilon(t) + B_i \left[I - \gamma(t) \right] K_i \left[\tilde{x}(t) + e(t) \right] \tag{3.50}$$

Using the fact that $\hat{x}(t_0) = \tilde{x}(t_0) = x_0$ such that $\varepsilon(t_0) = 0$, the solution of $\varepsilon(t)$ as defined in (3.50) may then be determined as

$$\varepsilon(t) = \int_{t_0}^{t} e^{A_i(t-t_0)} B_i \left[I - \gamma(\tau) \right] K_i \left[\tilde{x}(\tau) + e(\tau) \right] d\tau \qquad (3.51)$$

The norm of $\varepsilon(t)$ may then be computed and satisfies the following inequality:

$$\| \varepsilon(t) \| \leq \int_{t_0}^{t} \mu_i(\tau) \| I - \gamma(\tau) \| \| \tilde{x}(\tau) + e(\tau) \| d\tau$$

$$\leq \frac{1}{\kappa_1} \int_{t_0}^{t} \mu_i(\tau) \| I - \gamma(\tau) \| \| \tilde{x}(\tau) \| d\tau$$

$$\leq \frac{1}{\kappa_1} \int_{t_0}^{t} \left[\left(\mu_i(\tau) \| I - \gamma(\tau) \| \right) \left(\xi_i \| x_0 \| e^{\frac{1}{\kappa_1} \int_{t_0}^{t} \mu_i(s) \| I - \gamma(s) \| ds} \right) \right] d\tau$$

$$\leq E(\gamma, \tau) \| x_0 \| \qquad (3.52)$$

where we have define the function $\mathcal{E}(\gamma, \tau)$ below

$$E(\gamma, \tau) = \frac{\xi_i}{\kappa_i} \int_{t_0}^{t} \mu_i(\tau) \| I - \gamma(\tau) \| e^{\frac{1}{\kappa_1} \int_{t_0}^{t} \mu_i(s) \| I - \gamma(s) \| ds} d\tau \qquad (3.53)$$

with a property that $\mathcal{E}(\gamma, \tau) \neq 0$ whenever $\gamma(t) \neq I$ for all $t \in [t_0, t_1]$.

Using the derived bounds on both the trajectories of the compromised ET-SLS and their deviation from the optimal trajectories, we now arrive at determining a scheme for detecting the presence of cyber intrusion on the actuators of the ET-SLS. The proposed scheme is essentially formulated in the form of a detection intrusion function, which can be used as a reference for determining the presence of intrusion on the system actuator. Theorem 3.2 formally defines such a function and describes its use in the proposed intrusion detection scheme.

Theorem 3.2

Consider the ET-SLS in (3.14) with the feedback control law in (3.7). For a finite time duration $\delta > 0$, define the intrusion detection function $\pi(t)$ below:

$$\pi(t) = V_i \left(\tilde{x}(t - \delta) \right) - V_i \left(\tilde{x}(t) \right) - \int_{t-\delta}^{t} \left(\tilde{x}(\tau) Q_i \tilde{x}(\tau) + \rho \hat{u}_i^T(\tau) R_i \hat{u}_i(\tau) \right) d\tau \qquad (3.54)$$

Then, the system is under cyber intrusion if and only if $\pi(t) \neq 0$. In particular, for any integrable injected intrusion signal, the loss of optimality on the system performance which occurs under such an intrusion is bounded from above in the following form:

$$\| \bar{\pi}(t) \| \leq \mathcal{B}(\gamma, \tau) \| x_0 \|^2, \tag{3.55}$$

where the function $\mathcal{B}(\gamma, \tau)$ for any $\hat{\beta}_i = \| e^{-\beta(t-t_0)/2} \|$ is given by

$$\mathcal{B}(\gamma, \tau) = \| P_i \| \| \mathrm{E}(\gamma, t) \|^2 + 2\hat{\beta}_i \| P_i \| \| \mathrm{E}(\gamma, t) \|$$

$$+ \int_{t-\delta}^{t} \left(\| \Omega \| \| \mathrm{E}(\gamma, \tau) \|^2 + 2\hat{\beta}_i \| \mathrm{E}(\gamma, \tau) \| \| \Omega \| \right) d\tau. \tag{3.56}$$

Proof 3.3

Consider a finite time duration $\delta > 0$. Then from (3.29), we may write

$$V_i\big(x(t-\delta)\big) = x^T(t-\delta) P_i x(t-\delta)$$

$$= \min_u \int_{t-\delta}^{t} \left(x^T(\tau) Q_i x(\tau) + \rho u_i(\tau) R u(\tau) \right) d\tau + V_i(t)$$

$$= \min_u \int_{t-\delta}^{t} \left(x^T(\tau) Q_i x(\tau) + \rho u_i(\tau) R u(\tau) \right) d\tau + x^T(t) P_i x(t) \tag{3.57}$$

Let $\hat{u}(t)$ be the (optimal) control signal in the absence of attack and $\tilde{u}(t)$ be the control signal of the compromised system under the influence of attack. Then, the relationship between the accumulated costs that result from using each control signal will satisfy

$$\int_{t-\delta}^{t} \left(x^T Q_i x + \rho \hat{u}_i R_i \hat{u}_i \right) d\tau = \min_{u_i} \int_{t-\delta}^{t} \left(x^T Q_i x + \rho u_i R_i u \right) d\tau$$

$$\leq \int_{t-\delta}^{t} \left(x^T Q_i x + \rho \tilde{u}_i R_i \tilde{u} \right) d\tau \tag{3.58}$$

Based on the above relationship, one may infer that an intrusion is present whenever there is a nonzero difference between the two costs on both sides of (3.58).

Next, let us estimate the optimality loss on system performance when intrusion is present on the system. To this end, fix an operational period $[t_0, t_1]$ of the ET-SLS of duration $\delta > 0$. Then, under the optimal control law (3.7), the signal (3.54) becomes

$$\pi(t) = \tilde{x}^T(t_0)P_i\tilde{x}(t_0) - \tilde{x}^T(t_1)P_i\tilde{x}(t_1) - \int_{t-\delta}^{t}\left(\tilde{x}(\tau)Q_i\tilde{x}(\tau) + \rho\hat{u}_i^T(\tau)R_i\hat{u}_i(\tau)\right)d\tau$$

$$= \tilde{x}^T(t_0)P_i\tilde{x}(t_0) - \tilde{x}^T(t_1)P_i\tilde{x}(t_1) - \int_{t-\delta}^{t}\left(\tilde{x}(\tau)\left[Q_i + 4\rho P_iB_iR_i^{-1}B_i^TP_i\right]\tilde{x}(\tau)\right)d\tau$$

$$= \tilde{x}^T(t_0)P_i\tilde{x}(t_0) - \tilde{x}^T(t_1)P_i\tilde{x}(t_1) - \int_{t-\delta}^{t}\tilde{x}^T(\tau)\Omega\tilde{x}(\tau)d\tau \tag{3.59}$$

where we have defined $\Omega = Q_i + 4\rho P_iB_iR_i^{-1}B_i^TP_i$. Using the definition of the compromised system trajectories $\tilde{x}(t) = \varepsilon(t) + \hat{x}(t)$, then (3.59) can be rewritten as follows:

$$\pi(t) = \tilde{x}^T(t_0)P_i\tilde{x}(t_0) - \left[\varepsilon(t_1) + \hat{x}(t_1)\right]^T P_i\left[\varepsilon(t_1) + \hat{x}(t_1)\right]$$

$$- \int_{t-\delta}^{t}\left[\varepsilon(\tau) + \hat{x}(\tau)\right]^T \Omega\left[\varepsilon(\tau) + \hat{x}(\tau)\right]d\tau$$

$$= \tilde{x}^T(t_0)P_i\tilde{x}(t_0) - \hat{x}^T(t_1)P_i\hat{x}(t_1) - \int_{t-\delta}^{t}\hat{x}^T(\tau)\Omega\hat{x}d\tau - \varepsilon^T(t_1)P_i\varepsilon(t_1) - \varepsilon^T(t_1)P_i\hat{x}(t_1)$$

$$- \hat{x}^T(t_1)P_i\varepsilon(t_1) - \int_{t-\delta}^{t}\left[\varepsilon^T(\tau)\Omega\varepsilon(\tau) + \varepsilon^T(\tau)\Omega\hat{x}(\tau) + \hat{x}^T(\tau)\Omega\varepsilon(\tau)\right]d\tau \tag{3.60}$$

Now note on the right-hand side of (3.60) that the first three terms correspond to the integral HJB equation of the optimal cost. In this regard, the remaining terms essentially denote the optimality loss $\bar{\pi}(t)$ caused by the actuator intrusion, i.e.

$$\bar{\pi}(t) = -\left[\varepsilon^T(t_1)P_i\varepsilon(t_1) + \varepsilon^T(t_1)P_i\hat{x}(t_1) + \hat{x}^T(t_1)P_i\varepsilon(t_1)\right]$$

$$- \int_{t-\delta}^{t}\left[\varepsilon^T(\tau)\Omega\varepsilon(\tau) + \varepsilon^T(\tau)\Omega\hat{x}(\tau) + \hat{x}^T(\tau)\Omega\varepsilon(\tau)\right]d\tau \tag{3.61}$$

The norm of such a loss term is given as

$$\|\bar{\pi}(t)\| \leq \|\varepsilon^T(t_1)\|\,\|P_i\|\,\|\varepsilon(t_1)\| + \|\varepsilon^T(t_1)\|\,\|P_i\|\,\|\hat{x}(t_1)\| + \|\hat{x}^T(t_1)\|\,\|P_i\|\,\|\varepsilon(t_1)\|$$

$$+ \int_{t-\delta}^{t}\left[\|\varepsilon^T(\tau)\|\,\|\Omega\|\,\|\varepsilon(\tau)\| + \|\varepsilon^T(\tau)\|\,\|\Omega\|\,\|\hat{x}(\tau)\| + \|\hat{x}^T(\tau)\|\,\|\Omega\|\,\|\varepsilon(\tau)\|\right]d\tau$$

$$\leq \|P_i\|\,\|\varepsilon(t_1)\|^2 + 2\|\varepsilon(t_1)\|\,\|P_i\|\,\|\hat{x}(t_1)\|$$

$$+ \int_{t-\delta}^{t}\left[\|\Omega\|\,\|\varepsilon(\tau)\|^2 + 2\|\varepsilon(\tau)\|\,\|\Omega\|\,\|\hat{x}(\tau)\|\right]d\tau \tag{3.62}$$

By the conditions in (3.52) and (3.40), as well as noting that $\| \hat{x}(t) \| \leq \hat{\beta}_i \| x_0 \|$, where $\hat{\beta}_i$ is defined as in the theorem, it can be shown that (3.62) simplifies to (3.55).

It can be seen from (3.62) that the accumulated cost is zero whenever $\gamma(t) = I$ (i.e. no intrusion is present on the ET-SLS). This suggests that the presence of intrusion on the system can be identified/detected based on the deviation of the system's cost function from its nominal optimal value.

3.4.2 An MTD Control Scheme to Mitigate Cyber Attack

First, note from (3.13) that the probability $P_i > 0$ that each mode $i \in \mathcal{I}$ is active is greater than zero. This implies that there exists a final time t_f^*, sufficiently long enough after the initial time t_0, until which the system has been switched through all available modes. Then under the assumption that the attacker is not able to compromise all of the system actuators at once (cf. Assumption 3.1), the following result on the stability of the system in the presence of actuator intrusion may be stated. The proof of this theorem resembles that of [24, Theorem 4] and is thus omitted from this chapter due to space limitation.

Theorem 3.3

Consider the ET-SLS (3.10) with the set of stabilizing control gains $\mathcal{K}$ under the ET control update rule in (3.11). Let $\mathcal{K}_c \subset \mathcal{K}$ with $\mathcal{K} \setminus \mathcal{K}_c \neq \varnothing$ denoting the set of feedback gains which correspond to system actuators that are compromised by cyber attack. Now assume that the system detects an intrusion on the ith actuator/mode. Consider a strategy for MTD which after such a detection (i) takes out the corresponding ith mode from the queue of candidate actuating modes in the next switching instance and then (ii) chooses to switch to one of the remaining actuating modes in the queue with the best performance. Then, the closed-loop ET-SLS is asymptotically stable in the presence of actuator attack.

Note from the proposed MTD strategy in Theorem 3.3 that, whenever an intrusion is detected, the ET-SLS will switch its controller to one of the available candidate actuating modes available for the system. In particular, such a switch will be prioritized towards an actuating mode, which (compared to other modes in the queue) can guarantee the attainment of the best performance index. Thus, for an appropriate implementation of Theorem 3.3, one should ensure that each mode in the available set of candidate actuating modes is capable of achieving the attainable best performance. In the next section of this chapter, we propose the use of a metaheuristic approach based on PSO algorithm to generate such a set of candidate actuating modes.

3.5 Optimization of the Proposed MTD Control Scheme

This section describes the use of a PSO algorithm to optimize the implementation of the MTD scheme as proposed in Section 3.4.2. More specifically, this section uses the PSO algorithm to construct a set of candidate actuating modes that will be used in the

MTD strategy described in Theorem 3.3. In particular, given a set of candidate actuating modes, each of which implements the control law as defined in (3.7), the PSO algorithm is used to optimize feedback gain with respect to closed-loop system's cost function in (3.6). Thus, the proposed PSO algorithm is essentially used to tune the feedback gain of each candidate actuating mode.

3.5.1 Basic Algorithm of PSO

The PSO algorithm is a metaheuristic approach within the class of population-based stochastic optimization method. The working principle underlying the PSO algorithm is motivated by the observed phenomena on the collective behaviors of a group of animals which often create a formation (e.g. birds flocking or fish schooling) when searching for foods/shelter or avoiding predators. One important observation in such collective behaviors is that they are often produced through simple and local interaction/communication among the members of the group. Akin to such a principle of collective animal behavior, the PSO algorithm utilizes a collection/swarm of fictitious particles to explore the feasible search region of an optimization problem and uses their dynamics to search for the optimal solution of the corresponding problem. More specifically, the PSO algorithm begins with assigning such swarm of particles to visit and explore different regions of the search space. During each of such visits, every swarm will generate a memory of information about its own best position (quantified as the personal best, $\bar{\mathcal{P}}$) as well as its overall neighbors' best positions (quantified as the global best, $\bar{\mathcal{G}}$). After several iterations, such swarm of particles will be able to visit and thoroughly examine different regions of candidate optimal solutions and then finally determine the optimal solution of the optimization problem [10,15,25,35,42].

In the following, we briefly recall the basic idea in the PSO algorithm. The algorithm essentially starts with the determination of swarm of particles with an initial population of P and an objective function $F : \mathbb{R}^Q \to \mathbb{R}$ to be optimized on a Q-dimensional space of search region. To describe the dynamics of each particle in the swarm over the search region coordinates, consider the indices $z_1 = 1, 2, \ldots, P$ and $z_2 = 1, 2, \ldots, Q$, which correspond to the particle population and dimensional coordinate of the search region, respectively. Let us denote the z_1-th particle's position and velocity vectors with respect to the z_2-th coordinate as u_{z_1, z_2} and v_{z_1, z_2}, respectively. Then, based on the PSO algorithm, the position and velocity vectors of each particle are updated iteratively such that their evolution at the $(k+1)$th iteration satisfies the following equation [15,25,32]:

$$v_{z_1, z_2}^{k+1} = w^k v_{z_1, z_2}^k + c_1 r_1 \left(\bar{\mathcal{P}}_{z_1, z_2}^k - u_{z_1, z_2}^k \right) + c_2 r_2 \left(\bar{\mathcal{G}}_{z_2}^k - u_{z_1, z_2}^k \right),$$

$$u_{z_1, z_2}^{k+1} = u_{z_1, z_2}^k + v_{z_1, z_2}^{k+1}. \tag{3.63}$$

In Equation (3.63), the *inertial weight parameter* w^k controls each particle's prior momentum (velocity) contribution on the next iteration's velocity, while the *acceleration coefficients* c_1 and c_2 determine the learning rates of each particle. Furthermore, the uniformly distributed random numbers r_1 and r_2 with values $[0, 1]$ are used to control the stochasticity of each particle's dynamics. Figure 3.1 illustrates the vectorial progression of variables in (3.63) at each algorithm's iteration.

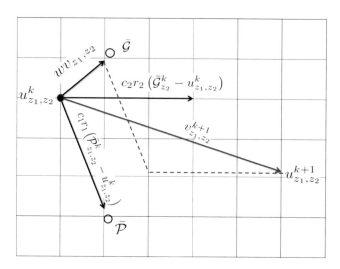

FIGURE 3.1
Illustration of vectorial progression in PSO iteration in Equation (3.63).

For every value of k in the iteration of (3.63), each particle evaluates the objective function $F(u_{z_1,z_2})$ to determine (i) its best positional value $\bar{\mathcal{P}}$, which corresponds to the optimum objective function, and (ii) the best neighboring particle (among others in the swarm) position $\bar{\mathcal{G}}$, which has the optimum objective function. The iteration of (3.63) then continues until convergence (i.e. no significant change on each particle's position/velocity occurs) or the maximum iteration number is reached. Note in accordance with Ref. [34] that the PSO iteration in (3.63) is guaranteed to be stable so long as the conditions $\sum_{i=1,2} c_i r_i < 4$ and $\frac{1}{2}\left(\sum_{i=1,2} c_i r_i\right) - 1 < w < 1$ are fulfilled. In practice, some heuristics are often used to choose the parameters of (3.63), including $v^0 = 0.1u^0, w \in [0.4, 0.9], c_1, c_2 \in \lfloor 1, 2.05 \rfloor, \mathcal{I} \in \lfloor 500, 10^6 \rfloor$, and $P = [10, 500]$.

3.5.2 PSO-Based LQR Tuning

The proposed PSO-based tuning for the used LQR feedback control gain is designed through the emulation of the eigenstructure assignment method [2,8,16]. A tuning method was motivated by the fact that the characteristics of closed-loop system's dynamic responses are closely related to the system eigenstructure (i.e. eigenvalues and eigenvectors). More specifically, the speed and shape of closed-loop system's responses are influenced by the system's eigenvalues and eigenvectors, respectively. The eigenstructure assignment methods thus refer to control design strategies, which aim at achieving a particular dynamic response characteristic of a closed-loop system through appropriate assignment of the corresponding eigenvalues and eigenvectors [2,7]. It has been well known that while the LQR-based control design method has the advantage of producing closed-loop system with guaranteed optimality and robustness, the method does not provide the flexibility of assigning the eigenstructure of the closed-loop system. In this regard, the emulation of the eigenstructure assignment method during the tuning process of the proposed LQR controller is expected to result in a closed-loop system that is not only robust but also captures certain dynamic response characteristics [8].

The objects in the tuning process of the used LQR controller are matrices Q and R in the cost function (3.6) [17]. Note that the value of the feedback gain (3.7) is influenced by these matrices. In particular, we assume in this chapter that each of such matrices has a diagonal form and then choose such diagonal elements as the particles in the PSO algorithm. Now, let A_{CL} be the matrix which corresponds to the closed-loop system dynamics (3.10) under the feedback gain in (3.7). Using ψ and Ψ^r (respectively Ψ^ℓ), respectively, to denote the eigenvalues and the right (respectively left) eigenvector of A_{CL}, we have the following relationship about the eigenstructure of matrix A_{CL} [2,16]:

$$A_{\mathrm{CL}}\, \psi_i = \psi_i\, \Psi_i^r$$
$$\left(\Psi_i^\ell\right)^T A_{\mathrm{CL}} = \psi_i \left(\Psi_i^\ell\right)^T \tag{3.64}$$

The objective of the proposed PSO-based LQR tuning is to minimize the condition numbers of the eigenvalue of A_{CL}. Recall that such numbers measure the matrix's eigenvalues sensitivity to perturbations or uncertainties on the system. Given the relationship in (3.64), the condition number η_i of ψ_i is defined as [14,22]:

$$\eta_i = \frac{\|\Psi_i^\ell\|_2 \|\Psi_i^r\|_2}{\left|\left(\Psi_i^\ell\right)^T \Psi_i^r\right|} \tag{3.65}$$

For a set of random initial values of the diagonal elements of matrices Q and R (i.e. the particle populations), the PSO algorithm will then seek to iteratively minimize a fitness function $\mathcal{F}$ of the form:

$$\mathcal{F} := \left|\sum_i \eta_i\right| \tag{3.66}$$

It can be seen that such a fitness function essentially is the aggregate of the cosine angles between the left and right eigenvectors of the closed-loop system, which results from the use of feedback gain that is produced at every iteration by each particle in the population. The minimum value of $\mathcal{F}$ essentially corresponds to a situation where the closed-loop system poles are well separated from each other on the left half side of the complex plane and produce commonly desired dynamic responses of closed-loop control systems. The search for such a minimum will be conducted by the PSO algorithm iteratively until convergence or maximum iteration number is reached.

3.6 Simulation Example

This section presents an illustrative example on the use of the previously described PSO algorithm to construct a set of candidate actuating modes for implementing the proposed MTD control method. The considered system model is given as follows:

TABLE 3.1

PSO Results

Mode	Q	R	Optimal Cost
1	$\begin{pmatrix} 100.00 & 0 \\ 0 & 35.97 \end{pmatrix}$	0.459	5.55×10^{-17}
2	$\begin{pmatrix} 100.00 & 0 \\ 0 & 14.44 \end{pmatrix}$	0.649	9.88×10^{-15}
3	$\begin{pmatrix} 61.76 & 0 \\ 0 & 50.20 \end{pmatrix}$	0.7029	1.17×10^{-5}

$$\dot{x}(t) = Ax(t) + B_i u(t), \quad x(0) = x_0, \, i = 1, 2, 3 \tag{3.67}$$

with the following state matrix and candidate actuator mode matrix/vector:

$$A = \begin{bmatrix} -1 & 1 \\ -1 & 1 \end{bmatrix}, \quad B_1 = \begin{bmatrix} 1 & 1 \\ 2 & 1 \end{bmatrix}, \quad B_2 = \begin{bmatrix} 1 \\ 2 \end{bmatrix}, \quad B_3 = \begin{bmatrix} 2 & 0 \\ 1 & 2 \end{bmatrix} \tag{3.68}$$

It can be shown that each pair (A, B_i) in the above model is stabilizable. The attack on each of such actuating mode is assumed to take the form of (normally distributed) random number within finite time length.

For each of the available candidate actuating mode, we first implemented the PSO algorithm from Section 3.5 to construct the optimal LQR gain. The results from PSO algorithm are summarized in Table 3.1, and the corresponding iterative progression of each mode's optimal cost is depicted in Figure 3.2. It can be seen in the table and the figure that mode 1 has the best performance, followed by mode 2 and then mode 3. The corresponding feedback gain K_i for each actuating mode is given as follows:

$$K_1 = \begin{bmatrix} 3.0135 & 8.7092 \\ 13.4847 & -0.8810 \end{bmatrix}, \quad K_2 = \begin{bmatrix} 10.1307 & 3.1403 \end{bmatrix},$$

$$K_3 = \begin{bmatrix} 8.7017 & 2.1559 \\ -2.1221 & 8.5561 \end{bmatrix} \tag{3.69}$$

Following the MTD scheme in Theorem 3.3, we then simulate the system dynamics for 70 time steps under possible attacks on the system actuator. We set three instances of attacks for the system, namely, during $t = \lfloor 10, 15 \rfloor, t = \lfloor 25, 32 \rfloor$ and $t = [50, 55]$. The actual trajectories of the closed-loop system's state variables are depicted in Figure 3.3 as solid lines. This figure also overlays the actual state trajectories with the reference trajectories (dashed lines) to illustrate the deviation which occurs due to the presence of actuator attacks. It can be seen in this figure that the closed-loop system remains stable under the presence of such attacks. This thus illustrates the effectiveness of the proposed MTD scheme with PSO-based controller tuning in maintaining the system stability and performance in the presence of cyber attacks.

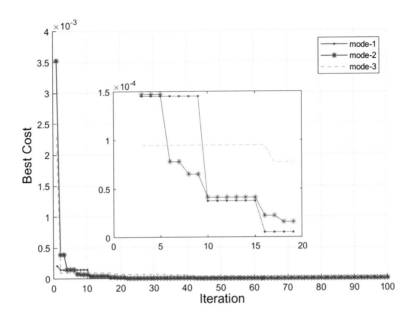

FIGURE 3.2
Comparison of iterative optimal cost progression for all modes.

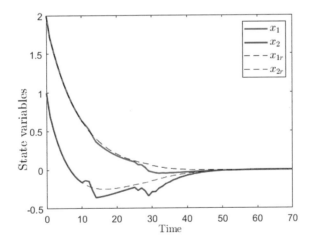

FIGURE 3.3
Trajectories of the closed-loop system state variables.

3.7 Conclusions and Future Scope

This chapter has presented a MTD control scheme to detect and mitigate cyber attacks from malicious adversaries on a CPS. The proposed scheme essentially utilizes a switching stabilizing controllers (SSC) strategy under an ET control update scheme to complicate external adversaries' attempts in delivering their attacks. To ensure the optimal performance of each controller that is used in the proposed MTD scheme, a metaheuristic approach for controller tuning based on PSO is also proposed. Simulation results are also presented to illustrate the effectiveness of the proposed MTD scheme with a PSO-based controller tuning method in maintaining the CPS stability and performance in the presence of cyber attacks. Future works will be directed towards extending the developed method to CPS with nonlinear dynamics.

Acknowledgments

This work was supported by Ministry of Research and Technology—National Research & Innovation Agency (Kemenristek-BRIN) of the Republic of Indonesia under the fundamental research scheme (PDUPT), 2021. The author gratefully acknowledges the partial financial support for this research work from the Institute for Research and Community Service (LPPM) of Parahyangan Catholic University, Indonesia.

References

1. Anders Ahlén et al. Toward wireless control in industrial process automation: A case study at a paper mill. *IEEE Control Systems Magazine*, 39(5):36–57, 2019.
2. Albert N. Andry, Eliezer Y. Shapiro, and Jonathan C. Chung. Eigenstructure assignment for linear systems. *IEEE Transactions on Aerospace and Electronic Systems*, 5:711–729, 1983.
3. Doug Britton. 3 reasons why moving target defense must be a priority. URL: https://gcn.com/articles/2019/06/10/moving-target-defense.aspx (accessed on 30 January 2021), 2019.
4. Mike Burshteyn. Moving target defense: State of the field in 2018? URL: https://blog.cryptomove.com/moving-target-defense-state-of-the-field-in-2018-a3a580668959 (accessed on 30 January 2021), 2018.
5. Alvaro Cardenas et al. Challenges for securing cyber physical systems. *In Proceedings of WKSH - Future Directions in Cyber-Physical Systems Security*, vol. 5, 2009, Newark, NJ.
6. Manuel Cheminod, Luca Durante, and Adriano Valenzano. Review of security issues in industrial networks. *IEEE Transactions on Industrial Informatics*, 9(1):277–293, 2012.
7. Jae Weon Choi. A simultaneous assignment methodology of right/left eigenstructures. *IEEE Transactions on Aerospace and Electronic Systems*, 34(2):625–634, 1998.
8. Jae Weon Choi and Young Bong Seo. LQR design with eigenstructure assignment capability [and application to aircraft flight control]. *IEEE Transactions on Aerospace and Electronic Systems*, 35(2):700–708, 1999.
9. Michelle S. Chong, Henrik Sandberg, and André M.H. Teixeira. A tutorial introduction to security and privacy for cyber-physical systems. *In Proceedings of 17th European Control Conference*, pp. 968–978, 2019, Naples, Italy.
10. Carlos A. Coello et al. Handling multiple objectives with particle swarm optimization. *IEEE Transactions on Evolutionary Computation*, 8(3):256–279, 2004.
11. Trustworthy Cyberspace. Trustworthy cyberspace: Strategic plan for the federal cybersecurity research and development program. *National Science & Technology Council*, 2011.

12. DHS. Moving target defense. URL: https://www.dhs.gov/science-and-technology/csd-mtd (accessed on 30 January 2021), 2013.

13. Derui Ding et al. A survey on security control and attack detection for industrial cyber-physical systems. *Neurocomputing*, 275:1674–1683, 2018.

14. Guangren R. Duan. Parametric eigenstructure assignment via output feedback based on singular value decompositions. *IEE Proceedings-Control Theory and Applications*, 150(1):93–100, 2003.

15. Soren Ebbesen, Pascal Kiwitz, and Lino Guzzella. A generic particle swarm optimization matlab function. *In Proceedings of 2012 American Control Conference*, pp. 1519–1524, 2012, Montréal, Canada.

16. Mil Fahmy and John O'reilly. On eigenstructure assignment in linear multivariable systems. *IEEE Transactions on Automatic Control*, 27(3):690–693, 1982.

17. Javad Hamidi. Control system design using particle swarm optimization (PSO). *International Journal of Soft Computing and Engineering*, 1(6):116–119, 2012.

18. Wilhelmus P.M.H. Heemels, Karl Henrik Johansson, and Paulo Tabuada. An introduction to event-triggered and self-triggered control. *In Proceedings of 51st IEEE Conference on Decision and Control*, pp. 3270–3285, 2012, Maui, Hawaii.

19. Joao P. Hespanha and A. Stephen Morse. Stability of switched systems with average dwell-time. In *Proceedings of 38th IEEE Conference on Decision and Control*, pp. 2655–2660, 1999, Phoenix, AZ.

20. Yan Hu et al. A survey of intrusion detection on industrial control systems. *International Journal of Distributed Sensor Networks*, 14(8): 1550147718794615, 2018.

21. Abdulmalik Humayed, Jingqian Lin, Fengjun Li, and Bo Luo. Cyber-physical systems security: A survey. *IEEE Internet of Things Journal*, 4(6):1802–1831, 2017.

22. Ralph K. Cavin III and Shankar P. Bhattacharyya. Robust and well-conditioned eigenstructure assignment via sylvester's equation. *Optimal Control Applications and Methods*, 4(3):205–212, 1983.

23. Sushil Jajodia et al. *Moving Target Defense: Creating Asymmetric Uncertainty for Cyber Threats*. Springer, Berlin/Heidelberg, Germany, 2011.

24. Aris Kanellopoulos and Kyriakos G. Vamvoudakis. A moving target defense control framework for cyber-physical systems. *IEEE Transactions on Automatic Control*, 65(3):1029–1043, 2020.

25. James Kennedy. Particle swarm optimization. *In Encyclopedia of Machine Learning*, pp. 760–766. Springer, Berlin/Heidelberg, Germany, 2011.

26. Kyoung-Dae Kim and Panganamala R. Kumar. Cyber–physical systems: A perspective at the centennial. *Proceedings of the IEEE*, 100(Special Centennial Issue):1287–1308, 2012.

27. Edward A. Lee. Cyber physical systems: Design challenges. *In Proceedings of 11th IEEE International Symposium on Real-Time Computing*, pp. 363–369, 2008.

28. Cheng Lei et al. Moving target defense techniques: A survey. *Security and Communication Networks*, 2018(Article ID 3759626)):1–25, 2018.

29. Michael Lemmon. Event-triggered feedback in control, estimation, and optimization. In *Networked Control Systems*, pp. 293–358. Springer, Berlin/Heidelberg, Germany, 2010.

30. Yuriy Zacchia Lun et al. State of the art of cyber-physical systems security: An automatic control perspective. *Journal of Systems and Software*, 149:174–216, 2019.

31. Nicolas Marchand, Sylvain Durand, and Jose Fermi Guerrero Castellanos. A general formula for event-based stabilization of nonlinear systems. *IEEE Transactions on Automatic Control*, 58(5):1332–1337, 2012.

32. Federico Marini and Beata Walczak. Particle swarm optimization (PSO): A tutorial. *Chemometrics and Intelligent Laboratory Systems*, 149:153–165, 2015.

33. Hamed Okhravi, Thomas Hobson, David Bigelow, and William Streilein. Finding focus in the blur of moving-target techniques. *IEEE Security & Privacy*, 12(2):16–26, 2013.

34. Ruben E. Perez and Kamran Behdinan. Particle swarm approach for structural design optimization. *Computers & Structures*, 85(19–20):1579–1588, 2007.
35. Gregorio Toscano Pulido and Carlos A. Coello. A constraint-handling mechanism for particle swarm optimization. *In IEEE Congress on Evolutionary Computation*, pp. 1396–1403, 2004, Portland, OR.
36. Paulo Tabuada. Event-triggered real-time scheduling of stabilizing control tasks. *IEEE Transactions on Automatic Control*, 52(9):1680–1685, 2007.
37. Tua A. Tamba. A pso-based moving target defense control optimization scheme. *In Proceedings of SICE International Symposium on Control Systems*, 2021, Tokyo, Japan (virtual).
38. Tua A. Tamba, Bin Hu, and Yul Y. Nazaruddin. On event-triggered implementation of moving target defense control. *In Proceedings of 21st IFAC World Congress*, 2020, Berlin, Germany (virtual).
39. Cliff Wang and Zhuo Lu. Cyber deception: Overview and the road ahead. *IEEE Security & Privacy*, 16(2):80–85, 2018.
40. Mark Yampolskiy et al. Systematic analysis of cyber -attacks on cps-evaluating applicability of DFD-based approach. *In Proceedings of 5th International Symposium on Resilient Control Systems*, pp. 55–62, 2012, Salt Lake City, UT.
41. Mark Yampolskiy et al. Taxonomy for description of cross-domain attacks on CPS. *In Proceedings of 2nd ACM International Conference on High Confidence Networked Systems*, pp. 135–142, 2013, Philadelphia, PA.
42. Yudong Zhang, Shuihua Wang, and Genlin Ji. A comprehensive survey on particle swarm optimization algorithm and its applications. *Mathematical Problems in Engineering*, 2015 (Article ID 931256): 1–38, 2015.
43. Rui Zhuang, Scott A. DeLoach, and Xinming Ou. Towards a theory of moving target defense. *In Proceedings of 1st ACM Workshop on Moving Target Defense*, pp. 31–34, Scottsdale, AZ.

4

Application of Salp Swarm Algorithm to Solve Constrained Optimization Problems with Dynamic Penalty Approach in Real-Life Problems

Omkar Kulkarni and G. M. Kakandikar
Dr. Vishwanath Karad MIT World Peace University

V. M. Nandedkar
Shri Guru Gobind Singhji Institute of Engineering & Technology

CONTENTS

4.1 Introduction

The sheet metal-forming technique plays a very vital role in the production of most body components used in the automobile sector [1]. The metallic-forming process is a technique/tool, which is regularly applied to fabricate hollow sheet metal components with high drawing ratios or complicated shapes. In sheet metal-forming process, a blank sheet is subjected to plastic deformation with the use of forming tools to verify the designed shape [2]. During this method, the blank sheet is probably going to develop defects if the method parameters don't seem to be designated properly. Therefore, it's vital to optimize the method parameters to avoid defects within the components and also to make an effort to reduce production price [3,4]. To determine the optimum values of the method parameters, it's essential to seek out their influence on the deformation behavior and formability of the sheet. Optimization of the process parameters, such as friction coefficient, blank holder force (BHF), and radius on die (Rd), can be acquired

DOI: 10.1201/9781003143505-4

based on their significance on the metal-forming characteristics [5–7]. In this paper, the optimization is carried out with the help of salp swarm algorithm (SSA). The component selected for the case study is heat shield of an engine. The material used is Al-coated CRCA (cold-rolled close-annealed). With the assistance of metal-forming simulation packages, we are able to establish the critical problem areas, and also, solutions can be validated with the help of computers instead of costly work operations of manufacturing. Metal-forming simulation also plays a vital role at the prior stages of product and tool design stage to choose various optimum parameters. In this paper, simulation is carried out with the help of Forming Suite software.

In the recent few years, metaheuristics algorithms have evolved and are used for the optimization purpose of the problems. The application of these algorithms extends to solve the real-life problems also. There are many bio- and nature-inspired algorithms, which are also known as optimization techniques such as evolutionary algorithms (EAs), swarm intelligence (SI), etc. There are also SI techniques such as particle swarm optimization (PSO) [8], cuckoo search (CS) algorithm [9], gray wolf optimizer (GWO) [10,11], artificial bee colony (ABC) algorithm [12], firefly algorithm (FA) [13], ant colony optimization (ACO) [14], grasshopper optimization algorithm (GOA) [15,16] etc. The EAs include genetic algorithm (GA) [17], differential evolution (DE) [18], etc. There are also many social-inspired algorithms such as cohort intelligence (CI) [19], ideology algorithm (IA) [20] etc.

Salps have their place to the own group of Salpidae, and their body is cylindrical in shape. The tissues of salps are similar to those of jelly fishes. The structure of the body of salps is similar to that of salps which use the pumping of water through body as propulsion to move forward. One of the most exciting behaviors of salps is their swarming behavior. At the depth of oceans, these salps get formed into swarms, which are called as salp chain. The main purpose of this behavior isn't very clear; however, a couple of researchers trust the performance for achieving higher locomotion, the usage of speedy coordinated locations and foraging behavior [21]. The salps are divided into two groups: leaders and followers. The salps present at the start of the chain are known as the leaders who lead the group, and the remaining are known as followers. As the leader is in best position when leading the group in search of food, this position represents optimum solution of the problem in search space.

In this paper, the process parameters selected for optimization are friction coefficient (μ), BHF and Rd. The results of these three important variables on the thinning of sheet are studied. The more the thinning, the more the wrinkling tendency. So the objective of the optimization is to maximize the thinning of component. To optimize the parameters, the SSA is used. The problem to be optimized or to be solved has constraints to be considered; hence, the SSA is modified to solve this problem, with dynamic penalty approach.

4.2 Framework of Salp Swarm Algorithm

The SSA is actually a recent technique to resolve the optimization issues. The idea of the SSA is based on the swarming and hunting of food behavior of the salps. The salps live in the deep ocean and are basically divided into two groups, namely, leader and followers. The salps always form a chain where in each salp is placed one behind the other. In order to fulfill the purpose of locomotion, they push out water forcibly so that they can get thrust. The salp present at the front is known as the leader, while others are follow-ers. The leader leads the swarm to the food location and others follow the obtaining of food source is obtaining the optimum solution, and the position of the salps or swarm represents the variables in the problem [21].

The position of salps is stored in n-dimensional array, where n is the number of unknown variables. This is how the position is stored in two-dimensional array. The equation for the same is presented as follows, which resembles the leader.

$$x_j^1 = \begin{cases} f_j + c_1\left((ub_j - lb_j)c_2 + lb_j\right) & c_3 \geq 0 \\ f_j - c_1\left((ub_j - lb_j)c_2 + lb_j\right) & c_3 < 0 \end{cases} \tag{4.1}$$

where x_j^1 is the position of leader salp in the j dimension, f_j represents the food source, lb_j and ub_j are the upper and lower bounds of the variables, and c_1, c_2, c_3 are the numbers randomly.

Equation (4.1) represents the front-runner or leader, which updated its location with respect to the food source. The coefficient c_1 is the utmost important variable of SSA, which helps in exploration and exploitation, which is defined by

$$c_1 = 2e^{-\left(\frac{l}{L}\right)^2} \tag{4.2}$$

Here, l is the instant iteration and L is the maximum number of iterations.

The follower salps are also following the leader, which also updates their positions, which are defined by

$$x_j^i = \frac{1}{2}\left(x_j^i + x_j^{i-1}\right) \tag{4.3}$$

The working of SSA is illustrated in pseudocode shown in Figure 4.1.

Initialize the salp population xi (i = 1, 2, ..., n) considering ub and lb
while *(end condition is not satisfied)*
Calculate the fitness of each search agent (salp)
F=the best search agent
Update c_1 by Eq. (2)
    for *each salp (xi)*
      if *(i==1)*
        Update the position of the leading salp by Eq. (1)
      else
        Update the position of the follower salp by Eq. (3)
      end
    end
    Amend the salps based on the upper and lower bounds of variables
end
return F

FIGURE 4.1
Pseudocode of SSA.

It ought to be cited that the food supply will be up-to-date throughout optimization for the reason that the chain of salp is very probable to discover a higher answer via exploring and exploiting the search space around it. The simulations shown above are that the chain of salp modeled is capable of chasing and shifting food source. Hence, the chain of salp has the ability to move closer to the food source most useful that changes over the path of iterations. However, the planned salp chain model and SSA algorithmic rule are powerful in fixing improvement issues; some remarks are listed as follows:

- SSA stores the solution, which is best obtained to this point, and applies it to the source of food variable used, so it by no means gets even lost supposing the whole population declines.
- The positions of the main salp in SSA are updated with reference to the food supply handiest, that is, the quality answer is gained up to now, so the leader continually explores and exploits the distance around it.
- The position of the follower salps is updated in SSA with respect to each other, in order that they flow slowly in the direction of the leading salp.
- Steady moves of follower salps prevents the SSA ending in local optima.
- Variable c_1 is reduced slowly over the route of iterations, so the SSA algorithm initially explores the search area and then exploits it.
- SSA has the best one key controlling parameter (c_1).
- SSA algorithm is easy and clean to apply.

4.3 Dynamic Penalty Approach

In general, the optimization problem is formulated as follows:

$$\text{Minimize } f(x) = f(x_i, \ldots, x_n) \tag{4.4}$$

Subjected to

$$g_i(x) \leq 0, \, i = 1, \ldots, n$$

$$h_j(x) = 0, \quad j = 1, \ldots, m$$

$$\text{Where } \varphi_i^{\text{lower}} \leq x_i \leq \varphi_i^{\text{upper}}, \, i = 1, \ldots, n$$

In the context of the SSA, the variables used as $x = (x_1, \ldots, x_n)$ are the positions of the salps. φ is the sampling interval with lower and upper bounds. $g_i(x)$ and $h_j(x)$ are the inequality and the equality constraints for the optimization problem [19].

The dynamic penalty approach is a powerful constraint handling technique, which is used in the problem solving of the main constrained problem mentioned in this paper.

The dynamic penalty is an approach that is used to solve the constrained optimization problems. This penalty approach converts the constrained optimization problem to unconstrained optimization problem by penalizing the function, which is not a feasible solution. The formulation of the dynamic penalty approach is given as follows:

$$f_q(x) = f(x) + \sum_{i=1}^{n} \left((q_k)^\alpha \times S \right) \times (g_i(x))^\beta \quad (4.5)$$

where α and β are the constant integers.

In this constraint handling dynamic penalty approach, at the initial stages the penalty applied is less in value. Over the iterations, the number of iterations proceeding the value of penalty applied to the nonfeasible solution also increases. The iteration value or the number of current iteration number q_k is directly used in the penalty value, which is also multiplied by Spenalty constant value; this results in a higher penalty value addition. The parameters α and β are very important, which should be tuned properly to get the best result. In the initial trials of the SSA, these parameters were selected. The $g_i(x)$ is the constraint value, which is not satisfied and is penalized.

4.4 Proposed Methodology

The experiments are performed with Taguchi design of experiments (DOE), including three input variables and performance parameter as uniform thickness distribution.

Analysis is carried out using analysis software, and thickness distribution of the product is carried out before optimization.

Mathematical model is developed using linear regression analysis, and constraints are set up.

Optimization problem is formulated, and with extensive literature survey, the lower and upper bounds have been decided after that the SSA is applied. The optimized process parameters are obtained, an analysis is performed on the optimized product, and the results are compared.

4.5 Design of Experiments

DOE strategies permit designers to determine concurrently the individual parameter and interactive effects of many factors that could have an effect on the output outcomes in any layout. DOE also provides a full insight of interplay between layout elements; therefore, it allows to turn any trendy layout into a robust one. Simply, DOE enables to pinpoint the sensitive components and touchy areas in designs that cause issues in output. Designers are then capable of fixing these problems and bringing strong and higher output designs prior actual shop floor work.

TABLE 4.1

Control Factors with Levels

Factors	Level 1	Level 2	Level 3	Units
Blank holder force	105	107	109	KN
Radius on die	2.5	5.25	8	mm
Coefficient of friction	0.005	0.0775	0.15	--

4.5.1 Selection of Orthogonal Array

The experiments are done based on Taguchi method. The orthogonal array is designed to get the maximum of the results from the minimum of experiments, and the experiments are designed on this basis level.

The control factors in this experiment are considered as BHF, Rd and coefficient of friction (μ). These factors are the most important parameters in the formation of sheet metal. Based on these numbers of parameters, the L9 orthogonal array is used. According to Table 4.1, there are three levels of each parameter selected.

4.5.2 Experimental Data

The orthogonal array is defined in L9 respect. Nine experiments were defined, and with respect to the same experiments, the thickness is measured over different areas. The average thickness is measured and is considered for the analysis. Further, the finite element analysis is performed for both before and after the optimization of parameters.

The experimental data is presented in Table 4.2.

Expt. no.	BHF	Rd	μ	Thickness
1.	105	2.5	0.005	0.3945
2.	105	5.25	0.0775	0.345
3.	105	8	0.15	0.348
4.	107	2.5	0.0775	0.4715
5.	107	5.25	0.15	0.4995
6.	107	8	0.005	0.4405
7.	109	2.5	0.15	0.551
8.	109	5.25	0.005	0.511
9.	109	8	0.0775	0.456

TABLE 4.2

Result Table

Parameters	Optimum Result from Experiments	Optimum Results from the Algorithm
BHF	109	108.99
Rd	2.5	2.5
μ	0.15	0.14
Thickness	0.505	0.559

4.5.3 Problem Formulation

The regression analysis is performed, and as the nature of problem states that the linear regression analysis is done, Equation 4.6 for the thickness is formed. The analysis is performed using Minitab 15. The larger the strategy used, the better the thickness because the thickness distribution becomes thinner in many cases. In the stated problem, there are constraints to be considered as for the Rd. This constraint is very important to be considered as it affects the thickness distribution on corners and the overall thickness of the drawn component.

The equation obtained from the regression analysis is as follows:

$$\text{Thickness} = -3.347 + 0.03588 \text{ BHF} - 0.01045 \text{ Rd} + 0.121\,\mu \qquad (4.6)$$

Subjected to

$$2.5 < \text{Rd} < 8$$

$$\text{BHF} = \frac{\pi}{4}\left(d_0^2 + 2r\right)^2 * P$$

$$\text{Rd} = 0.035\left[50 + (d_0 - d_1)\sqrt{s_0}\right]$$

$$\text{BHF} = \text{Blank holder force}$$

$$\text{Rd} = \text{Radius on die}$$

$$\mu = \text{Coefficient of friction}$$

$$P = \text{Pressure applied, i.e., } 2.5 \text{ N/mm}^2$$

$$d_0 = \text{Blank diameter}$$

$$r = \text{Radius of blank}$$

4.6 Results and Discussion

In the experimentation, the parameters considered are BHF, Rd and coefficient of friction. The response variable is thickness. The thickness distribution in the formed component is getting thinner in many areas as shown in Figure 4.2. The blank thickness is 0.6 mm. The red area shown in Figure 4.2 indicates the thinning of the sheet metal on

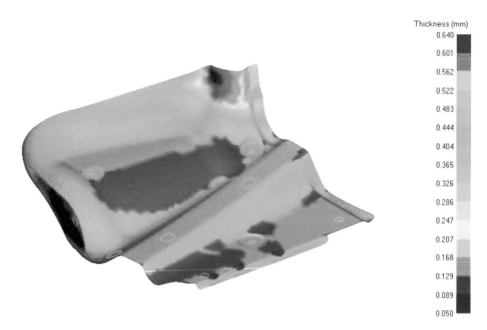

FIGURE 4.2
Thickness distribution of original component.

the corners and edges. On the basis of these results, Taguchi methodology was adopted and L9 orthogonal array was selected. Then, the problem statement was defined by regression analysis, and the problem was formulated. This formulated problem was then integrated with the SSA, and the constraints in the problem are also considered. Then, the stated problem was then solved by using the SSA, and the optimum results are obtained. The SSA algorithm gives the optimum parameter settings so that the thickness which should be achieved should be uniform and maximum. This problem is solved as maximization problem. The problem was solved on desktop with i5 processor and 4Gb RAM. MATLAB version 19 was used to solve the problem. The other software packages used for the analysis and DOE are Minitab 15 and Forming Suite. The finite element analysis (FEA) results are obtained from the Forming Suite. Figure 4.3 indicates the optimum parameter setting finite element analysis.

In Figure 4.3, it can be observed that the distribution of the thickness is most even and excessive thinning is reduced. In Table 4.2, it is observed that the parameters obtained from the algorithm are optimized and the thickness is close to 0.6 mm, which is the blank thickness.

4.7 Conclusion

This paper aims at the optimization of the parameters of the rectangular component, which is performed by the SSA. It is observed that the obtained parameters from the SSA algorithm are giving more thickness value; this means that the parameters obtained

FIGURE 4.3
Thickness distribution after optimization.

are more optimized. The value obtained for the thickness with the optimized result is 0.559 mm, and that without the algorithm is 0.505 mm so the increase in thickness is 10.69%. This increase of 10.69% reduces the thinning area and the uniform distribution of the thickness. This ensures that the algorithm gives the optimized results, which are proved with the help of finite element analysis also.

References

1. M. Nalbant, H. Gokkaya, G. Sur, "Application of Taguchi method in the optimization of cutting parameters for surface roughness in turning", *Materials*, 28(4), 1379–1385, 2007.
2. R.A. Fisher, *Statistical Methods for Research Worker*, London: Oliver & Boyd, 1925.
3. R. K. Roy, *A Primer on the Taguchi Method*, New York: Van Nostrand Reinhold, 1990.
4. T.-P. Dao, S.-C. Huang, "Study on optimization of process parameters for hydromechanical deep drawing of trapezoid cup", *Journal of Engineering Technology and Education*, 8(1), 53–73, 2011.
5. P. P. Date, "Sheet metal formability", *Proceedings of the Training Programme of Sheet Metal Forming Research Association at Indian Institute of Technology*, Mumbai, 2005.
6. K. Lang, *Hand Book of Metal Forming*. New York: McGraw Hill Publications, pp. 20.3–20.7, 1985.
7. R. Pierce, *Sheet Metal Forming*. Bristol Philadelphia and New York: Adam Hilgar, pp. 5–15, 1991.
8. N. K. Kulkarni, S. Patekar, T. Bhoskar, O. Kulkarni, G. M. Kakandikar, V. M. Nandedkar, "Particle swarm optimization applications to mechanical engineering: A review", *Materials Today: Proceedings*, 2, 2631–2639, 2015.

9. A. S. Joshi, O. Kulkarni, G. M. Kakandikar, V. M. Nandedkar, "Cuckoo search optimization-a review", *Materials Today: Proceedings*, 4, 7262–7269, 2017.

10. S. Patekar, G. M. Kakandikar, V. M. Nandedkar, "Numerical modelling and BIO inspired optimization of thinning in automotive sealing cover: Grey wolf approach", 2016/12/16.

11. O. Kulkarni, S. Kulkarni, "Process parameter optimization in WEDM by Grey Wolf optimizer", *Materials Today: Proceedings*, 5, 4402–4412, 2018.

12. X. Li, G. Yang, "Artificial bee colony algorithm with memory", *Applied Soft Computing*, 41, 362–372, 2016.

13. G. M. Kakandikar, O. Kulkarni, S. Patekar, T. Bhoskar, "Optimising fracture in automotive tail cap by firefly algorithm", *International Journal of Swarm Intelligence (IJSI)*, 5(1), 2020. doi: 10.1504/IJSI.2020.10027781.

14. O. Kulkarni, S. Patekar, T. Bhoskar, N. Kulkarni, G. M. Kakandikar, V. M. Nandedkar, "Ant colony optimization and its applications in mechanical engineering review", *Industrial Engineering Journal*, 9, 41–45, 2016.

15. A. G. Neve, G. M. Kakandikar, O. Kulkarni, "Application of grasshopper optimization algorithm for constrained and unconstrained test functions", *International Journal of Swarm Intelligence and Evolutionary Computation*, 6, 165, 2017.

16. A. G. Neve, G. M. Kakandikar, O. Kulkarni, V. M. Nandedkar, "Optimization of railway Bogie Snubber spring with grasshopper algorithm", *Data Engineering and Communication Technology, Advances in Intelligent Systems and Computing*, 926–937, 2020.

17. T. Bhoskar, O. K. Kulkarni, N. K. Kulkarni, S. L. Patekar, G. M. Kakandikar, V. M. Nandedkar, "Genetic algorithm and its applications to mechanical engineering: A review", *Materials Today: Proceedings*, 2, 2624–2630, 2015.

18. L. Zhang, X. Xu, C. Zhou, M. Ma, Z. Yu, "An improved differential evolution algorithm for optimization problems", in *Advances in Computer Science, Intelligent System and Environment, Advances in Intelligent and Soft Computing*, vol. 104. Berlin, Heidelberg: Springer, pp. 233–238, 2020.

19. O. Kulkarni, N. Kulkarni, A. J. Kulkarni, G. Kakandikar, "Constrained cohort intelligence using static and dynamic penalty function approach for mechanical components design", *International Journal of Parallel, Emergent and Distributed Systems*, 33(6), 570–588, 2018, Taylor & Francis.

20. T. T. Huan, A. J. Kulkarni, J. Kanesan, et al., "Ideology algorithm: A socio-inspired optimization methodology", *Neural Computing and Applications*, 28, 845–876, 2017. doi: 10.1007/s00521-016-2379-4.

21. S. Mirjalili, A. H. Gandomi, S. Z. Mirjalili, S. Saremi, H. Faris, S. M. Mirjalili, "Salp Swarm algorithm: A bio-inspired optimizer for engineering design problems", *Advances in Engineering Software*, 114, 163–191, 2017. doi: 10.1016/j.advengsoft.2017.07.002.

5

Optimization of Robot Path Planning Using Advanced Optimization Techniques

R. V. Rao and S. Patel

S.V. National Institute of Technology

CONTENTS

DOI: 10.1201/9781003143505-5

Notation

Abbreviations	Full form
γ and δ	Controlling parameters
$(\text{Dist})_{\text{C-OB}}$	Euclidean distance between the best candidate solution and the detected obstacle
$(\text{Dist})_{\text{C-G}}$	Euclidean distance between the best candidate and the goal
$(\text{Dist})_{\text{R-OB}}$	Distance between the location of robot and nearby obstacle
f_i	Objective function
k_1 and k_2	Controlling parameters
$O_{k,j,i}$	Value of the kth variable during the ith iteration for the jth candidate
$O_{k,\text{best},i}$	Value of the variable j for the best candidate
$O_{k,\text{worst},i}$	Value of the variable k for the worst candidate
$O'_{k,j,i}$	Updated value of $O_{k,j,i}$
$r_{1,k,i}$	First random number for the kth variable during the ith iteration in the range
$r_{2,k,i}$	Second random number for the kth variable during the ith iteration in the range
x_{OB} and y_{OB}	Positional coordinates of the detected obstacle
x_{COB} and y_{COB}	Positional coordinates of the best candidate solution
x_G and y_G	Positional coordinates of the goal
x_O and y_O	Positional coordinates of the nearest obstacle
x_c and y_c	Positional coordinates of the candidate

5.1 Introduction

In the past, the use of a robot path planning was restricted in the food processing plant and electrical industry only. The use of a mobile robot is increasing day by day in real life. Recently, the use of robot path planning is increased in the fields of space technology, medical, entertainment industry, agricultural, mining, defense technology, and many more. The task of navigation is not easy, and it requires lots of information. The robot used in robot path planning is equipped with many high-tech gadgets and sensors, which are helping to create a local map of the workspace, locate the position of obstacles,

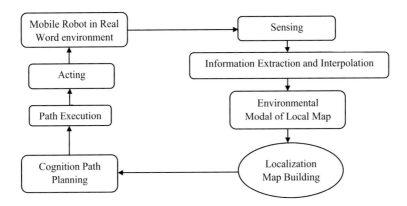

FIGURE 5.1
Flow diagram for mobile robot navigation (Patle et al. 2018).

and ensure smoothness of path and safety of the robot. Smooth and safe path obtained from the origin of the robot to the destination (by detecting and avoiding the obstacles) is the vital role of any optimization technique. The most significant step in robot navigation is the selection of the optimization technique. Based on the optimization technique, robot moves the next position in the workspace. The safety of the robot and the smoothness of the path highly depend on the optimization technique. At present, many scientists try to develop new optimization techniques to navigate the robot in the complex workspace.

The steps required in the navigation of the mobile robot are presented in Figure 5.1.

A robot is equipped with many sensors like infrared, proximity, vision, etc. These sensors help to sense the position of the obstacles in the workspace. Based on information obtained by sensors, the robot analyzes the data collected from sensors and makes one environmental model of a local map. Now, the robot knows the position, type, and shape of obstacles present in the workspace. The next step is to produce the path from the origin to the destination by avoiding the obstacles present in the workspace. This job is done by intelligent optimization techniques.

5.1.1 Navigational Methodology Used for Mobile Robot Path Planning

Many navigation methodologies have been applied by the researcher in mobile robot path planning. The scientist classifies navigation methodology into two approaches, i.e., reactive approach and classical approach.

5.1.1.1 Classical Approaches for Mobile Robot Navigation

In the past, robot navigation problems were solved by classical approaches. These approaches do not require any intelligence technique to navigate the robot to the destination. These approaches are easy to understand and easy to implement in the workspace. By using these approaches to the robot navigation, it is identified that either optimum path would be obtained, or path would be unfeasible. The main problem of these approaches is stuck at local optimum point, high computational time, and failure to quick response; hence, these approaches are less favorable for on-time implementation.

Presently, these approaches are obsolete for a mobile robot path planning problem. Artificial potential field, cell decomposition, and roadmap approach are some of the classical approaches.

5.1.1.2 Reactive Approaches for Mobile Robot Navigation

Recently, many reactive approaches have emerged such as shuffled frog leaping algorithm, cuckoo search (CS) algorithm, neural network, firefly algorithm (FA), genetic algorithm (GA), ant colony optimization (ACO), artificial bee colony (ABC), particle swarm optimization (PSO), and other advanced optimization algorithms. These approaches are very effective to solve complex navigation problems. These approaches can handle multi-robots, the dynamic behavior of obstacles, and moving goal problems in the workspace. It is observed that most of the researchers applied reactive approaches to mobile robot navigation over classical approaches.

5.1.2 Classification of Navigation Strategy

There are two navigation strategies to navigate the robot from the origin to the destination in the workspace, i.e., local navigation strategy and global navigation strategy. These strategies are based on the information required for defining the workspace. An obstacle location, the origin point of the path, and destination are known to the robot in the global navigation strategy, i.e., completely known workspace. Prior information of the workspace is not necessary to navigate the robot from the origin to the destination in local navigation strategy, i.e., unknown workspace. In the local navigation strategy, the robot moves straight toward the destination in the workspace until an obstacle is detected near the robot. Global navigation techniques cannot handle the dynamic environment. The location of the target position, obstacle position, shape, and size are continuously changing with respect to time in a dynamic environment problem. Global navigation strategy requires prior information, and it does not change after implementing a global strategy. Reactive approaches work based on local navigation strategy as they are more intelligent and find an optimum path between the origin and the destination.

5.2 Literature Review

Liang et al. (2013) applied a bacterial foraging approach in the navigation of a mobile robot. In this model, an optimal feasible path from the origin (start point) and the destination (goal) in the workspace is obtained. The workspace is surrounded by stationary obstacles. The authors had considered two case studies with various types of obstacles present in a stationary environment to estimate the performance of the proposed approach.

Mohanty and Parhi (2013) applied a CS-based approach in the application for robot navigation. The authors had considered static and dynamic environments with dissimilar obstacles. The authors had used a global navigation strategy to obtain a feasible path in robot navigation. The authors had proposed the Euclidean distance-based fitness function to navigate the robot between the origin and the goal in the static workspace.

Depending upon the fitness function value, the robot avoided the obstacles and moved in the direction of the target point. Several simulation results were presented to demonstrate the potential of the algorithm.

Mohanty and Parhi (2014) proposed a probabilistic invasive weed optimization (IWO) algorithm to optimize the robot path problem. The authors had applied this algorithm in a completely unknown and partially known workspace. The authors had formulated a fitness function considering two criteria, namely, target seeking behavior and obstacle avoidance behavior. Only a static environment was considered, i.e., static obstacle. Based on the objective function value, the robot had navigated toward the target. An optimal path was generated, and the results were compared with the results of the other advanced optimization techniques.

Das et al. (2016b) proposed an improved gravitational search algorithm (IGSA) in multi-robot path planning. They had considered a dynamic environment in which the position, shape, and size of the obstacles are changed with respect to time. A robot path planning methodology was developed with the help of IGSA to obtain the new position of mobile robots from the previous position of robots. The authors had considered seven obstacles and five robots in a cultivated environment to verify the performance of the proposed algorithm in the field of robot navigation. Finally, the authors had compared IGSA solutions with the solution given by other optimization techniques. The authors had used five Khepera-2 robots in the stationary environment for experimental purpose.

Liang and Lee (2015) applied an efficient artificial bee colony (EABC) algorithm for solving the problem of multiple robot navigation in the same environment. The authors had considered three objective functions for target, obstacles, and robots for collision avoidance. The first objective function was to obtain a minimum path from the origin to the destination. The second objective function was to avoid the robot collision, and the third objective function was to avoid the stationary obstacles existing in the workspace.

Mohanty and Parhi (2016) used a CS algorithm for the navigation of a mobile robot in a static and partially known environment. In this environment, the positions of the obstacles were fixed. Two objective functions were formulated: the first was the distance between the goal and the robot, which needs to be minimized, and the second one was the distance between the robot and obstacles, which needs to be maximized. Finally, the authors had formed a combined fitness function, which was minimized. The fitness function fulfilled the condition of obstacle avoidance and goal-achieving behavior of the robot that was present in an unknown environment. Depending upon the fitness function value, the robot avoided the obstacles and moved towards the target point.

Das et al. (2016a,b) applied an improved particle swarm optimization (IPSO) and differentially perturbed velocity (DV) algorithm as a hybrid algorithm to obtain an optimal path for the multi-robots. The authors had considered multiple robots in the same environment but the goal and the starting position of each robot were different. The authors had applied a hybrid algorithm to minimize the total time and path length of all robots to their respective target positions. The energy factor was also considered along with the path length and time in the proposed scheme. Finally, the optimal and feasible path was generated with the help of the proposed algorithm. In the experimental study, the authors had used Khepera-2 robot in a static environment.

Zeng et al. (2016) proposed an ant colony algorithm with the free step length concept to move the robot toward the target. The authors had considered a grid workspace to demonstrate the proposed algorithm. In the grid environment, the workspace is divided into small rectangular or square boxes. The dimension of these boxes depends upon the size of the robot and the obstacle present in the workspace. The main drawback of the grid environment is that only rectangular- or square-shaped obstacles are considered in the workspace. If the shape and the size of the robot or the obstacle change, then we have to change the grid size according to the obstacles and the robot. It can be observed that the grid-type workspace could not handle the dynamic problem of robot navigation.

Kim et al. (2017) suggested a GA to obtain the optimized path with minimum navigation time for an autonomous vehicle under the environment loads such as current, wave, and wind. Initially, the path problems were solved without considering the environment load. The authors had considered 10 and 20 m radius around the obstacles for the safety of unmanned aerial vehicle (UAV). Finally, three objective functions, namely, obstacle avoidance, reaching the target point, and minimizing travel time, were considered. For the simulation purpose, the positions of the obstacles were generated randomly in MATLAB.

Pandey and Parhi (2017) applied fuzzy-wind-driven optimization (FWDO) to obtain the minimum and feasible path for robot navigation. The authors had defined intermediate points between the origin and the destination in the workspace, and the robot must pass through these intermediate points during navigation. The authors used Khepera-3 robot for the experimental study with the proposed algorithm in the static and the dynamic obstacles.

Patle et al. (2018a) attempted optimization in the field of robot path planning using FA. The authors proposed a local navigation strategy to obtain a collision-free path from the origin to the destination. In the simulation, five different workspaces with different static and dynamic obstacles were considered to validate the proposed algorithm. Finally, they designed an effective controller for the generation of a collision-free path in an uncertain environment. The constraints like path smoothness and path safety were also considered.

Patle et al. (2018b) applied the probability distribution-based fuzzy logic for robot navigation in a cultivated environment. The authors had studied the geometric constraints present in the workspace and produced six fuzzy parameters. Certain values of these six fuzzy parameters were assigned, and a probabilistic fuzzy logic controller was tested by using Khepera-2 mobile robot. Finally, the authors had simulated the code in MATLAB to get an optimum result in path length and simulation time.

Orozco-Rosas et al. (2019) proposed evolutionary artificial potential field (memEAPF) approach for the mobile robot path planning. They used GA and artificial potential field method to obtain safe and optimized path in complex environment. The proposed approach had compared with artificial potential field-based robot path planning methods. They had applied this method in dynamic environment.

Lopez et al. (2020) applied the real-time method (BCM-DO) to include the restrictions imposed by dynamic objects. The researchers had formulated one optimization function. They had formulated the problem as one of the constrained optimizations in the velocity space of the robot. They tested this system in the simulation finally applied to a hotel assistant robot (BellBot).

It has been observed from the literature review that the researcher had used advanced optimization techniques such as FA, GA, wind-driven optimization (WDO), ACO, EABC, PSO, IGSA, CS, etc. However, these advanced optimization techniques need proper tuning of their algorithm-specific parameters in addition to tuning of common controlling parameters. Improper tuning of parameters leads to inferior results. Hence, there is need to apply algorithm-specific parameter-less optimization techniques. In this dissertation, recently developed algorithm-specific parameter-less algorithms such as Jaya and Rao algorithms are used for optimization robot path planning.

5.3 Optimization Algorithms

There are many methods available to solve the problems of optimization. Some of them are inspired by the processes happening in nature. These methods generally start with an initial set of variables and then move towards the global optimum value (minimum or maximum) of the objective function. The population-based algorithms can be classified into two groups: swarm intelligence (SI)-based algorithms and evolutionary algorithms (EA). Some of the well-known SI-based algorithms are shushed frog leaping (SFL), fire fly (FF) algorithm, ABC, PSO, ACO, etc. The group of EAs are evolution programming (EP), evolution strategy (ES), GA, differential evolution (DE), artificial immune algorithm (AIA), bacteria foraging optimization (BFO), etc.

All the SI- and evolutionary-based algorithms are probabilistic algorithms and require common controlling parameters such as the number of generations, population size, elite size, etc. Also, these algorithms involve parameters that are specific to the algorithms. The performance of the mentioned algorithms depends upon the proper tuning of the algorithm-specific parameters. The improper tuning of algorithm-specific parameters can lead to trapping of the solution at local optima or increased computational efforts. Considering this issue, the teaching learning-based optimization (TLBO) algorithm was introduced by Rao (2011), which does not require any algorithm-specific parameters. The TLBO algorithm requires only common control parameters like the number of generations and population size, and because of this fact, it has gained interest and wide acceptance among the optimization researchers.

Another algorithm-specific parameter-less algorithm named as Jaya has been proposed by Rao (2016). The Jaya algorithm contains only one phase, and it is comparatively simpler to apply. This algorithm is based on the concept that the solution obtained for a given problem should move towards the best solution and should avoid the worst solution. The Jaya algorithm requires only common control parameters and does not require any algorithm-specific control parameters.

5.3.1 Jaya Algorithm

Let $Z(x)$ be the objective function to be optimized. Let 'p' be the number of design variables and 'q' be the number of candidate solutions (i.e., population size, $j=1, 2, ..., q$) at any iteration i. From the entire candidate solutions, the best value of objective function $Z(x)$ (i.e., $Z(x)_{best}$) obtained with best candidate is termed as 'best', and the worst value of $Z(x)$ (i.e., $Z(x)_{worst}$) obtained by the worst candidate is termed as 'worst'. Figure 5.2 shows the flowchart of the Jaya algorithm.

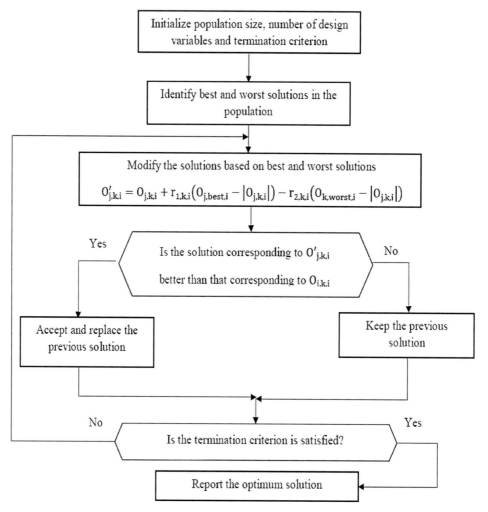

FIGURE 5.2
Flow chart of Jaya algorithm (Rao, 2016).

The value of the kth variable during the ith iteration for the jth candidate is termed as $O_{k,j,i}$, and this value is modified as per Equation (5.1).

$$O'_{k,j,i} = O_{k,j,i} + r_{1,k,i}\left(O_{k,best,i} - |O_{k,j,i}|\right) - r_{2,k,i}\left(O_{k,worst,i} - |O_{k,j,i}|\right) \tag{5.1}$$

where $O_{k,best,i}$ is the value of the variable j for the best candidate and $O_{k,worst,i}$ is the value of the variable k for the worst candidate. $O'_{k,j,i}$ is the updated value of $O_{k,j,i}$, and $r_{1,k,i}$ and $r_{2,k,i}$ are the two random numbers for the kth variable during the ith iteration in the range [0, 1]. The term $r_{1,k,i}(O_{k,best,i} - |O_{k,j,i}|)$ indicates the tendency of the solution to move closer to the best solution, and the term $r_{2,k,i}(O_{k,worst,i} - |O_{k,j,i}|)$ indicates the tendency of the solution to avoid the worst solution. $O'_{k,j,i}$ is accepted if it gives a better function value. All the accepted function values at the end of the iteration are maintained, and these values become the input to the next iteration. In the proposed algorithm, the solution obtained for a given problem is moving towards the best solution and avoiding the worst solution. The random numbers r_1 and r_2 take care of good exploration of the search space.

5.3.2 Rao Algorithms

Rao algorithms have been developed by Rao (2020). These are also algorithm-specific parameter-less algorithms.

Let $Z(x)$ be the objective function to be optimized. Let 'p' be the number of design variables and 'q' be the number of candidate solutions (i.e., population size, $j=1, 2, ..., q$) at any iteration i. From the entire candidate solutions, the best value of objective function $Z(x)$ (i.e., $Z(x)_{best}$) obtained with best candidate is termed as 'best' and the worst value of $Z(x)$ (i.e., $Z(x)_{worst}$) obtained by the worst candidate is termed as 'worst'. If $O_{k,j,i}$ is the value of the kth variable for the jth candidate during the ith iteration, then this value is modified as per the following equations:

$$O'_{k,j,i} = O_{k,j,i} + r_{1,k,i}\left(O_{k,\text{best},i} - O_{k,\text{worst},i}\right), \tag{5.2}$$

$$O'_{k,j,i} = O_{k,j,i} + r_{1,k,i}\left(O_{k,\text{best},i} - O_{k,\text{worst},i}\right) + r_{2,k,i}\left(\left|O_{k,j,i} \text{ or } O_{k,l,i}\right| - \left|O_{k,l,i} \text{ or } O_{k,j,i}\right|\right), \tag{5.3}$$

$$O'_{k,j,i} = O_{k,j,i} + r_{1,k,i}\left(O_{k,\text{best},i} - \left|O_{k,\text{worst},i}\right|\right) + r_{2,k,i}\left(\left|O_{k,j,i} \text{ or } O_{k,l,i}\right| - \left(O_{k,l,i} \text{ or } O_{k,j,i}\right)\right), \tag{5.4}$$

where $O_{k,\text{best},i}$ is the value of the variable k for the *best* candidate and $O_{k,\text{worst},i}$ is the value of the variable k for the *worst* candidate during the ith iteration and $O'_{k,j,i}$ is the updated value of $O_{k,j,i}$, $r_{1,k,i}$ and $r_{2,k,i}$ are the two random numbers for the kth variable during the ith iteration in the range [0, 1].

Figure 5.3 shows the flowchart of Rao algorithms.

In Equations (5.3) and (5.4), the term $O_{k,j,i}$ or $O_{k,l,i}$ indicates that the candidate solution j is compared with any randomly picked candidate solution l and the information is exchanged based on their fitness values. If the fitness value of jth solution is better than the fitness value of lth solution, then the term '$O_{k,j,i}$ or $O_{k,l,i}$' becomes $O_{k,j,i}$. On the other hand, if the fitness value of lth solution is better than the fitness value of jth solution, then the term '$O_{k,j,i}$ or $O_{k,l,i}$' becomes $O_{k,l,i}$. Similarly, if the fitness value of jth solution is better than the fitness value of lth solution, then the term '$O_{k,l,i}$ or $O_{k,j,i}$' becomes $O_{k,l,i}$. If the fitness value of lth solution is better than the fitness value of jth solution, then the term '$O_{k,l,i}$ or $O_{k,j,i}$' becomes $O_{k,j,i}$.

5.4 Applications of the Jaya and Rao Algorithms to Robot Path Planning Optimization

The aim of this dissertation work is to optimize the robot path by using Jaya and Rao algorithms and to compare the same with the results of the other algorithms to see if there are any improvements. The Jaya and Rao algorithms are coded in MATLAB 2015.

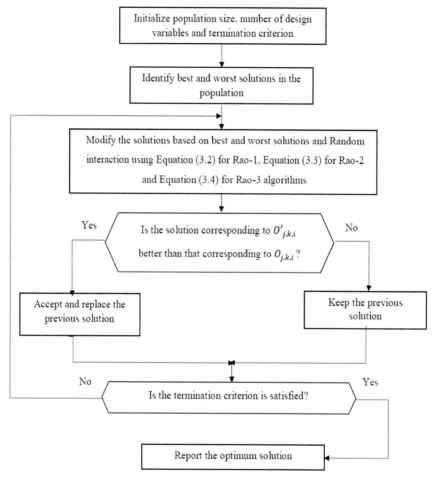

FIGURE 5.3
Flow chart of Rao algorithms (Rao, 2020).

The four case studies have been attempted for justification and demonstration of the Jaya and Rao algorithms.

5.4.1 Case Study 1

This case study was previously attempted by using GA (Wang et al., 2010), PSO (Mohamed et al. 2011), and CS algorithm (Mohanty and Parhi 2016). In this case study, a stationary workspace has been considered for the optimization of the robot path.

5.4.1.1 Objective Function Formulation for Case Study 1

Initially, the robot at the origin (start-a) in the workspace is shown in Figure 5.4, and the robot starts to move in a straight direction towards the destination until sensors detect

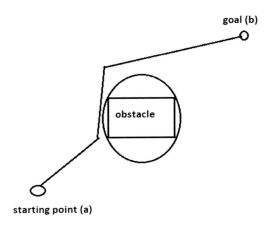

FIGURE 5.4
Robot path in the presence of an obstacle.

the obstacle nearby the robot. The robot traverses from the origin (start-a) to the destination (goal-b) by avoiding a static obstacle present in the workspace. The path obtained between the origin and the destination is optimal, desired, collision-free, and smooth for the robot to reach the destination.

To formulate the objective function, we need to consider two criteria, i.e., goal searching and obstacle avoidance.

5.4.1.2 Obstacle Avoidance Behavior for Case Study 1

The safety of the robot is an important aspect of robot navigation in a complex workspace. Obstacle avoidance behavior helps the robot to maintain a specific safe distance from the detected obstacle. The candidates generated during the navigation should have the safest distance from the detected obstacles. The Euclidean distance between the best candidate solution and the detected obstacle is calculated by the following equation:

$$(\text{Dist})_{\text{C-OB}} = \sqrt{(x_{\text{OB}} - x_{\text{COB}})^2 + (y_{\text{OB}} - y_{\text{COB}})^2} \qquad (5.5)$$

where x_{OB} and y_{OB} are the positional coordinates of the detected obstacle, and x_{COB} and y_{COB} are the positional coordinates of the best candidate solution.

5.4.1.3 Goal Searching Behavior for Case Study 1

Goal searching behavior is helping in acquiring the new position of the robot in such a way that the distance between the robot and the target should be minimum. The candidate solutions generated during the navigation should have a minimum distance from the goal. The Euclidean distance between the best candidate and the goal is calculated by the following equation:

$$(\text{Dist})_{\text{C-G}} = \sqrt{(x_{\text{G}} - x_{\text{COB}})^2 + (y_{\text{G}} - y_{\text{COB}})^2} \qquad (5.6)$$

where x_G and y_G are the positional coordinates of the goal, and x_{COB} and y_{COB} are the positional coordinates of the best candidate.

The final objective function is expressed as follows:

$$\text{Objective function}\left(f_i\right) = k_1 \frac{1}{\min_{\text{OBS}_j \in \text{OBS}_d}\left\|\left(\text{Dist}\right)_{\text{C-OB}}\right\|} + k_2\left\|\left(\text{Dist}\right)_{\text{C-G}}\right\| \qquad (5.7)$$

where k_1 and k_2 are the controlling parameters.

The workspace is considered with 'm' number of obstacles existing in the workspace, and we denote them as OBS_1, OBS_2, OBS_3, …, OBS_n. The positional coordinates of the obstacles are $(x_{\text{OBS1}}, y_{\text{OBS1}})$, $(x_{\text{OBS2}}, y_{\text{OBS2}})$, $(x_{\text{OBS3}}, y_{\text{OBS3}})$, …, $(x_{\text{OBSn}}, y_{\text{OBSn}})$, respectively. From Equation (5.7), we can understand that when the robot is nearer to target, the value of $\left\|\left(\text{Dist}\right)_{\text{C-G}}\right\|$ will be reduced. When the robot moves far from the obstacles, the value of $\min_{\text{OBS}_j \in \text{OBS}_d}\left\|\left(\text{Dist}\right)_{\text{C-OB}}\right\|$ will be more. We need to combine the two objective functions for minimization. From the above discussion, we can understand that $\min_{\text{OBS}_j \in \text{OBS}_d}\left\|\left(\text{Dist}\right)_{\text{C-OB}}\right\|$ is the maximization function (keeping the robot away from the obstacle). Hence, to form the combined objective function for minimization, we need to consider the reciprocal of $\min_{\text{OBS}_j \in \text{OBS}_d}\left\|\left(\text{Dist}\right)_{\text{C-OB}}\right\|$.

From Equation (5.7), we can observe that the robot path length and smoothness depend on the controlling parameters k_1 and k_2. When k_1 is large, the robot considers $\min_{\text{OBS}_j \in \text{OBS}_d}\left\|\left(\text{Dist}\right)_{\text{C-OB}}\right\|$ as a prime objective and the optimization algorithms plot trajectory in such a way that robot is always far from the obstacle. If k_1 is small, the optimization algorithms plot trajectory in such a way that the robot may hit the obstacle. On the other hand, when k_2 is large, the robot considers $\left\|\left(\text{Dist}\right)_{\text{C-G}}\right\|$ as a prime objective and the optimization algorithms try to plot a shorter path. In this work, controlling parameters are chosen by the various trials in the workspace.

In the fixed environment, the location, shape, and size of the obstacles do not change after the workspace is generated. Initially, the robot does not know about the location, shape, and size of the obstacles. The robot has a specific range of sensing in the workspace. When the robot starts traveling towards the goal, the robot must sense the unknown obstacles within the sensing range. In this case study, the sensing range is chosen as 2 cm around the location of the robot. Two examples are demonstrated by generating obstacles of dissimilar shapes and various sizes in the workspace.

5.4.1.4 Case Study 1: Example 1

This problem was attempted by the researcher in the past using GA (Wang et al., 2010), PSO (Mohamed et al. 2011), and CS algorithm (Mohanty and Parhi 2016). The total number of function evaluation applied by the previous researchers was 3,000. Hence, to demonstrate and validate the Jaya and Rao algorithms, the same number of function evaluation is used for fair comparison. Hence, in the present work, various combinations of number of generations and population size are tried to get function evaluations of 3,000. Finally, the number of iterations of 100 and population size of 30 are set for obtaining optimum results. In this example, six number of obstacles are considered in the workspace.

- The specifications of example are:
  - Start point=(5, 95)
  - Goal point=(95, 5)
- Parameter bounds:
  - $0 \leq x \leq 100$
  - $0 \leq y \leq 100$

Figure 5.5 shows the optimum path from starting (5, 95) to goal (95, 5) by robot after the application of Jaya and Rao algorithms.

Jaya and Rao algorithms have calculated 47 points. Table 5.1 shows the coordinates obtained in example 1 of case study 1 from starting point (5, 95) to goal (95, 5). Rao-3 algorithm has calculated 47 optimum coordinates as shown in Table 5.1.

Table 5.2 presents the comparison of results of Jaya and Rao algorithms with CS algorithm, PSO, and GA.

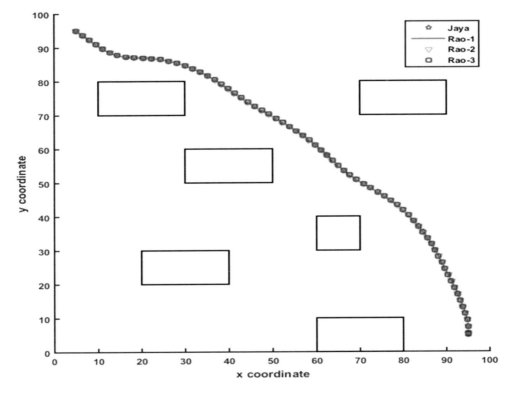

FIGURE 5.5
Path traveled by robot for example 1 of case study 1 by Jaya and Rao algorithms.

TABLE 5.1

Coordinates for Robot in Example 1 of Case Study

No.	X-Coordinate	Y-Coordinate	No.	X-Coordinate	Y-Coordinate
1	5.00	95.00	25	63.83	57.11
2	7.27	93.04	26	65.90	54.94
3	9.63	91.19	27	68.01	52.81
4	12.09	89.48	28	70.36	50.96
5	14.92	88.49	29	72.74	49.12
6	17.90	88.15	30	75.19	47.39
7	20.89	87.92	31	77.46	45.44
8	23.87	87.58	32	79.60	43.35
9	26.80	86.96	33	81.54	41.06
10	29.62	85.93	34	83.32	38.65
11	32.28	84.55	35	85.04	36.19
12	34.81	82.94	36	86.35	33.50
13	37.23	81.18	37	87.68	30.81
14	39.50	79.22	38	88.78	28.03
15	41.62	77.10	39	89.89	25.24
16	44.00	75.28	40	90.94	22.43
17	46.35	73.42	41	91.96	19.62
18	48.75	71.62	42	93.03	16.82
19	51.05	69.69	43	93.84	13.93
20	53.37	67.80	44	94.57	11.03
21	55.70	65.90	45	95.04	8.06
22	57.85	63.81	46	95.02	5.06
23	59.92	61.64	47	95.00	5.00
24	61.93	59.43			

TABLE 5.2

Results Comparison for Example 1 of Case Study 1

Method	Total Path Length (cm)
Genetic algorithm (Wang et al. 2010)	171.4
Particle swarm optimization (Mohamed et al. 2011)	147.6
Cuckoo search algorithm (Mohanty and Parhi 2016)	145.25
Jaya algorithm	135.2107
Rao-1 algorithm	135.1213
Rao-2 algorithm	135.0509
Rao-3 algorithm	**134.9616**
Bold value shows best result obtained by the algorithm	

It can be seen that Rao-3 algorithm has produced a minimum total path length of 134.9616 cm. The performance of the algorithms may be arranged in the following sequence:

Rao-3- Rao-2-Rao-1 – Jaya - cuckoo search algorithm – PSO - GA.

5.4.1.5 Case Study 1: Example 2

This problem was attempted by the researchers in the past using particle swarm optimization (Mohamed et al. 2011) and CS algorithm (Mohanty and Parhi 2016). The total number of function evaluations used by the previous researchers was 3,000. Hence, to demonstrate and validate the Jaya and Rao algorithms, the same number of function evaluation is used for fair comparison. Hence, in the present work, various combinations of number of generations and population size are tried to get function evaluations of 3,000. Finally, the number of iterations of 100 and population size of 30 are set for obtaining optimum results. In this example, three number of obstacles are considered in the workspace.

- The specifications of example are:
    - Start point=(0, 0)
    - Goal point=(95, 80)
- Parameter bounds:
    - $0 \leq x \leq 100$
    - $0 \leq y \leq 100$

Figure 5.6 shows the optimum path from starting (0, 0) to goal (95, 80) by robot after the application of Jaya and Rao algorithms.

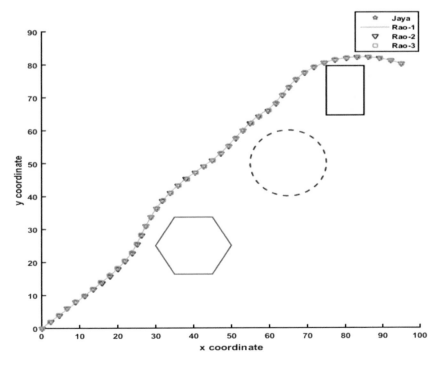

FIGURE 5.6
Path traveled by robot for example 2 of case study 1 by Jaya and Rao algorithms.

TABLE 5.3

Coordinates for Robot in Example 2 of Case Study 1

No.	X-Coordinate	Y-Coordinate	No.	X-Coordinate	Y-Coordinate
1	0.00	0.00	16	48.59	55.89
2	3.86	3.18	17	51.71	59.79
3	7.60	6.49	18	55.09	63.47
4	11.27	9.88	19	58.96	66.64
5	14.87	13.35	20	62.09	70.53
6	18.44	16.85	21	65.24	74.42
7	21.59	20.73	22	69.01	77.70
8	23.94	25.14	23	73.38	80.12
9	25.86	29.76	24	78.18	81.52
10	28.07	34.24	25	83.15	82.09
11	30.86	38.39	26	88.14	81.84
12	34.11	42.19	27	93.04	80.82
13	37.74	45.63	28	95.00	80.00
14	41.69	48.69	29	95.00	80.00
15	45.41	52.02			

TABLE 5.4

Results Comparison for Example 2 of Case Study 1

Method	Total Path Length (cm)
Particle swarm optimization (Mohamed et al. 2011)	145.26
Cuckoo search algorithm (Mohanty and Parhi 2016)	145.01
Jaya algorithm	**132.0981**
Rao-1 algorithm	132.5449
Rao-2 algorithm	132.4979
Rao-3 algorithm	132.2857
Bold value shows best result obtained by the algorithm	

Jaya and Rao algorithms have calculated 29 points. Table 5.3 shows the coordinates obtained in example 2 of case study 1 from starting point (0, 0) to goal (95, 80). Jaya algorithm has calculated 29 optimum coordinates as shown in Table 5.3.

Table 5.4 presents the comparison of results of Jaya and Rao algorithms with those of CS algorithm and PSO.

It can be seen that Jaya algorithm has produced a minimum total path length of 132.0981 cm. The performance of the algorithms may be arranged in the following sequence:

Jaya - Rao-3- Rao-2-Rao-1 - cuckoo search algorithm - PSO.

5.4.2 Case Study 2

This case study was previously attempted by Patle et al. (2018a) using a FA and Joshi et al. (2011) using neuro-fuzzy approach. In this case study, a stationary workspace has been considered for the optimization of the robot path.

5.4.2.1 Objective Function Formulation for Case Study 2

In formulation of the objective function, several challenges have to be considered, which include the detection of the obstacle, avoiding random solutions, feasible and smooth trajectory, etc. Initially, the robot at the starting position (x_1, y_1) tries to move in the direction of the target position (x_n, y_n) as shown in Figure 5.7. The Jaya and Rao algorithms produce the number of solutions in the circle of searching radius, and the best solution is chosen from the group of the solution generated.

5.4.2.2 Obstacle Avoidance Behavior for Case Study 2

Obstacle avoidance behavior helps the robot to maintain a specific safe distance from the detected obstacle. The best candidate is selected by using Euclidean distance shown by Equation (5.8).
  Then, Euclidean distance is

$$(\text{Dist})_{\text{C-OB}} = \sqrt{(x_O - x_c)^2 + (y_O - y_c)^2} \tag{5.8}$$

Let $(\text{Dist})_{\text{R-OB}}$ be the Euclidean distance between the robot and the nearest obstacle. x_c and y_c are the positional coordinates of the best candidate solution, and x_O and y_O are the positional coordinates of the nearest obstacle, respectively.
  Then, the distance between the location of robot and the nearby obstacle is presented as:

$$(\text{Dist})_{\text{R-OB}} = \sqrt{(x_O - x_c)^2 + (y_O - y_c)^2} \tag{5.9}$$

5.4.2.3 Goal Searching Behavior for Case Study 2

The main aim of optimization of robot path planning is to obtain the minimum path, and this is achieved by the goal searching behavior. This behavior helps to find the next best position of the robot, i.e., best candidate solution. The searching process is continued until the optimum path is generated.

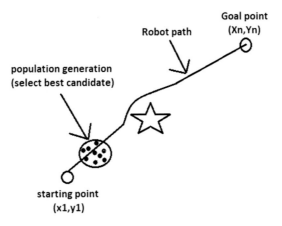

FIGURE 5.7
Robot path in the presence of an obstacle.

Let $(\text{Dist})_{C-G}$ be the Euclidean distance between the candidate solution and the goal. x_c and y_c are the positional coordinates of the candidate, and x_G and y_G are the coordinates of goal position, respectively. The distance between the candidate and the goal is calculated by Equation (5.10):

$$(\text{Dist})_{C-G} = \sqrt{\left(x_G - x_c\right)^2 + \left(y_g - y_c\right)^2} \tag{5.10}$$

From the above study, we can see that one objective is of maximization type and other objective is of minimization type. Hence, to combine the objective functions and to minimize the value of the combined objective function, we need reciprocal of maximization function.

Finally, the optimization problem is represented as follows:

$$\text{Fitness function}\left(f_i\right) = C_1 \frac{1}{\underset{O_j \in O_d}{\min} \left\|(\text{Dist})_{C\text{-OB}}\right\|} + C_2 \left\|(\text{Dist})_{C-G}\right\| \tag{5.11}$$

The workspace is considered with 'n' number of obstacles, which are represented as $O_1, O_2, O_3, O_4, \ldots, O_n$, and their coordinate positions are (x_{O1}, y_{O1}), (x_{O2}, y_{O2}), (x_{O3}, y_{O3}), $\ldots$ (x_{On}, y_{On}). The value of the first term and the second term of Equation (5.11) depends on the new position of the robot. When the new position of the robot is far from the obstacle, the value of the first term is reduced in the fitness function, and the second term will be reduced when the new position of the robot is closer to the target. From Equation (5.11), we can observe that the robot path length and smoothness depend on the controlling parameters C_1 and C_2. When C_1 is large, the robot considers $\underset{O_j \in O_d}{\min} \left\|(\text{Dist})_{C\text{-OB}}\right\|$ as a prime objective, and the optimization algorithms plot trajectory in such a way that robot is always far from the obstacle. If C_1 is small, the optimization algorithms plot trajectory in such a way that the robot may hit the obstacle. On the other hand, when C_2 is large, the robot considers $\left\|(\text{Dist})_{C-G}\right\|$ as a prime objective and the optimization algorithms try to plot a shorter path. In this work, controlling parameters are chosen by the various trials in the workspace.

In this case study, the author presented the application of a FA for local path planning in a different workspace. Initially, the robot knows the origin and the destination. When the robot starts moving, the robot must know the nature and position of the obstacles to obtain the path. The sensors provide information about size, position, shape, and distance between the robot and the obstacle to the robot. Five examples are demonstrated by generating obstacles of different shapes and sizes in the workspace. The first four examples are single robot moving in the workspace, and the fifth example is multirobots moving in the same workspace.

5.4.2.4 Case Study 2: Example 1

This problem was attempted by the researcher in the past using neuro-fuzzy approach (Joshi et al. 2011) and FA (Patle et al. 2018a). The total number of function evaluation used by the previous researchers was 5,000. Hence, to demonstrate and validate the Jaya and Rao algorithms, the same number of function evaluation is used for fair comparison. Hence, in the present work, various combinations of number of generations and population size are tried to get function evaluations of 5,000. Finally, the number of iterations of 100 and population size of 50 are set for obtaining optimum results. In this example, seven number of obstacles are considered in the workspace.

- The specifications of example are:
  - Start point=(20, 0)
  - Goal point=(25, 85)
- Parameter bounds:
  - $0 \leq x \leq 90$
  - $0 \leq y \leq 90$

Figure 5.8 shows the optimum path from starting (20, 0) to goal (25, 85) by robot after the application of Jaya and Rao algorithms.

Jaya and Rao algorithms have calculated 32 points. Table 5.5 shows the coordinates obtained in example 1 of case study 2 from starting point (20, 0) to goal (25, 85). Rao-2 algorithm has calculated 32 optimum coordinates as shown in Table 5.5.

Table 5.6 presents the comparison of results of Jaya and Rao algorithms with those of neuro-fuzzy and FAs.

It can be seen that Rao-2 algorithm has produced a minimum total path length of 154.5871 cm. The performance of the algorithms may be arranged in the following sequence:

Rao-2- Rao-3- Rao-1 – Jaya - firefly algorithm - neuro-fuzzy approach.

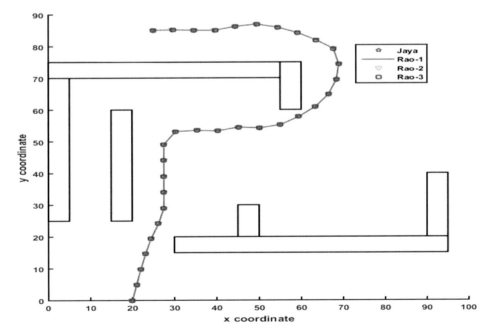

FIGURE 5.8
Path traveled by robot for example 1 of case study 2 by Jaya and Rao algorithms.

TABLE 5.5

Coordinates for Robot in Example 1 of Case Study 2

No.	X-Coordinate	Y-Coordinate	No.	X-Coordinate	Y-Coordinate
1	20.00	0.00	17	55.07	55.24
2	20.97	4.90	18	59.40	57.74
3	21.97	9.80	19	63.28	60.90
4	23.10	14.67	20	66.42	64.79
5	24.36	19.51	21	68.41	69.37
6	26.07	24.21	22	68.95	74.34
7	27.50	29.00	23	67.63	79.17
8	27.50	34.00	24	63.44	81.88
9	27.50	39.00	25	59.00	84.20
10	27.50	44.00	26	54.29	85.85
11	27.50	49.00	27	49.39	86.88
12	30.33	53.13	28	44.44	86.19
13	35.31	53.51	29	39.58	85.01
14	40.31	53.33	30	34.58	85.00
15	45.18	54.48	31	29.59	85.21
16	50.17	54.24	32	25.00	85.00

TABLE 5.6

Results Comparison for Example 1 of Case Study 2

Method	Total Path Length (cm)
Neuro-fuzzy (Joshi et al. 2011)	198
Firefly algorithm (Patle et al. 2018a)	186
Jaya algorithm	154.6636
Rao-1 algorithm	154.6353
Rao-2 algorithm	**154.5871**
Rao-3 algorithm	154.6000

Bold value shows best result obtained by the algorithm

5.4.2.5 Case Study 2: Example 2

This problem was attempted by the researcher in the past using fuzzy-neural network (Shi et al. 2009) and FA (Patle et al. 2018a). The total number of function evaluation used by the previous researchers was 5,000. Hence, to demonstrate and validate the Jaya and Rao algorithms, the same number of function evaluation is used for fair comparison. Hence, in the present work, various combinations of number of generations and population size are tried to get function evaluations of 5,000. Finally, the number of iterations of 100 and population size of 50 are set for obtaining optimum results. In this example, four number of obstacles are considered in the workspace.

- The specifications of example are:
  - Start point = (22, 5)
  - Goal point = (21, 26)

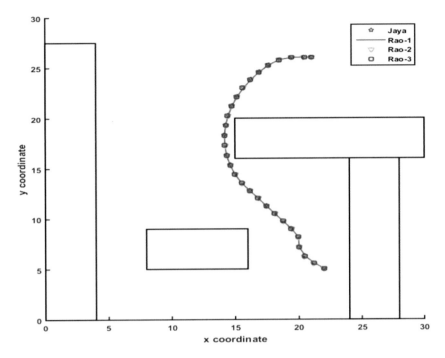

FIGURE 5.9
Path traveled by robot for example 2 of case study 2 by Jaya and Rao algorithms.

- Parameter bounds:
  - $5 \leq x \leq 26$
  - $5 \leq y \leq 26$

Figure 5.9 shows the optimum path from starting (22, 5) to goal (21, 26) by robot after the application of Jaya and Rao algorithms.

Jaya and Rao algorithms have calculated 29 points. Table 5.7 shows the coordinates obtained in example 2 of case study 2 from starting point (22, 5) to goal (21, 26). Rao-3 algorithm has calculated 29 optimum coordinates as shown in Table 5.7.

Table 5.8 presents the comparison of results of Jaya and Rao algorithms with those of fuzzy-neural network and FA.

It can be seen that Rao-3 algorithm has produced a minimum total path length of 137.2647 cm. The performance of the algorithms may be arranged in the following sequence:

Rao-3- Rao-1 – Jaya - Rao-2- firefly algorithm - fuzzy-neural network.

5.4.2.6 Case Study 2: Example 3

This problem was attempted by the researcher in the past using FA (Patle et al. 2018a). The total number of function evaluation used by the previous researchers was 5,000. Hence, to demonstrate and validate the Jaya and Rao algorithms, the same number of

TABLE 5.7

Coordinates for Robot in Example 2 for Case Study 2

No.	X-Coordinate	Y-Coordinate	No.	X-Coordinate	Y-Coordinate
1	22.00	5.00	16	14.25	17.31
2	21.20	5.60	17	14.23	18.31
3	20.47	6.28	18	14.31	19.31
4	20.02	7.18	19	14.53	20.29
5	19.96	8.17	20	14.82	21.24
6	19.40	9.00	21	15.23	22.15
7	18.75	9.76	22	15.72	23.03
8	18.09	10.52	23	16.30	23.84
9	17.44	11.27	24	16.96	24.59
10	16.80	12.03	25	17.73	25.23
11	16.14	12.79	26	18.56	25.78
12	15.51	13.57	27	19.54	26.00
13	15.00	14.42	28	20.54	26.00
14	14.64	15.36	29	21.00	26.00
15	14.38	16.32			

TABLE 5.8

Results Comparison for Example 2 of Case Study 2

Method	Total Path Length (cm)
Fuzzy-neural network (Shi et al. 2009)	180
Firefly algorithm (Patle et al. 2018a)	151.5
Jaya algorithm	137.649
Rao-1 algorithm	137.3625
Rao-2 algorithm	137.6914
Rao-3 algorithm	**137.2647**
Bold value shows best result obtained by the algorithm	

function evaluation is used for fair comparison. Hence, in the present work, various combinations of number of generations and population size are tried to get function evaluations of 5,000. Finally, the number of iterations of 100 and population size of 50 are set for obtaining optimum results. In this example, three number of obstacles are considered in the workspace.

- The specifications of example are:
  - Start point $= (32, 8)$
  - Goal point $= (78, 80)$
- Parameter bounds:
  - $0 \leq x \leq 85$
  - $0 \leq y \leq 85$

Figure 5.10 shows the optimum path from starting (32, 8) to goal (78, 80) by robot after the application of Jaya and Rao algorithms.

Jaya and Rao algorithms have calculated 50 points. Table 5.9 shows the coordinates obtained in example 3 of case study 2 from starting point (32, 8) to goal (78, 80). Rao-1 algorithm has calculated 50 optimum coordinates as shown in Table 5.9.

Table 5.10 presents the comparison of results of Jaya and Rao algorithms with that of FA.

It can be seen that Rao-1 algorithm has produced the minimum total path length of 179.4836 cm and simulation time of 8.466 s. The performance of the algorithms may be arranged in the following sequence:

Rao-1- Rao-2-Rao-3 – Jaya - firefly algorithm.

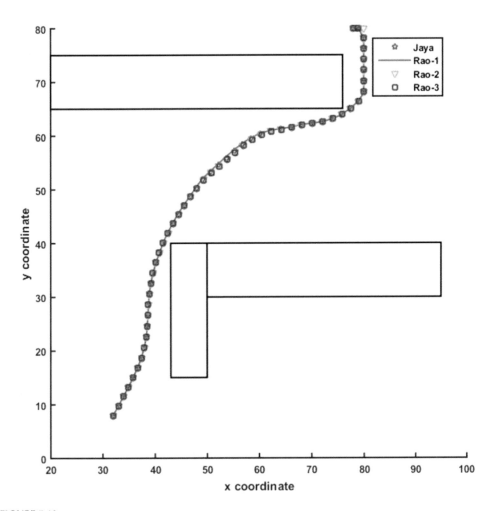

FIGURE 5.10
Path traveled by robot for example 3 of case study 2 by Jaya and Rao algorithms.

TABLE 5.9

Coordinates for Robot in Example 3 of Case Study 2

No.	X-Coordinate	Y-Coordinate	No.	X-Coordinate	Y-Coordinate
1	32.00	8.00	26	50.77	53.16
2	32.97	9.75	27	52.28	54.47
3	34.00	11.46	28	53.77	55.81
4	34.95	13.22	29	55.30	57.09
5	35.88	14.99	30	56.89	58.30
6	36.76	16.79	31	58.56	59.40
7	37.50	18.64	32	60.38	60.22
8	38.04	20.57	33	62.28	60.84
9	38.41	22.53	34	64.26	61.16
10	38.59	24.53	35	66.23	61.50
11	38.66	26.53	36	68.17	61.98
12	38.75	28.52	37	70.15	62.25
13	38.94	30.51	38	72.13	62.56
14	39.24	32.49	39	74.05	63.11
15	39.67	34.44	40	75.87	63.93
16	40.25	36.36	41	77.53	65.05
17	40.93	38.24	42	78.99	66.41
18	41.73	40.07	43	80.00	68.14
19	42.61	41.87	44	80.00	70.14
20	43.57	43.62	45	80.00	72.14
21	44.63	45.31	46	80.00	74.14
22	45.71	47.00	47	80.00	76.14
23	46.85	48.64	48	79.97	78.14
24	48.06	50.23	49	78.93	79.84
25	49.31	51.79	50	78.00	80.00

TABLE 5.10

Results Comparison for Example 3 of Case Study 2

Method	Total Path Length (cm)	Simulation Time(s)
Firefly algorithm (Patle et al, 2018a)	180.2	28.9
Jaya algorithm	179.695	16.143
Rao-1 algorithm	**179.4836**	**8.466**
Rao-2 algorithm	179.5817	10.526
Rao-3 algorithm	179.6115	10.896
Bold value shows best result obtained by the algorithm		

5.4.2.7 Case Study 2: Example 4

This problem was attempted by the researcher in the past using FA (Patle et al. 2018a). The total number of function evaluation used by the previous researchers was 5,000. Hence, to demonstrate and validate the Jaya and Rao algorithms, the same number of function evaluation is used for fair comparison. Hence, in the present work, various combinations of number of generations and population size are tried to get function evaluations of 5,000. Finally, the number of iterations of 100 and population size of 50 are set for obtaining optimum results. In this example, four number of obstacles are considered in the workspace.

- The specifications, for example, are:
  - Start point=(46, 84)
  - Goal point=(100, 28)
- Parameter bounds:
  - $0 \leq x \leq 100$
  - $0 \leq y \leq 100$

Figure 5.11 shows the optimum path from starting (46, 84) to goal (100, 28) by robot after the application of Jaya and Rao algorithms.

Jaya and Rao algorithms have calculated 39 points. Table 5.11 shows the coordinates obtained in example 4 of case study 2 from starting point (46, 84) to goal (100, 28). Rao-1 algorithm has calculated 39 optimum coordinates as shown in Table 5.11.

Table 5.12 presents the comparison of results of Jaya and Rao algorithms with that of FA.

It can be seen that Rao-1 algorithm has produced a minimum total path length of 210.9485 cm and a simulation time of 13.716 s. The performance of the algorithms may be arranged in the following sequence:

Rao-1- Rao-3-Rao-2 – Jaya - firefly algorithm.

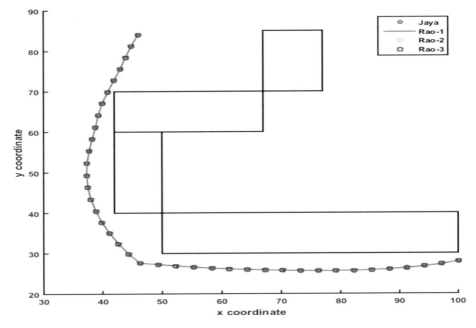

FIGURE 5.11
Path traveled by robot for example 4 of case study 2 by Jaya and Rao algorithms.

TABLE 5.11

Coordinates for Robot in Example 4 of Case Study 2

No.	X-Coordinate	Y-Coordinate	No.	X-Coordinate	Y-Coordinate
1	46.00	84.00	21	46.37	27.58
2	44.92	81.20	22	49.34	27.18
3	43.97	78.36	23	52.32	26.82
4	43.07	75.50	24	55.30	26.50
5	42.01	72.69	25	58.29	26.26
6	40.98	69.87	26	61.29	26.07
7	40.06	67.02	27	64.28	25.94
8	39.39	64.09	28	67.28	25.82
9	38.91	61.13	29	70.28	25.72
10	38.31	58.19	30	73.28	25.63
11	37.77	55.24	31	76.28	25.59
12	37.40	52.26	32	79.28	25.60
13	37.33	49.26	33	82.28	25.67
14	37.56	46.27	34	85.27	25.82
15	38.10	43.32	35	88.26	26.05
16	38.88	40.42	36	91.25	26.35
17	39.93	37.62	37	94.22	26.75
18	41.18	34.89	38	97.17	27.28
19	42.65	32.27	39	100.00	28.00
20	44.37	29.82			

TABLE 5.12

Results Comparison for Example 4 of Case Study 2

Method	Total Path Length (cm)	Simulation Time(s)
Firefly algorithm (Patle et al, 2018a)	253.9	32.500
Jaya algorithm	211.01967	23.865
Rao-1 algorithm	**210.9485**	**13.716**
Rao-2 algorithm	210.9572	19.873
Rao-3 algorithm	210.9526	17.488
Bold value shows best result obtained by the algorithm		

5.4.2.8 Case Study 2: Example 5

This problem was attempted by the researcher in the past using FA (Patle et al. 2018a). The total number of function evaluation used by the previous researchers was 5,000. Hence, to demonstrate and validate the Jaya and Rao algorithms, the same number of function evaluation is used for fair comparison. Hence, in the present work, various combinations of number of generations and population size are tried to get function evaluations of 5,000. Finally, the number of iterations of 100 and population size of 50 are set for obtaining optimum results. In this example, 14 number of obstacles are considered in the workspace.

- The specifications, for example, are:

(1) Robot 1	(2) Robot 2	(3) Robot 3
• Start point=(94,22)	• Start point=(7,41)	• Start point=(35,97)
• Goal point=(97,97)	• Goal point=(97,97)	• Goal point=(97,97)

- Parameter bounds:
  - $0 \le x \le 100$
  - $0 \le y \le 100$

Figure 5.12 shows the optimum path traveled by three robots whose starting points are different but the goal is the same in the given environment after the application of Jaya and Rao algorithms.

Table 5.13 presents the path length comparison of results of Jaya and Rao algorithms with that of FA.

It can be seen that Rao-1 algorithm has produced the minimum total path length of 162.3084 cm for robot 2 and Rao-3 algorithm has produced the minimum total path length of 138.2142 cm for robot 3. The performance of the algorithms for the minimum total path length may be arranged in the following sequence:

Sequence of algorithms' performance for robot 1:

Firefly algorithm - Rao-3-Rao-2 – Jaya - Rao1

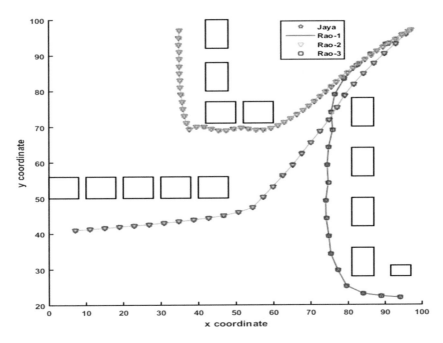

FIGURE 5.12
Path traveled by robot for example 5 of case study 2 by Jaya and Rao algorithms.

TABLE 5.13

Path Length Comparison in Example 5 of Case Study 2

Robots	Total Path Length (cm)				
	Firefly Algorithm (Patle et al. 2018a)	Jaya Algorithm	Rao-1 Algorithm	Rao-2 Algorithm	Rao-3 Algorithm
Robot 1	**129.74**	130.3758	130.3990	130.3600	130.3541
Robot 2	181.81	162.3331	**162.3084**	162.3320	162.3135
Robot 3	159.76	138.3030	138.2541	138.2280	**138.2142**

Bold value shows best result obtained by the algorithm

TABLE 5.14

Simulation Time Comparison in Example 5 of Case Study 2

Robots	Total Simulation Time (s)				
	Firefly Algorithm (Patle et al. 2018a)	Jaya Algorithm	Rao-1 Algorithm	Rao-2 Algorithm	Rao-3 Algorithm
Robot 1	20.88	13.4282	**10.201**	11.215	10.857
Robot 2	30.10	20.5021	**11.022**	14.823	14.3961
Robot 3	25.19	18.2056	**11.039**	12.532	12.711

Bold value shows best result obtained by the algorithm

Sequence of algorithms' performance for robot 2:

Rao-1- Rao-3-Rao-2 – Jaya -firefly algorithm

Sequence of algorithms' performance for robot 3:

Rao-3- Rao-2-Rao-1 – Jaya - firefly algorithm

Table 5.14 presents the simulation time comparison of results of Jaya and Rao algorithms with that of FA.

It can be seen that Rao-1 algorithm has required a minimum time of 10.201, 11.022, and 11.039 s for the navigation of robot 1, robot 2, and robot 3, respectively. The performance of the algorithms for simulation time may be arranged in the following sequence:

Sequence of algorithms' performance for robot 1:

Rao-1- Rao-3-Rao-2 – Jaya -firefly algorithm

Sequence of algorithms' performance for robot 2:

Rao-1- Rao-3-Rao-2 – Jaya -firefly algorithm

Sequence of algorithms' performance for robot 3:

Rao-1- Rao-2-Rao-3 – Jaya -firefly algorithm

5.4.3 Case Study 3

This case study was previously attempted by Patle et al. (2018b) using a probability and fuzzy logic (PFL) algorithm. In this case study, a stationary workspace has been considered for the optimization of the robot path.

5.4.3.1 Objective Function Formulation for Case Study 3

The formation of objective function is based on the Euclidean distance. In the formation of objective function, different criteria such as obstacle detection, smoothness in robot path, etc are to be considered.

5.4.3.2 Obstacle Avoidance Behavior for Case Study 3

The basic motive of obstacle avoidance is to make sure that the safest distance is available between the robot and the obstacle present in the workspace. In this motive, we consider the distance between the center of the obstacle and the center of the robot. This is needed to be maximum to make sure safety of the robot path. The Euclidean distance between the robot and the nearest obstacle is calculated by the following equation:

$$(EDist)_{R\text{-OB}} = \sqrt{(x_{OB} - x_{ROB})^2 + (y_{OB} - y_{ROB})^2} \tag{5.12}$$

where x_{OB} and y_{OB} are the positional coordinates of the detected obstacle, and x_{ROB} and y_{ROB} are the positional coordinates of the robot.

5.4.3.3 Target-Seeking Behavior for Case Study 3

We have to minimize the total path length cover by robot without touching any obstacle in workspace; i.e., we needed to decrease the length of path trajectory between the robot and the target.

Then, the distance between the candidate and the goal is calculated as:

$$(EDist)_{C\text{-G}} = \sqrt{(x_R - x_G)^2 + (y_R - y_G)^2} \tag{5.13}$$

Let x_R and y_R be the positional coordinates of the candidate, and x_G and y_G be the positional coordinates of goal, respectively.

Finally, the objective function is given below:

$$\text{Objective function}\,(f_i) = \gamma \frac{1}{\min_{OB_j \in OB_d} \left\| (EDist)_{R\text{-OB}} \right\|} + \delta \left\| (EDist)_{N\text{-G}} \right\| \tag{5.14}$$

where γ and δ are the controlling parameters.

The workspace is considered with 'n' number of obstacles, which are represented as $O_1, O_2, O_3, O_4, ..., O_n$, and their coordinate positions are (x_{O1}, y_{O1}), (x_{O2}, y_{O2}), (x_{O3}, y_{O3}), ..., (x_{On}, y_{On}). From the objective function equation, it can be clearly understood that path length, smoothness, and turning angle of trajectory depend on γ and δ. From the above study, we can see that one objective is of maximization type and the other objective is of minimization type. Hence, to combine the objective functions and to minimize the value of the combined objective function, we need reciprocal of maximization function. In this work, controlling parameters are chosen by doing various trials in the workspace.

The robot must know the position and nature of the obstacle to achieve the appropriate movement in the workspace and generate a feasible path from the starting to the target. For any robot, path planning, map building, and location of the robot in the workspace are the requirements for navigation. In this case study, the author had considered a PFL rule to obtain the position of obstacle in the workspace. The sensors like infrared sensors, proximity sensors, and ultrasonic sensors are a very powerful tool to detect the obstacle. Three examples are demonstrated by generating obstacles of different shapes and sizes in the workspace. The first example is about the single robot moving in the workspace, and the second and third examples are multi-robots moving in the same workspace.

5.4.3.4 Case Study 3: Example 1

This problem was attempted by the researcher in the past using PFL algorithm (Patle et al. 2018b). The total number of function evaluation applied by the previous researchers was 3,000. Hence, to demonstrate and validate the Jaya and Rao algorithms, the same number of function evaluation is used for fair comparison. Hence, in the present work, various combinations of number of generations and population size are tried to get function evaluations of 3,000. Finally, the number of iterations of 100 and population size of 30 are set for obtaining optimum results. In this example, two number of obstacles are considered in the workspace.

- The specifications, for example, are:
  - Start point = (10, 10)
  - Goal point = (82, 82)
- Parameter bounds:
  - $10 \leq x \leq 82$
  - $10 \leq y \leq 82$

Figure 5.13 shows the optimum path from starting (10, 10) to goal (82, 82) by robot after the application of Jaya and Rao algorithms.

Jaya and Rao algorithms have calculated 60 points. Table 5.15 shows the coordinates obtained in example 1 of case study 3 from starting point (10, 10) to goal (82, 82). Rao-2 algorithm has calculated 60 optimum coordinates as shown in Table 5.15.

Table 5.16 presents the comparison of results of Jaya and Rao algorithms with that of PFL.

It can be observed that Rao-2 algorithm has produced the minimum total path length of 178.920 cm. The performance of the algorithms may be arranged in the following sequence:

Rao-2-Rao-1-Rao-3 – Jaya -probability and fuzzy logic.

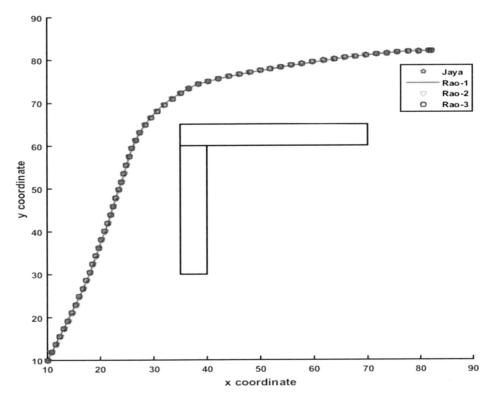

FIGURE 5.13
Path traveled by robot for example 1 of case study 3 by Jaya and Rao algorithm.

TABLE 5.15

Coordinates for Robot in Example 1 of Case Study 3

No.	X-Coordinate	Y-Coordinate	No.	X-Coordinate	Y-Coordinate
1	10.00	10.00	31	29.62	66.48
2	10.78	11.84	32	30.79	68.10
3	11.56	13.68	33	32.09	69.61
4	12.36	15.51	34	33.57	70.96
5	13.15	17.35	35	35.10	72.24
6	13.93	19.19	36	36.73	73.39
7	14.67	21.05	37	38.46	74.40
8	15.36	22.93	38	40.38	74.97
9	16.05	24.80	39	42.28	75.57
10	16.73	26.68	40	44.23	76.04
11	17.37	28.57	41	46.16	76.54
12	17.99	30.47	42	48.10	77.04
13	18.59	32.38	43	50.06	77.45

(*Continued*)

TABLE 5.15 (*Continued*)

Coordinates for Robot in Example 1 of Case Study 3

No.	X-Coordinate	Y-Coordinate	No.	X-Coordinate	Y-Coordinate
14	19.13	34.31	44	52.01	77.87
15	19.65	36.24	45	53.97	78.29
16	20.17	38.17	46	55.92	78.70
17	20.78	40.07	47	57.88	79.12
18	21.36	41.99	48	59.85	79.46
19	21.86	43.92	49	61.82	79.80
20	22.37	45.85	50	63.79	80.14
21	22.87	47.79	51	65.77	80.46
22	23.46	49.70	52	67.74	80.77
23	24.09	51.60	53	69.72	81.06
24	24.56	53.54	54	71.70	81.34
25	24.99	55.50	55	73.69	81.58
26	25.47	57.44	56	75.68	81.76
27	26.05	59.35	57	77.67	81.90
28	26.72	61.24	58	79.67	82.00
29	27.54	63.06	59	81.67	82.01
30	28.55	64.78	60	82.00	82.00

TABLE 5.16

Results Comparison for Example 1 of Case Study 3

Method	Total Path Length (cm)
Probability and fuzzy logic (PFL) (Patle et al. 2018b)	187.1
Jaya algorithm	178.9939
Rao-1 algorithm	178.9777
Rao-2 algorithm	**178.920**
Rao-3 algorithm	178.989
Bold value shows best result obtained by the algorithm	

5.4.3.5 Case Study 3: Example 2

This problem was attempted by the researcher in the past using PFL algorithm (Patle et al. 2018b). The total number of function evaluation applied by the previous researchers was 3,000. Hence, to demonstrate and validate the Jaya and Rao algorithms, the same number of function evaluation is used for fair comparison. Hence, in the present work, various combinations of number of generation and population size are tried to get function evaluations of 3,000. Finally, the number of iterations of 100 and population size of 30 are set for obtaining optimum results. In this example, one obstacle is considered in the workspace.

- The specifications, for example, are:

(1) Robot 1	(2) Robot 2
• Start point=(22, 22)	• Start point=(82, 22)
• Goal point=(92, 92)	• Goal point=(12, 92)

- Parameter bounds:
  - $10 \leq x \leq 95$
  - $10 \leq y \leq 95$

Figure 5.14 shows the optimum path traveled by two robots whose starting points and goal are different after the application of Jaya and Rao algorithms.

Jaya and Rao algorithms have calculated 55 points for robot 1 and robot 2. Table 5.17 shows the coordinates obtained in example 2 of case study 3. Rao-1 algorithm has calculated 55 optimum coordinates as shown in Table 5.17.

Table 5.18 presents the comparison of results of Jaya and Rao algorithms with that of PFL.

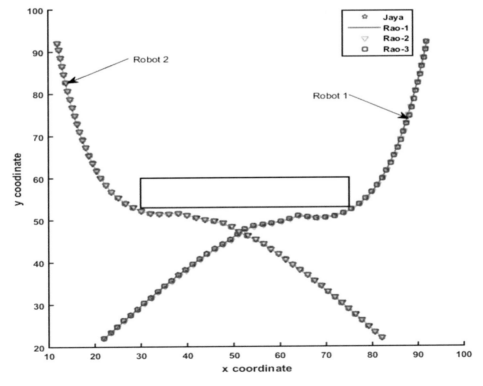

FIGURE 5.14
Path traveled by robots for example 2 of case study 3 by Jaya and Rao algorithms.

TABLE 5.17

Coordinates for Robot 1 and Robot 2 in Example 2 of Case Study 3

No.	Robot 1 X Coordinate	Robot 1 Y Coordinate	Robot 2 X Coordinate	Robot 2 Y Coordinate	No.	Robot 1 X Coordinate	Robot 1 Y Coordinate	Robot 2 X Coordinate	Robot 2 Y Coordinate
1	22.00	22.00	82.00	22.00	29	68.11	50.48	36.04	51.51
2	23.44	23.39	80.55	23.38	30	70.10	50.56	34.04	51.45
3	24.89	24.76	79.10	24.75	31	72.06	50.97	32.06	51.70
4	26.29	26.19	77.66	26.13	32	73.93	51.68	30.13	52.24
5	27.72	27.59	76.20	27.51	33	75.69	52.63	28.32	53.09
6	29.20	28.93	74.75	28.88	34	77.32	53.79	26.67	54.22
7	30.68	30.27	73.28	30.23	35	78.73	55.20	25.14	55.50
8	32.15	31.63	71.85	31.63	36	80.08	56.67	23.75	56.94
9	33.62	32.98	70.38	32.98	37	81.24	58.30	22.47	58.48
10	35.12	34.31	68.88	34.30	38	82.25	60.03	21.40	60.16
11	36.61	35.64	67.35	35.58	39	83.22	61.78	20.47	61.93
12	38.13	36.93	65.81	36.86	40	84.14	63.55	19.50	63.68
13	39.67	38.21	64.23	38.08	41	84.92	65.39	18.64	65.49
14	41.21	39.48	62.61	39.26	42	85.67	67.25	17.91	67.35
15	42.78	40.72	61.07	40.54	43	86.29	69.15	17.27	69.24
16	44.40	41.89	59.52	41.80	44	86.89	71.05	16.60	71.13
17	46.05	43.02	57.91	42.99	45	87.47	72.97	16.08	73.06
18	47.69	44.16	56.30	44.17	46	87.99	74.90	15.59	75.00
19	49.39	45.21	54.62	45.26	47	88.56	76.81	15.06	76.93
20	50.94	46.48	52.87	46.22	48	89.06	78.75	14.61	78.87
21	52.67	47.49	51.10	47.14	49	89.56	80.69	14.21	80.83
22	54.50	48.29	49.50	48.34	50	90.07	82.62	13.78	82.78
23	56.44	48.77	47.68	49.17	51	90.52	84.57	13.34	84.73
24	58.40	49.12	45.76	49.71	52	90.91	86.53	12.94	86.69
25	60.36	49.56	43.80	50.11	53	91.34	88.48	12.58	88.66
26	62.27	50.12	41.85	50.55	54	91.74	90.44	12.21	90.63
27	64.13	50.88	40.00	51.32	55	92.00	92.00	12.00	92.00
28	66.11	50.65	38.03	51.68					

TABLE 5.18

Results Comparison for Example 2 of Case Study 3

	Total Path Length (cm)	
Method	Robot 1	Robot 2
Probability and fuzzy logic (Patle et al. 2018b)	171.5	186.7
Jaya algorithm	170.9039	185.2219
Rao-1 algorithm	**170.6962**	**185.0888**
Rao-2 algorithm	170.7479	185.479
Rao-3 algorithm	170.6991	185.2112
Bold value shows best result obtained by the algorithm		

It can be seen that Rao-1 algorithm has produced the minimum total path length of 170.6962 cm for robot 1 and 185.0888 cm for robot 2. The performance of the algorithms may be arranged in the following sequence:

Sequence of algorithms' performance for robot 1:

Rao-1- Rao-3-Rao-2 – Jaya -probability and fuzzy logic

Sequence of algorithms' performance for robot 2:

Rao-1- Rao-3 –Jaya - Rao-2 – probability and fuzzy logic

5.4.3.6 Case Study 3: Example 3

This problem was attempted by the researcher in the past using PFL algorithm (Patle et al. 2018b). The total number of function evaluation applied by the previous researchers was 3,000. Hence, to demonstrate and validate the Jaya and Rao algorithms, the same number of function evaluation is used for fair comparison. Hence, in the present work, various combinations of number of generations and population size are tried to get function evaluations of 3,000. Finally, the number of iterations of 100 and population size of 30 are set for obtaining optimum results. In this example, 13 number of obstacles are considered in the workspace.

- The specifications, for example, are:

(1) Robot 1	(2) Robot 2
• Start point=(10,90)	• Start point=(10,90)
• Goal point=(67,20)	• Goal point=(67,65)

- Parameter bounds:
  - $10 \leq x \leq 90$
  - $10 \leq y \leq 90$

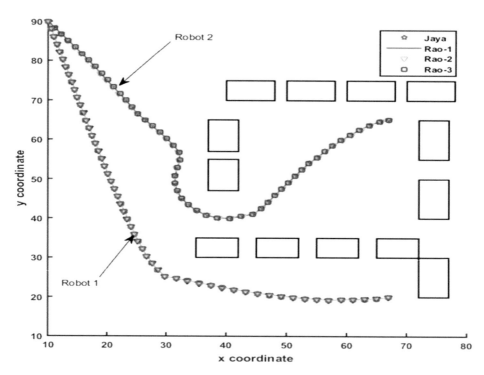

FIGURE 5.15
Path traveled by robots for example 3 of case study 3 by Jaya and Rao algorithms.

Figure 5.15 shows the optimum path traveled by two robots whose starting points are the same and the goals are different after the application of Jaya and Rao algorithms.

Jaya and Rao algorithms have calculated 54 points for robot 1 and 49 points for robot 2. Table 5.19 shows the coordinates obtained in example 3 of case study 3 for robot 1 and robot 2. Rao-2 algorithm has calculated 60 optimum coordinates for robot 1, and Rao-1 algorithm has calculated 49 optimum coordinates for robot 1, as shown in Table 5.19.

Table 5.20 presents the comparison of results of Jaya and Rao algorithms with that of PFL.

It can be seen that Rao-2 algorithm has produced a minimum total path length of 163.0174 cm for robot 1 and Rao-1 algorithm has produced a minimum total path length of 95.9419 cm for robot 2. The performance of the algorithms may be arranged in the following sequence:

Sequence of algorithms' performance for robot 1:

 Rao-2 – Jaya - Rao-3 – Rao-1-probability and fuzzy logic

Sequence of algorithms' performance for robot 2:

 Rao-1- Rao-3-Rao-2 – Jaya -probability and fuzzy logic

TABLE 5.19

Coordinates for Robot 1 and Robot 2 in Example 3 of Case Study 3

No.	Robot 1		Robot 2		No.	Robot 1		Robot 2	
	X Coordinate	Y Coordinate	X Coordinate	Y Coordinate		X Coordinate	Y Coordinate	X Coordinate	Y Coordinate
1	10.00	90.00	10.00	90.00	28	24.01	37.87	38.57	44.28
2	10.50	88.06	11.15	88.37	29	24.61	35.96	40.08	42.97
3	10.94	86.11	12.29	86.72	30	25.25	34.07	42.05	42.65
4	11.44	84.18	13.45	85.09	31	25.95	32.19	44.04	42.45
5	11.97	82.25	14.58	83.44	32	26.72	30.35	46.04	42.48
6	12.49	80.32	15.65	81.75	33	27.62	28.56	48.04	42.55
7	12.97	78.38	16.72	80.06	34	28.55	26.79	50.02	42.84
8	13.46	76.44	17.77	78.36	35	29.78	25.22	51.96	43.34
9	13.94	74.50	18.93	76.73	36	31.70	24.68	53.80	44.12
10	14.42	72.56	20.00	75.05	37	33.62	24.10	55.52	45.13
11	14.88	70.61	21.10	73.37	38	35.54	23.55	57.12	46.33
12	15.43	68.69	22.18	71.69	39	37.45	22.96	58.63	47.65
13	15.98	66.77	23.21	69.98	40	39.39	22.48	59.98	49.12
14	16.48	64.83	24.25	68.27	41	41.32	21.95	61.21	50.70
15	16.98	62.89	25.22	66.52	42	43.26	21.48	62.37	52.32
16	17.48	60.96	26.25	64.81	43	45.21	21.00	62.71	54.29
17	18.03	59.04	27.31	63.11	44	47.16	20.59	63.03	56.27
18	18.56	57.11	28.42	61.45	45	49.13	20.25	63.37	58.24
19	19.07	55.17	29.42	59.72	46	51.11	19.92	64.59	59.82
20	19.62	53.25	30.41	57.99	47	53.08	19.62	65.90	61.33
21	20.14	51.32	31.57	56.36	48	55.07	19.41	66.96	63.03
22	20.68	49.40	32.51	54.60	49	57.07	19.26	67.00	65.00
23	21.26	47.48	33.51	52.87	50	59.07	19.32	N/A	N/A
24	21.77	45.55	34.41	51.08	51	61.06	19.40	N/A	N/A
25	22.34	43.63	35.32	49.30	52	63.06	19.47	N/A	N/A
26	22.87	41.70	36.25	47.53	53	65.06	19.62	N/A	N/A
27	23.41	39.78	37.33	45.85	54	67.00	20.00	N/A	N/A

TABLE 5.20

Results Comparison for Example 3 of Case Study 3

	Total Path Length (cm)	
Method	Robot 1	Robot 2
Probability and fuzzy logic (Patle et al. 2018b)	164.82	123.41
Jaya algorithm	163.0311	96.0586
Rao-1 algorithm	163.1076	**95.9419**
Rao-2 algorithm	**163.0174**	95.9523
Rao-3 algorithm	163.0856	95.9512

Bold value shows best result obtained by the algorithm

5.4.4 Case Study 4

This case study was previously attempted by Cen et al. (2008) using ACO, Liang et al. (2013) using BFO, and Mohanty and Parthi (2014) using an IWO. In this case study, a stationary workspace has been considered for the optimization of the robot path.

5.4.4.1 Objective Function Formulation for Case Study 4

$$\text{Objective function}\left(f_i\right) = k_1 \frac{1}{\min_{\text{OBS}_j \in \text{OBS}_d}\left\|(\text{Dist})_{\text{C-OB}}\right\|} + k_2 \left\|(\text{Dist})_{C-G}\right\| \qquad (5.15)$$

The workspace is considered with 'm' number of obstacles that are existing in the workspace, and we denote them as OBS_1, OBS_2, OBS_3, ..., OBS_n. The positional coordinates of the obstacles are $(x_{\text{OBS1}}, y_{\text{OBS1}})$, $(x_{\text{OBS2}}, y_{\text{OBS2}})$, $(x_{\text{OBS3}}, y_{\text{OBS3}})$, ..., $(x_{\text{OBS}n-1}, y_{\text{OBS}n-1})$, and $(x_{\text{OBS}n}, y_{\text{OBS}n})$, respectively. From Equation (5.15), it can be clearly understood that path length, smoothness, and turning angle of trajectory depend on k_1 and k_2. We can see that one objective is of maximization type, and the other objective is of minimization type. Hence, to combine the objective functions and to minimize the value of the combined objective function, we need reciprocal of maximization function. In this work, controlling parameters are chosen by doing various trials in the workspace.

In this case study, the robot knows about stating positional coordinates, target positional coordinates, and searching radius. To obtain a new position in the workspace, the navigation algorithm produces a different number of solutions within the circle of searching radius. Initially, the robot moves in a straight direction to the destination until the sensor on the robot senses an obstacle. When the robot detects an obstacle within the search radius, the navigation algorithm will be started to obtain the best next position for the robot, and the robot reaches the next position to avoid obstacles successfully. Two examples are demonstrated by generating obstacles of different shapes and sizes in the workspace.

5.4.4.2 Case Study 4: Example 1

This problem was attempted by the researcher in the past using BFO (Liang et al. 2013) and IWO (Mohanty and Parthi 2014). The total number of function evaluation used by the previous researchers was 6,000. Hence, to demonstrate and validate the Jaya and Rao algorithms, the same number of function evaluation is used for fair comparison. Hence, in the present work, various combinations of number of generations and population size are tried to get function evaluations of 6,000. Finally, the number of iterations of 300 and population size of 20 are set for obtaining optimum results. In this example, six number of obstacles are considered in the workspace.

- The specifications, for example, are:
  - Start point=(0, 0)
  - Goal point=(25, 25)

- Parameter bounds:
  - $0 \leq x \leq 25$
  - $0 \leq y \leq 25$

Figure 5.16 shows the optimum path from starting (0, 0) to goal (25, 25) by robot after the application of Jaya and Rao algorithms.

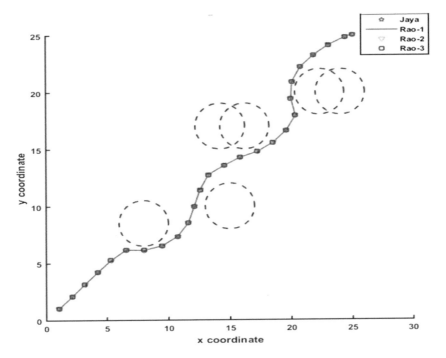

FIGURE 5.16
Path traveled by robot for example 1 of case study 4 by Jaya and Rao algorithms.

TABLE 5.21

Coordinates for Robot in Example 1 for Case Study 4

No.	X-Coordinate	Y-Coordinate	No.	X-Coordinate	Y-Coordinate
1	0.00	0.00	15	14.33	13.47
2	1.07	1.05	16	15.66	14.15
3	2.12	2.12	17	17.10	14.57
4	3.18	3.18	18	18.42	15.29
5	4.24	4.24	19	19.52	16.31
6	5.32	5.28	20	20.29	17.60
7	6.63	6.02	21	20.05	19.07
8	8.12	5.98	22	19.87	20.56
9	9.58	6.35	23	20.43	21.95
10	10.86	7.13	24	21.44	23.06
11	11.69	8.38	25	22.66	23.93
12	12.06	9.83	26	23.98	24.65
13	12.43	11.29	27	25.00	25.00
14	13.08	12.64			

TABLE 5.22

Results Comparison for Example 1 of Case Study 4

Method	Total Path Length (cm)
Bacteria foraging optimization (Liang et al. 2013)	192.21
Invasive weed optimization (Mohanty and Parhi 2014)	171.76
Jaya algorithm	145.7433
Rao-1 algorithm	145.7969
Rao-2 algorithm	**145.3899**
Rao-3 algorithm	145.6957
Bold value shows best result obtained by the algorithm	

Jaya and Rao algorithms have calculated 27 points. Table 5.21 shows the coordinates obtained in example 1 of case study 4 from starting point (0, 0) to goal (25, 25). Rao-2 algorithm has calculated 27 optimum coordinates as shown in Table 5.21.

Table 5.22 presents the comparison of results of Jaya and Rao algorithms with those of BFO and IWO.

It can be seen that Rao-2 algorithm has produced a minimum total path length of 145.3899 cm. The performance of the algorithms may be arranged in the following sequence:

Rao-2- Rao-3 –Jaya - Rao-1-invasive weed optimization -bacteria foraging optimization.

5.4.4.3 Case Study 4: Example 2

This problem was attempted by the researcher in the past using ACO (Cen et al. 2008) and IWO (Mohanty and Parthi 2014). The total number of function evaluation used by the previous researchers was 6,000. Hence, to demonstrate and validate the Jaya and Rao algorithms, the same number of function evaluation is used for fair comparison. Hence, in the present work, various combinations of number of generations and population size are tried to get function evaluations of 6,000. Finally, the number of iterations of 300 and population size of 20 are set for obtaining optimum results. In this example, two number of obstacles are considered in the workspace.

- The specifications, for example, are:
  - Start point=(0, 0)
  - Goal point=(23, 23)
- Parameter bounds:
  - $0 \leq x \leq 23$
  - $0 \leq y \leq 23$

Figure 5.17 shows the optimum path from starting (0, 0) to goal (23, 23) by robot after the application of Jaya and Rao algorithms.

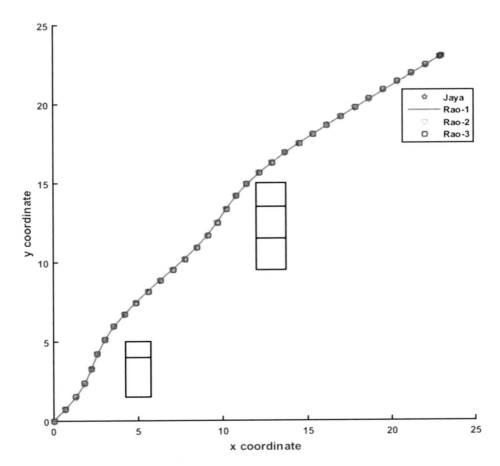

FIGURE 5.17
Path traveled by robot for example 2 of case study 4 by Jaya and Rao algorithms.

TABLE 5.23

Coordinates for Robot in Example 2 of Case Study 4

No.	X-Coordinate	Y-Coordinate	No.	X-Coordinate	Y-Coordinate
1	0.00	0.00	19	10.63	14.27
2	0.68	0.74	20	11.28	15.02
3	1.28	1.53	21	12.00	15.72
4	1.77	2.41	22	12.76	16.37
5	2.11	3.34	23	13.54	16.99
6	2.45	4.29	24	14.36	17.57
7	2.87	5.19	25	15.19	18.13
8	3.42	6.03	26	16.01	18.70
9	4.04	6.82	27	16.86	19.23
10	4.71	7.56	28	17.69	19.78
11	5.44	8.24	29	18.54	20.31
12	6.16	8.93	30	19.40	20.83
13	6.91	9.59	31	20.24	21.36
14	7.64	10.28	32	21.10	21.88
15	8.33	11.00	33	21.96	22.38
16	8.96	11.78	34	22.82	22.90
17	9.52	12.60	35	23.00	23.00
18	10.05	13.45			

TABLE 5.24

Results Comparison for Example 2 of Case Study 4

Method	Total Path Length (cm)
Ant colony optimization (Cen et al. 2008)	174.81
Invasive weed optimization (Mohanty and Parhi 2014)	169.07
Jaya algorithm	127.7678
Rao-1 algorithm	127.7759
Rao-2 algorithm	**127.7172**
Rao-3 algorithm	127.7619
Bold value shows best result obtained by the algorithm	

Jaya and Rao algorithms have calculated 35 points. Table 5.23 shows the coordinates obtained in example 2 of case study 4 from starting point (0, 0) to goal (23, 23). Rao-2 algorithm has calculated 35 optimum coordinates as shown in Table 5.23.

Table 5.24 presents the comparison of results of Jaya and Rao algorithms with those of ACO and IWO.

It can be seen that Rao-2 algorithm has produced a minimum total path length of 127.7172 cm. The performance of the algorithms may be arranged in the following sequence:

Rao-2- Rao-3 –Jaya - Rao-1-invasive weed optimization-ant colony optimization.

5.5 Conclusions

Robot path planning is a complex task, and advanced optimization tools are required to find the shortest and feasible paths. This dissertation work demonstrates the successful application of Jaya and Rao algorithms for the optimum robot path planning with avoidance of static obstacles present in the workspace.

Four case studies are presented, which have 12 examples. The performance of Jaya and Rao algorithms is studied in terms of path length, time required for simulation, and accuracy of the solution. Compared to the other algorithms, the Jaya and Rao algorithms are simple and easy to apply in robot path planning problems. In the first case study, the Jaya and Rao-2 algorithms have performed better than GA, PSO, and CS algorithm. In the second case study, Rao-1, Rao-2, and Rao-3 algorithms have performed better than the fuzzy-neural network and the FA. In the third case study, Rao-1 and Rao-2 algorithms have performed better than the PFL algorithm. In the fourth case study, the Rao-2 algorithm has performed better than the ACO, BFO, and IWO algorithm.

All the results obtained by Jaya and Rao algorithms in these four case studies are feasible. The Jaya and Rao algorithms are found useful for optimal robot path planning. In the present work, the Jaya and Rao algorithms are used to solve the navigation problems in static environment; i.e., obstacles are static. However, these algorithms may also be used to solve the navigation problems of dynamic environment, i.e., when the obstacles are dynamic.

References

Cen, Y., Song, C., Xie, N., & Wang, L. (2008). Path planning method for mobile robot based on ant colony optimization algorithm. *IEEE Conference on Industrial Electronics and Applications*, 298–301.

Das, P. K., Behera, H. S., & Panigrahi, B. K. (2016a). Intelligent-based multi-robot path planning inspired by improved classical Q-learning and improved particle swarm optimization with perturbed velocity. *Engineering Science and Technology, an International Journal*, 19(1), 651–669.

Das, P. K., Behera, H. S., Jena, P. K., & Panigrahi, B. K. (2016b). Multi-robot path planning in a dynamic environment using improved gravitational search algorithm. *Journal of Electrical System and Information Technology*, 3(2), 295–313.

Joshi, M. M., & Zaveri, M. A. (2011). Reactive navigation of autonomous mobile robot using Neuro-Fuzzy system. *International Journal of Robotics and Automation*, 2(3), 45–128.

Kim, H., Kim, S. H., Jeon, M., Kim, J., Song, S., & Paik, K. J. (2017). A study on path optimization method of an unmanned surface vehicle under environmental loads using genetic algorithm. *Ocean Engineering*, 142, 616–624.

Liang, J. H., & Lee, C. H. (2015). Efficient collision-free path-planning of multiple mobile robots system using efficient artificial bee colony algorithm. *Advances in Engineering Software*, 79, 47–56.

Liang, X., Li, L., Wu, J., & Chen, H. (2013). Mobile robot path planning based on adaptive bacterial foraging algorithm. *Journal of Central South University*, 20(12), 3391–3400.

Lopez, J., Pablo, S. V., Cacho, M. D., & Guillen, E. L. (2020). Obstacle avoidance in dynamic environments based on velocity space optimization. *Robotics and Autonomous Systems*, 131, 103569.

Mohamed, A. Z., Lee, S. H., & Hsu, H. Y. (2011). Autonomous mobile robot system concept based on PSO path planner and VSLAM. *In IEEE International Conference on Computer Science and Automation Engineering*, 4, 92–97.

Mohanty, P. K., & Parhi, D. R. (2013). Cuckoo search algorithm for the mobile robot navigation. *Lecture Notes in Computer Science*, 4, 527–536.

Mohanty, P. K., & Parhi, D. R. (2014). A new efficient optimal path planner for mobile robot based on invasive weed optimization algorithm. *Frontiers of Mechanical Engineering*, 9(4), 317–330.

Mohanty, P. K., & Parhi, D. R. (2016). Optimal path planning for a mobile robot using cuckoo search algorithm. *Journal of Experimental & Theoretical Artificial Intelligence*, 28(1–2), 35–52.

Orozco-Rosas, U., Montiel, O., & Sepúlveda, R. (2019). Mobile robot path planning using membrane evolutionary artificial potential field. *Applied Soft Computing*, 77, 236–251.

Pandey, A., & Parhi, D. R. (2017). Optimum path planning of mobile robot in unknown static and dynamic environments using Fuzzy-wind driven optimization algorithm. *Defence Technology*, 13(1), 47–58.

Patle, B. K., Pandey, A., Jagadeesh, A., & Parhi, D. R. (2018a). Path planning in uncertain environment by using firefly algorithm. *Defence Technology*, 14(6), 691–701.

Patle, B. K., Parhi, D. R. K., Jagadeesh, A., & Kashyap, S. K. (2018b). Application of probability to enhance the performance of Fuzzy based mobile robot navigation. *Applied Soft Computing*, 75, 265–283.

Patle, B. K., Babul, L. G., Pandey, P. D. R. K., & Jagadeesh, A. (2019). A review: On path planning strategies for navigation of mobile robot. *Defence Technology*, 15(4), 582–606.

Rao R. V. (2016). Jaya: A simple and new optimization algorithm for solving constrained and unconstrained optimization problems. *International Journal of Industrial Engineering Computations*, 7(1), 19–34.

Rao, R. V. (2020). Rao algorithms: Three metaphor-less simple algorithms for solving optimization problems. *International Journal of Industrial Engineering Computations*, 11(1), 107–130.

Rao, R. V. (2011). An elitist teaching-learning-based optimization algorithm for solving complex constrained optimization problems. *International Journal of Industrial Engineering Computations*, 3(4), 535–560.

Shi, W., Wang, K., & Yang, S. K., (2009). A Fuzzy-neural network approach to multi-sensor integration for obstacle avoidance of a mobile robot. *Intelligent Automation & Soft Computing*, 15(2), 289–301.

Wang, J., Zhang, Y., & Xia, L. (2010). Adaptive genetic algorithm enhancements for path planning of mobile robots. *International Conference on Measuring Technology and Mechatronics Automation*, 1, 416–419.

Zeng, M. R., Xi, L., & Xiao, A. M. (2016). The free step length ant colony algorithm in mobile robot path planning. *Advance Robotics*, 30(23), 1509–1514.

6

Semi-Empirical Modeling and Jaya Optimization of White Layer Thickness during Electrical Discharge Machining of NiTi Alloy

Mahendra Uttam Gaikwad and A. Krishnamoorthy

Sathyabama Institute of Science and Technology

Vijaykumar S. Jatti

D.Y. Patil College of Engineering

CONTENTS

Abbreviations

WLT: White layer thickness
EDM: Electrical discharge machining

6.1 Introduction

Surface integrity parameters reflect the condition or quality of machined surface, which not only helps in deciding the performance but also helps to define the life of machined

DOI: 10.1201/9781003143505-6

parts or components. Various parameters that fall under the category of surface integrity are surface roughness, white layer thickness (WLT), heat-affected zone, microhardness, residual stresses, etc. [1]. A WLT is considered to be one of the important surface integrity parameters in electrical discharge machining (EDM) process. As a well-known fact, EDM is a thermomechanical process where the workpiece is subjected to continuous heating and cooling phases. This phenomenon gives rise to the formation of a recast layer on the EDM-machined surface, which is also known as WLT. In order to investigate the WLT during EDM machining of various materials, efforts have been made by various researchers. Boujelbene et al. [2] investigated WLT during EDM machining of $^{50}CrV_4$ steel with the variation in discharge energy. They stated that increases in discharge energy make the workpiece surface rough, causing an increase in WLT. Also with an increase in discharge energy, more melting and recasting of material takes place, which may damage the tool surfaces, and correspondingly, the formation of microcracks on workpiece starts. P. Sharma et al. [3] investigated the formation of WLT during wire electrical discharge machining (WEDM) of Inconel 718 material for rough and trim operation. They reported that minimum WLT and surface damage take place in trim-cut operation in comparison with rough WEDM operation. Ultrafine discharge pulses are generated in trim operations, which can be easily flushed away by pressurized waves generated in the absence of plasma channel. This will lead to a reduction in the thickness of a white layer in WEDM process. Navas et al. [4] carried out a comparative investigation on residual stresses and surface integrity generated in AISI O1 tool steel during turning, grinding, and WEDM processes. They stated that WEDM is more detrimental to surface integrity, which will lead to the progression of the cracks on the machined surface, thus reducing the service life of machined parts. This is because WEDM forms tensile stresses at the machined parts and leads to the generation of surface superficial white layer below the tensile stress area. M. Gaikwad et al. [5] investigated the effect of process parameters (current, voltage, pulse-on time, and pulse-off time) on the formation of WLT during EDM machining of NiTi 60 alloys. They stated that current is the dominant parameter for the generation of WLT because an increase in current causes the machined surface to become rough and thereby increases the thickness of a white layer. In connection with predictive analysis through empirical modeling of surface integrity parameters like MRR, SR, and WLT, some researchers have placed their valuable suggestions – among them, N. K. Mandal et al. [6] carried out empirical modeling of surface roughness parameter during CNC milling operation by considering process variables such as speed, feed, and depth of cut and work material. They used Buckingham's Pi-theorem and generated models; the results obtained stated that model is reliant on workpiece material because of which variable coefficients and the power indexes were obtained. They had carried out a comparative analysis between the experimental and semi-empirical models; the results show that prediction based on a model for surface roughness of work is reasonably accurate. N. G. Patil and P. K. Brahmankar [7] determined MRR by adopting dimensional analysis method during wire-EDM machining of metal matrix composites by considering thermophysical properties of the workpiece and machining variables such as pulse-on time and average gap voltage. The results stated that the developed model shows close results with experimental one. R. Bobbili et al. [8] presented mathematical model using Buckingham's pie theorem and invested experimentally the MRR and surface roughness during wire-EDM machining of aluminum 7,017 materials. The variables of model considered were pulse-on time, flushing pressure, input power, thermal diffusivity, and latent heat of vaporization; the results predicted by this model show good agreement with experimental results. Also, an increase in pulse-on time causes an improvement in MRR and declination of surface finish. The machined surface shows the formation of crater surface, and the formation of craters increases with increasing current and pulse-on time. D. Poro's and S. Zaborski [9] carried out wire-EDM machining of titanium (Ti_6Al_4V) alloy by using coated and uncoated brass wires to determine the volumetric deficiency of cutting by semi-empirical modeling method. The semi-empirical modeling method was developed and the efficiency of wire-EDM machining decreases with an increase in thermal conduction and specific heat capacity of materials. They achieved the highest efficiency (17.75 mm³/min), which is 18% greater for the uncoated brass electrode. Optimization of surface integrity parameters such as WLT, surface roughness, MRR, etc. was carried out. B. Roy and A. Mandal [10] carried out WEDM machining of Nitinol-60 alloy to determine WLT by considering pulse-on time, gap voltage, and flow rate as machining parameters. In this analysis, the relationship between the machining parameters and output responses was developed using response surface methodology (RSM). They stated that WLT was also found to increase with the flow rate, which indicated that flushing was insufficient even at a maximum flow rate of 6 LPM and the molten material solidified time due to higher quenching rates. S. Dewangan et al. [11] investigated the influence of electrode materials (copper, brass, and graphite) on process parameters (current, pulse-on time, duty cycle, and polarity) and corresponding determines the effect of WLT formation during EDM machining of AISI P20 tool steel. With the use of Taguchi L18 orthogonal array method they reported, pulse current is the most dominating factor affecting WLT. They also reported that the maximum WLT was obtained with the use of graphite tool. Muthukumar et al. [12] used APSO optimization technique by considering P_{on}, P_{off}, gap voltage, and wire feed as input parameters to improve MRR, SR, and kerf width during wire-EDM machining of AISI D3 die steel. They developed the empirical model to predict output variables. The optimum results obtained showed that SR and kerf width values decreased with a rise in MRR. M. U. Gaikwad et al. [13] used Jaya modern optimization technique to improve MRR and surface roughness during EDM machining of NiTi alloy. They found an improvement in surface roughness (from 2.804 to 2.420 μm) by maintaining the current (5.2 A) at a lower rate, whereas an improvement in MRR (from 0.7409 to 0.758 mm³/min) is achieved by an increment in current (7.1 A). They also concluded that the selection of optimum process results in good quality of workpiece; machining period was largely based upon the application of optimization techniques. N. Sharma et al. [14] investigated the effect of process variables such as P_{on}, P_{off}, C, V, and wire tension on output factors such as MRR and SR during machining HSLA using brass wire as electrode. They developed a mathematical model with RSM technique and found that P_{on} and P_{off} are significant factors for MRR, and P_{off} and V are significant for SR. Takale et al. [15] implemented Taguchi method during wire-EDM machining of shape memory alloy ($Ti_{49.4}$-$Ni_{50.6}$) used for orthopedic purposes. They found the best operational setting for maximum MRR and minimum SR with voltage of 140, capacitance of 0.4 μF, and wire feed of 30 μm/s. Rao et al. [16] applied Jaya algorithms to optimize the MRR, taper angle, and dross formation rate during EDM machining of AISI 4340 steel; they obtained good results with the improvements in performance parameters.

6.1.1 Research Novelty

Very few investigations have been reported for empirical modeling and analysis during EDM machining of NiTi60 alloy. Few studies have been reported for empirical modeling and analysis for WLT as output variables during EDM machining of NiTi60 alloy. Also, very few researchers have carried out empirical modeling and analysis with the use of Die Sink EDM machine. Earlier researcher has carried out empirical modeling and analysis but few researchers have carried out a comparative investigation between experimental and empirical modeling. Such kind of comparative investigation helps the researcher for validation purpose.

6.2 Method and Material

6.2.1 Experimental Details

Experiments were conducted with the aid of Die Sink EDM machine. The various parts of EDM machine are shown in Figure 6.1. During this machining process, NiTi 60 alloy is selected as a workpiece, and copper is chosen as electrode material. Experimentation was conducted by considering three levels, and various process variables are shown in Table 6.1. The experimental WLT was evaluated using Taguchi design of experiments considering L9 orthogonal array method. The corresponding values of experimental WLT are mentioned in Table 6.1. After the experimental work, empirical modeling and Jaya optimization work were carried out, which are discussed in the next section.

FIGURE 6.1
Die Sink EDM machine.

TABLE 6.1

Process Variable and Operation Levels

Process Variables	Phase I	Phase II	Phase III
Vtg	40	55	80
Crt	4	6	8
P_{on}	20	40	60
P_{off}	5	7	9

6.2.2 Empirical Modeling

Empirical modeling was carried out with the aid of Buckingham's pie theorem, where it is possible to assemble all the variables occurring in the problem [17]. In this present model, thermophysical properties like density, thermal conductivity, and coefficient of thermal expansion along with machining variables like voltage, current, P_{on}, P_{off}, and duty factors are considered. The dimensions of these variables along with other details are mentioned in Table 6.2.

According to Buckingham's π theorem, the correlation of WLT with workpiece can be expressed as follows:

$$WLT = f(C, V, P_{on}, P_{off}, D, K, \alpha) \tag{6.1}$$

The dimensionless homogeneous equation has nine variables and five fundamental dimensions, namely, mass (*M*), length (*L*), time (*T*), current (*C*), and temperature (*θ*).

So the number of π will be $(n - m) = 9 - 5 = 4$ terms. The functional relationship for π terms can be written as follows:

$$g(\pi_1, \pi_2, \pi_3, \pi_4)$$

Further, each dimensionless π term can be calculated using Buckingham's π theorem as listed follows:

$$\pi_1 = C^{a1}, V^{b1}, \rho^{c1}, K^{d1}, \alpha^{e1}, WLT \tag{6.2}$$

TABLE 6.2

Machining Parameter and Dimensions

	Factors	Symbol	Unit	Dimensions
Output response	White layer thickness	Wlt	μm	*L*
Machining variables	Pulse-on time	P_{on}	μs	*T*
	Peak current	*c*	A	*I*
	Duty cycle	*D*	ms/ms	$M^0 L^0 T^0 I^0 \theta^0$
	Voltage	*V*	v	$ML^2/T^3 I$
Material properties	Thermal conductivity	*c*	Cal/mole°C	$ML/T^3 \theta^{-1}$
	Density	ρ	g/cm³	M/L^3
	Coefficient of thermal expansion	α	°C⁻¹	θ^{-1}

$$\pi_2 = C^{a2}, V^{b2}, \rho^{c2}, K^{d2}, \alpha^{e2}, P_{on} \tag{6.3}$$

$$\pi_3 = C^{a3}, V^{b3}, \rho^{c3}, K^{d3}, \alpha^{e3}, P_{off} \tag{6.4}$$

$$\pi_4 = C^{a4}, V^{b4}, \rho^{c4}, K^{d4}, \alpha^{e4}, D \tag{6.5}$$

The power of each variable was obtained and rearranged to generate the dimensionless form by equating both sides dimensionally as follows:

$$\pi_1 = \frac{CV\alpha \, WLT}{K} \tag{6.6}$$

$$\pi_2 = \frac{\rho^{0.33}, K^{1.66}, P_{on}}{C^{1.33}, V^{1.33}, \alpha^{1.66}} \tag{6.7}$$

$$\pi_3 = \frac{K^{1.66}, P_{off}}{C^{1.66}, V^{1.33}, \alpha^{1.66}, \rho^{0.33}} \tag{6.8}$$

$$\pi_4 = D \tag{6.9}$$

As π_1 is a function of other three π terms, so it can be written as:

$$\pi_1 = f(\pi_2, \pi_3, \pi_4) \tag{6.10}$$

With the help of the above equation, the material removal rate can be determined as follows:

$$WLT = A \, \frac{(K)}{CV\alpha} \, \frac{\left(\rho^{0.33}, K^{1.66}, P_{on}\right)^a}{C^{1.33}, V^{1.33}, \alpha^{1.66}} \, \frac{\left(K^{1.66}, P_{off}\right)^b}{C^{1.66}, V^{1.33}, \alpha^{1.66}, \rho^{0.33}} \, (D)^c \tag{6.11}$$

The unknown power and coefficient of empirical model of Equation (6.11) were determined using multiple linear regression analysis, which is represented in Table 6.3.

The WLT was further evaluated with the use of ImageJ software by using Equation (6.12), and the corresponding obtained values are mentioned in Table 6.4. It can be observed that the values of WLT obtained with the help of ImageJ software are very near to those of the experimental one.

$$WLT = \frac{Image \, area}{Image \, length} \tag{6.12}$$

TABLE 6.3

Coefficients of Model

Power and Coefficient	A	a	b	c
Results obtained	−15.658	−0.00704	0	39.815

TABLE 6.4

Evaluation of Experimental and Empirical WLT

Trial Nos.	Voltage (V)	Current (A)	Pulse-on (P_{on})	Pulse-off (P_{off})	Expt. WLT (µm)	Empirical WLT (µm)	WLT (ImageJ) (µm)
1	40	4	20	5	8	11.013	9
2	40	6	40	7	9	10.675	10.19
3	40	8	60	9	11	9.85	11.21
4	55	4	40	9	8	11.646	8.39
5	55	6	60	5	13	8.768	13.51
6	55	8	20	7	17	13.867	18.63
7	80	4	60	7	12	9.868	12.95
8	80	6	20	9	7	5.332	7.89
9	80	8	40	5	11	6.678	11.79

6.2.3 Jaya Optimization

Jaya optimization is used to determine the best and worst solutions from various numbers of solutions. If $f(X)$ is the objective function to be maximized or minimized, then the best and worst solutions are determined using Equation (6.13). Here, the problem is formulated to minimize the WLT during EDM machining.

$$X'_{j,k,I} = X_{j,k,I} + \left[r_{1,j,i}\left(X_{j,best,I} - \left|X_{j,k,i}\right|\right) \right] - \left[r_{2,j,i}\left(X_{j,worst,I} - \left|X_{j,k,i}\right|\right) \right] \tag{6.13}$$

where

$X_{j,best,I}$: value of j variable for the best solution.

$X_{j,worst,I}$: value of j variable for the worst solution.

$X'_{j,k,I}$: refined value of $X_{j,k,I}$,

$r_{1,j,i}$ and $r_{2,j,i}$: random numbers for the jth variable during the ith iteration having a range from 0 to 1.

$X'_{j,k,I}$ is accepted if it gives better function value.

During the iteration of Jaya optimization, all the accepted function values are maintained till the end of iteration and these values become the input to the next iteration [15]. An advanced optimization technique named as Jaya algorithm is implemented in this research work, with the objective to minimize the function of $F(WLT)$. Optimization was carried out with a population size of nine; process variables were voltage, current, P_{on}, and P_{off} with numbers of iteration of 10. The initial population is randomly created within the range of process parameters as mentioned in Equation (6.14), and the corresponding objective function values obtained are mentioned in Table 6.5. For minimization function, the lowest value is considered as best, while the highest value is considered as worst.

For minimization function, $F(WLT)$

$$= 6.43 + 0.0088 \text{ Vtg} + 1.583 \text{ Crt} + 0.0417 \, P_{on} - 1.000 \, P_{off}$$

Subjected to $40 \leq \text{Vtg} \leq 80, \ 4 \leq \text{Crt} \leq 8, \ 20 \leq P_{on} \leq 60, \ 5 \leq P_{off} \leq 9$ (6.14)

TABLE 6.5

Initial Population of WLT

Trials No	Voltage (V)	Current (A)	Pulse-on (P_{on})	Pulse-off (P_{off})	F(WLT) (μm)
1	40	4	20	5	8.878
2	40	6	40	7	10.886
3	40	8	60	9	12.889
4	55	4	40	9	5.886 (best)
5	55	6	60	5	13.989
6	55	8	20	7	13.663
7	80	4	60	7	8.868
8	80	6	20	9	8.664
9	80	8	40	5	16.120 (worst)

TABLE 6.6

Refined Values of the Variables and the Objective Function during 10th Generation

Trials No	Voltage (V)	Current (A)	Pulse-on (P_{on})	Pulse-off (P_{off})	F(WLT) (μm)
1	40	4.5	33	9	4.887
2	56	6.1	39	5.5	4.846
3	64	5.9	43	7.2	4.846
4	63	7.8	56	6.8	4.846
5	51	4	20	8.2	4.846 (best)
6	62	5	45	5	4.846
7	80	6.9	59	7	4.846
8	72	7.1	51	5.5	4.846
9	67	5.2	35	8.9	4.846

It can be observed from Table 6.5 that for minimizing WLT, the best solution value is 5.886 (trial 4th) and the worst solution value is 16.120 (trial 9th); now the objective is to improve the best solution values for which 10 iterations were carried out. By assuming random values of input process variables within their working range, the objective function values seem to be improving as mentioned in Table 6.6; hence, the objective of Jaya algorithm moving toward the best is satisfied. It can be observed from Table 6.6 that for minimizing WLT, the best solution value is 4.846 (trial 5th).

6.2.4 Convergence Analysis for WLT

Figure 6.2 shows the convergence graph for WLT, which shows that the implementation of Jaya optimization algorithm will help to predict WLT EDM machining of NiTi alloy. It is observed from Figure 6.2 that the value of WLT goes on decreasing from 4.887 to 4.846 μm in just two number of iterations. After the second iteration, the plot seems to be straight line or the values of WLT remain constant.

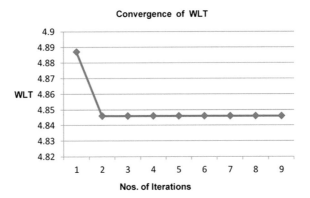

FIGURE 6.2
Convergence graph for WLT.

6.3 Results and Discussions

6.3.1 Comparative Analysis of WLT

The developed model is different from the earlier studies, which have been reported for empirical modeling and analysis for WLT during EDM machining of NiTi60 alloy. The predicted WLT was determined by using empirical model Equation (6.11); the values of the model are mentioned in Table 6.4. The comparative result obtained for WLT by experimental and empirical models shows a close agreement between them, which can be observed from Table 6.4 and Figure 6.3. It is observed from Table 6.4 that current is the dominant parameter for the generation of WLT because an increase in current causes the machined surface to become rough and thereby increases the thickness of a white layer [5]. This phenomenon can also be evident from SEM images as shown in Figure 6.4; for trial no. 6 with 8 A current, the maximum value of WLT was found to be 17 μm where large globules, large voids, and surface cracks are generated on the machined surface. Due to the presence of such high value of WLT, the life of EDM-machined part will be affected [18].

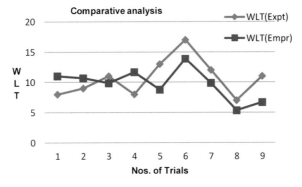

FIGURE 6.3
Comparative analysis of WLT.

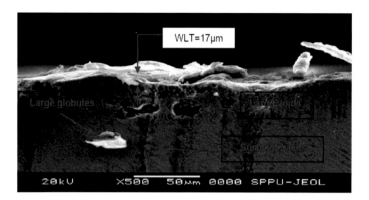

FIGURE 6.4
Trial no: 6: voltage of 55 V, current of 8 A, pulse-on time of 20 μs, pulse-off time of 7 μs.

6.3.2 WLT Evaluation Using ImageJ Software

Using ImageJ software and Equation (6.12), the WLT was calculated as shown in Figure 6.5a and b, respectively. For trial no. 1, it is observed from Figure 6.5a that the WLT area was found to be 2286.41 μm², whereas from Figure 6.5b the length of WLT was found to be 254.027 μm. By substituting these values into Equation (6.12), the WLT obtained with ImageJ software was 9 μm, which was close to the experimental value as mentioned in Table 6.4. Similarly, WLT for the remaining trials can also be evaluated.

6.3.3 Optimum Parameter Setting Using Jaya Technique

Using Jaya optimization technique, a reduction in WLT was achieved using optimum parameters setting. The result obtained showed that WLT was optimized up to 4.846 μm with the optimum parameters setting of voltage of 55 V, current of 4 A, pulse-on time of 20 μs, and pulse-off time of 8.2 μs, respectively. Table 6.7 shows the same; hence, moving toward the best solution condition is satisfied.

6.4 Conclusions

- The comparative investigation between experimental and empirical models shows a close agreement between them. This states that an empirical model generated for WLT will be applicable for the prediction of result for the given set of input variables.
- The Jaya optimization is meant for moving towards the best solution; this condition is likely to be satisfied during EDM machining of NiTi 60. Where it was possible to optimize of WLT up to 4.846 μm with the optimum parameter settings of voltage of 55 V, current of 4 A, pulse-on time of 20 μs, and pulse-off time of 8.2 μs, respectively.

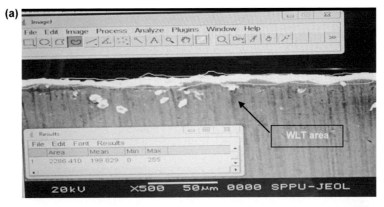

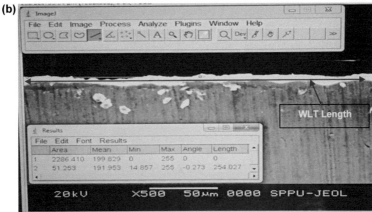

FIGURE 6.5
(a) Evaluation of WLT area through ImageJ software and (b) evaluation of WLT length through ImageJ software.

TABLE 6.7

Optimum Combinations

Objective	EDM Parameters				Optimum Result
	Voltage	Current	Pulse-on	Pulse-off	
Min. WLT	51	4	20	8.2	4.846

- In comparison with EDM parameters, current seems to be the most dominating parameter by an increase in WLT, and correspondingly, degradation in quality of machined specimen takes place. This is likely to be evident with SEM images.

- ImageJ software is the other means, which provides information regarding surface quality. An effort has been made with the use of this software where the area and length of WLT were determined.

- With respect to the current research, similar analysis can also be carried out by varying the EDM process parameters for different materials.

References

1. S. Tripathy, D. K. Tripathy (2017). *Optimization of Process Parameters and Investigation on Surface Characteristics during EDM and Powder Mixed EDM*. Springer, Singapore, vol. 385–391.
2. M. Boujelbene, E. Bayraktar, W. Tebni, S. Ben Salem (2009). Influence of machining parameters on surface integrity in electrical discharge machining. *International Journal of Material Science and Engineering* 37, 110–116.
3. P. Sharma, A. Tripathy, N. Sahoo (2018). Evaluation of surface integrity of WEDM processed Inconel 718 for jet engine application. *IOP Conference Series: Materials Science and Engineering* 323, 012019.
4. V. G. Navas, I. Ferreres, J. A. Maranon, C. G. Rosales, G. J. Sevillano (2008). Electro discharge machining (EDM) versus hard turning and grinding-comparison of residual stresses and surface integrity generated in AISI 01 tool steel. *Journal of Materials Processing Technology* 195, 186–194.
5. M. U. Gaikwad, A. Krishnamoorthy, V. S. Jatti (2020). Investigation on effect of process parameter on surface integrity during electrical discharge machining of NiTi60. *Multidiscipline Modeling in Materials and Structures*, 16(6), 1385–1394.
6. N. K. Mandal, N. K. Singh, U. C. Kumar, V. Kumar (2016). Semi-empirical modeling of surface roughness in CNC end milling. *International Journal of Mechatronics, Electrical and Computer Technology* 6(22), 3099–3109.
7. N. G. Patil, P. K. Brahmankar (2010). Determination of material removal rate in wire electro-discharge machining of metal matrix composites using dimensional analysis. *The International Journal of Advanced Manufacturing Technology* 51, 599–610.
8. R. Bobbili, V. Madhu, A. K. Gogia (2015). Modelling and analysis of material removal rate and surface roughness in wire-cut EDM of armour materials. *Engineering Science and Technology, an International Journal* 37(4), 1–5.
9. D. Poro's, S. Zaborski (2009). Semi-empirical model of efficiency of wire electrical discharge machining of hard-to-machine materials. *Journal of Materials Processing Technology* 209, 1247–1253.
10. B. K. Roy, A. Mandal (2019). Surface integrity analysis of nitinol-60 shape memory alloy in WEDM. *Materials and Manufacturing Processes* 34(10), 1091–1102.
11. S. Dewangan, C. K. Biswas, S. Gangopadhyay (2014). Influence of different tool electrode materials on EDMed surface integrity of AISI P20 tool steel. *Materials and Manufacturing Processes* 29, 1387–1394.
12. V. Muthukumar, A. Suresh Babu, R. Venkatasamy, N. Senthil Kumars (2015). An accelerated particle swarm optimization algorithm on parametric optimization of WEDM of die-steel. *Journal of the Institution of Engineers (India): Series C* 96(1), 49–56.
13. M. U. Gaikwad, A. Krishnamoorthy, V. S. Jatti (2019). Process parameters optimization using Jaya algorithm during EDM machining of Niti60 alloy. *International Journal of Scientific &Technology Research* 8(11), 1168–1174.
14. N. Sharma, R. Khanna, R. D. Gupta (2013). Multi quality characteristics of WEDM process parameters with RSM. *International Conference on Design and Manufacturing* 64, 710–719.
15. A. M. Takale, N. K. Chougule, P. H. Selmokar, M. G. Gawari. (2018). Multi-response optimization of micro-WEDM process parameters of $Ti_{49.4}$-$Ni_{50.6}$ shape memory alloy for orthopedic implant application. *Advanced Materials Research* 1150, 1–21.
16. R. V. Rao, D. P. Rai, J. Ramkumar, J. Balic (2016). A new multiobjective Jaya algorithm for optimization of modern machining processes. *Advances in Production Engineering & Management Journals* 11(40), 271–286. doi: 10.14743/apem2016.4.226.
17. E. Krause (2004). *Fluid Mechanics*, Springer, Berlin, Heidelberg, New York.
18. R. Subramanian, K. Marimuthu, M. Sakthivel (2013). Study of crack formation and re-solidified layer in EDM process on $T_{90}Mn_2W_{50}Cr_{45}$ tool steel. *Materials and Manufacturing Processes* 28(6), 664–669.

7

Analysis of Convolution Neural Network Architectures and Their Applications in Industry 4.0

Gaurav Bansod, Shardul Khandekar, and Soumya Khurana

Pune Institute of Computer Technology

CONTENTS

DOI: 10.1201/9781003143505-7

7.1 Introduction

In the past decade, the use of convolution neural networks (CNNs) has increased significantly to perform tasks that are fundamental to computer vision and image processing such as object detection, image classification and segmentation, and video processing, which are performed with more than human-level accuracy by the CNNs. Nowadays, CNNs are not only used for feature extraction but also used for feature generation, called generative modeling, to give the state-of-the-art results. This ability of CNNs enables them to be ideal for applications that are specific to Industry 4.0. Industry 4.0 refers to the evolution of smart machines, which is heavily dependent on interconnectivity, automation, machine learning, and deep learning, which provide it real-time decision-making capability.

Through this chapter, we have provided a detailed description of various CNN architectures and their applications in Industry 4.0. We have given an overview of each architecture, which includes its structure, advantages, and limitations. CNN architectures like GoogLeNet [3], YOLO (you only look once) [8], ResNets [5], and generative adversarial networks (GANs) [9] are discussed in this chapter. These architectures in general consist of various components, viz., convolution layers, pooling layers, normalization layers, fully connected layers, padding layers, loss and activation functions.

- **Convolution layer:** This layer performs the convolution operation on the input image based on the specified filter size and strides to reduce the dimension of the input as shown in Figure 7.1. It extracts features using small-sized filters from the input data and still preserves the relationship between the adjacent pixels.

- **Pooling layer**: This layer is used to extract the most prominent features from a particular area of the input image as shown in Figure 7.2, which drastically reduces the size of the image and trainable parameters. It does this in two ways, namely, maximum and average pooling. Maximum pooling extracts pixels with the largest raw value, while average pooling extracts a pixel value based on the average of the pixels in the area.

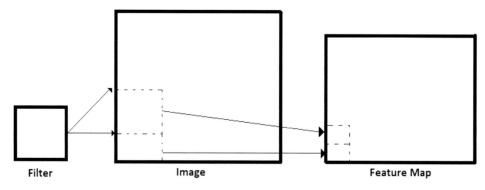

Filter   Image   Feature Map

FIGURE 7.1
The filter performing a convolution operation with the input image to form a feature map.

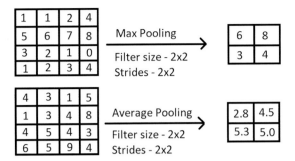

FIGURE 7.2
Maximum pooling and average pooling being performed to extract predominant features. (Source: Machine Curve [18].)

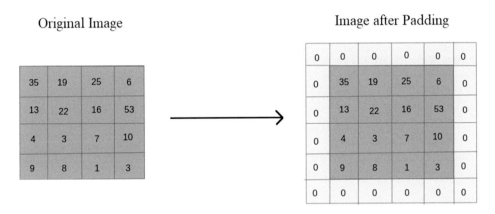

FIGURE 7.3
Image structure before and after padding.

- **Padding layer**: This layer is used to increase the size of the input to the required size by adding redundant pixels to the boundary of the image, which is described by Figure 7.3. This is done to extract the features present at the boundary of the image so as to retain any valuable information.

  The output size of the image after the convolution operation can be determined by

$$W_2 = \frac{W_1 - F + 2P}{S} + 1 \tag{7.1}$$

  where
  W_1 = Shape of the input vector (width or height)
  P = Shape of the padding layer
  S = Size of the strides
  W_2 = Shape of the output vector.

- **Normalization layer**: This layer is used to assign initial random weights and configurations to the input of each layer. Batch normalization is a type of normalization that is used to standardize the input to each layer, which reduces the number of iterations and stabilizes the learning process. For any layer, with d-dimensional input $x = (x_1, x_2, x_{3,...} x_d)$, the normalization with expectation and variance for a batch B is defined as

$$\hat{x}_k = \frac{x_k - E[x_k]_B}{\sqrt{\mathrm{Var}[x_k]_B}} \tag{7.2}$$

where
$E[x_k]$ = Weighted mean of x
$\mathrm{Var}[x_k]$ = Variance of x

- **Fully connected layer**: These layers serve as the output layers of most CNN architectures. The input to the fully connected layer is the output of the last convolution or pooling layer, which is flattened and fed to the fully connected layer whose details are shown in Figure 7.4. The main goal of these layers is to classify the input based on the given training labels.

- **Loss and activation functions:** Loss functions like mean squared error which is a regression loss function, binary cross-entropy for binary classification, and multi-class cross-entropy for multi-class classification are used to determine the difference between the actual value and the predicted value, which in turn can be used to update the weight matrix. Activation functions are used to determine whether a node in the network will be activated or not, which is eventually relevant to the model's final prediction. Most commonly used activation functions like sigmoid, tanh, rectified linear unit (ReLu), and leaky ReLu are shown in Figure 7.5.

CNNs were originally used to extract features from small, low-resolution images. The first CNN was used to recognize handwritten and machine-printed letters and digits. Today, CNNs have evolved and are being used for localization, segmentation, classification of images as well as for generating high-resolution images without the need for human intervention and annotations. To perform all these various tasks, different CNN architectures were proposed over the years, as illustrated in Figure 7.6, with each of

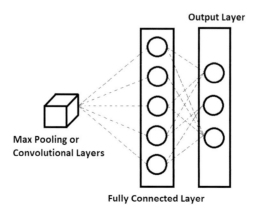

FIGURE 7.4
Fully connected layers present at the end of a CNN architecture.

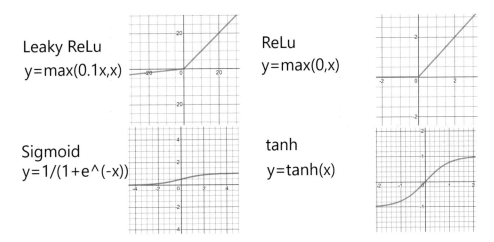

FIGURE 7.5
Various activation functions used in CNN architecture. (Source: A basic overview of CNN [19].)

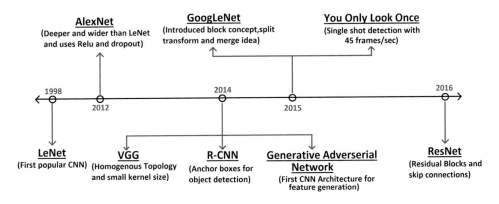

FIGURE 7.6
CNN architecture that brought major architectural innovations over the period of 1998–2016.

them having their strengths like lower training time and faster and more accurate prediction, and weaknesses like heavy dependency on powerful hardware. In this paper, we will be focusing on a few of the important underlying CNN architectures, their inception, and probable industrial use cases.

7.2 Evolution of Convolution Neural Network Architectures

7.2.1 LeNet

LeNet, also known as lenet-5, is a CNN structure, which was first introduced by Y. LeCun [1], a French computer scientist, in 1998. This neural network architecture was initially used for handwritten and machine-printed character recognition in documents.

The LeNet architecture needs less computing power and requires less memory to store its weights due to its small and simple architecture.

7.2.1.1 Architecture Description

The LeNet architecture consists of a total of seven layers, including convolution layer, pooling layer, and fully connected layer. The first layer is an input layer, which takes an input in the form of a grayscale image with dimensions 32×32 pixels. The grayscale images used as an input have their pixel values normalized between -0.1 and 1.175 to ensure that their mean is 0 and standard deviation is 1. The subsequent layers of the LeNet architecture are described below and shown in Figure 7.7.

- The first layer is a convolution layer, which consists of six filters, each filter of size (5×5) and (1×1) strides with 'same' padding to map an output of dimension (28×28).
- The second layer is an average pooling layer with (2×2) strides to have an output with a dimension of (14×14).
- The third layer is a convolution layer, which is similar to the first layer with the exception that this layer has 16 filters and 'valid' padding.
- The fourth layer is an average pooling layer, which is similar to the second layer with (2×2) strides and an output of 16 (5×5) feature graphs.
- The fifth layer is a convolution layer with 120 filters, each of size (5×5) and an output size of (1×1). This is defined as a convolution layer rather than a fully connected layer because if the input to the first layer is larger than a (32×32) image, this layer will not produce an output with dimensions of (1×1).
- The sixth layer, i.e., the last layer, is a fully connected layer with 84 feature graphs as an output.

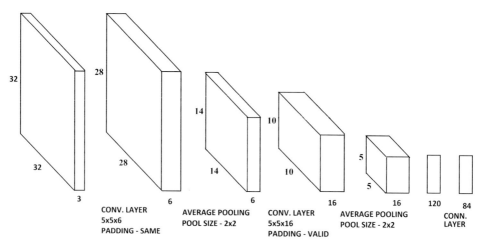

FIGURE 7.7
The layer structure and input and output size for LeNet.

7.2.1.2 Limitations

LeNet was designed in the late 1990s for the purpose of character recognition, but it cannot be used for a variety of other use cases, such as extracting complex features from high-resolution images because of its smaller number of layers. Another limitation is that it tends to over-fit and has no built-in mechanism to avoid this due to the lack of dropout layers.

7.2.2 AlexNet

AlexNet is a CNN that was created by A. Krizhevsky [2], a Ukrainian computer scientist. The major upgrade that this neural network architecture brought was that more layers and subsequently more depth were instrumental for higher performance, which was made possible by the availability of graphics processing units for the purpose of training the network. Due to the availability of better hardware, this network enhanced the performance when validated across multiple image databases.

7.2.2.1 Architecture Description

AlexNet consists of eight layers, five convolution layers, and three fully connected layers with the first layer taking an RGB image with dimensions 224×224 pixels as an input. AlexNet has an architecture, which is similar to that of LeNet with the improvement of having more layers, more filters per layer, and stacked convolution layers. Some of the new methodologies that AlexNet implemented were as follows:

- **Use of ReLu:** At the time, activation functions like tanh ($f(x) = \tanh(x)$) and sigmoid ($f(x) = (1+e^{-x})^{-1}$) were used to predict the output of neurons given the input feature vector x. These functions were saturating nonlinearity and hence required longer training time as compared to non-saturation nonlinearities like rectified linear unit (ReLu) $f(x) = \max(x, 0)$.
- **Use of local response normalization (LRN):** One of the important features of ReLu is that it does not require the inputs to be normalized to prevent the model from becoming saturated. But normalization can help to improve the overall efficiency of the model. LRN is used to ensure lateral inhibition that is the process of inhibiting the activity of neighboring neurons. There are two types of LRN, namely, inter-channel and intra-channel. The one that is used in AlexNet is inter-channel and is defined as follows:

$$b_{x,y}^i = a_{x,y}^i \left/ \left(k + \alpha \sum_{j=\max(0,i-n/2)}^{\min(N-1,i+n/2)} \left(a_{x,y}^i \right)^2 \right)^\beta \right. \tag{7.3}$$

where i is the output of ith filter, $a(x, y)$ and $b(x, y)$ refer to the pixel value intensity at location (x, y) before and after normalization, N refers to the total number of channels, and β, α, k, n are hyperparameters.

- **Use of overlapping pooling:** Pooling is used to represent the most dominant feature or pixel value inside the kernel of size ($f \times f$). In the traditional pooling, the

TABLE 7.1

Structure of AlexNet

Type	Filter Size	Strides	Padding	Number of Filters/Nodes	Output Shape
Convolution	11×11	4×4	-	96	$55 \times 55 \times 96$
Max pooling	3×3	2×2	-	-	$27 \times 27 \times 96$
Convolution	5×5	-	2	256	$27 \times 27 \times 256$
Max pooling	3×3	2×2	-	-	$13 \times 13 \times 256$
Convolution	3×3	-	1	384	$13 \times 13 \times 384$
Convolution	3×3	-	1	384	$13 \times 13 \times 384$
Convolution	3×3	-	1	256	$13 \times 13 \times 256$
Max pooling	3×3	2×2	-	-	$6 \times 6 \times 256$
Dense	-	-	-	4,096	$1 \times 1 \times 4,096$
Dense	-	-	-	4,096	$1 \times 1 \times 4,096$
Dense (softmax)	-	-	-	1,000	$1 \times 1 \times 1,000$

filter size was the same as stride size, which is also called maximum pooling but in overlapping pooling, the stride size is less than the filter size, and this tends to give a smoother transition effect for detecting various features. This also makes the model less prone to over-fitting. Table 7.1 shows the architecture of AlexNet.

7.2.2.2 Limitations

The total number of parameters that AlexNet has is 60 million, and this results in over-fitting. The other drawback is that due to a very deep architecture, it was computationally expensive at the time and the time required for the model to converge during training was very high.

7.2.3 GoogLeNet

GoogLeNet (or Inception v1) was introduced by researchers at Google [3] in 2014. GoogLeNet significantly reduced error rate for image classification as compared to its predecessors like AlexNet. The architecture introduced new methods like (1 × 1) convolution, global average pooling, and inception module, which were not present in other state-of-the-art architectures of its time.

7.2.3.1 Architecture Description

GoogLeNet consists of 27 layers, including the convolution and pooling layers. Additional features such as 1 × 1 convolution, global average pooling, and inception module enabled it to create a deeper architecture, which was able to extract complex features.

- **1 × 1 Convolution:** The concept of (1 × 1) convolution was introduced by M. Lin et al. in their paper Network in Network [21]. (1 × 1) convolution is the convolution operation performed by a filter of size (1 × 1), and it will convolve with the entire image to reduce the dimensionality of the output and also to introduce nonlinearity with the help of activation functions. This was used in GoogLeNet to reduce the number of dimensions before carrying out the computationally expensive (3×3) and (5×5) convolutions.

- **Global average pooling:** This is used to extract a single feature from each layer, reducing the dimension of the image from (height $\times$ width $\times$ depth) to $(1 \times 1 \times$ depth). This results in a reduction of total parameters, as visualized in Figure 7.8a, and minimizes over-fitting.

- **Inception module:** The main purpose of the inception block was to create sparsely connected convolution layers as compared to fully connected layers so as to decrease the number of computations and prevent the model from over-fitting. It initially applies (1×1) convolution, which is followed by (3×3), (5×5) convolution and (3×3) maximum pooling. The final output for the module is obtained by parallel stacking the output of the corresponding filters as shown in Figure 7.8b (Table 7.2).

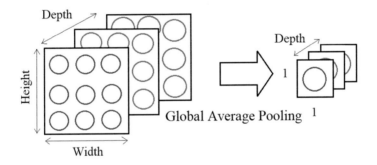

FIGURE 7.8a
A visualization of global average pooling.

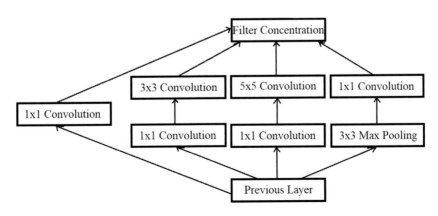

FIGURE 7.8b
A visualization of global average pooling.

TABLE 7.2

Structure of GoogLeNet

Layers	Filter Size	Strides	Output Size	# 1×1	# 3×3 Reduce	# 3×3	# 5×5 Reduce	# 5×5	# Pool proj
Convolution	7×7	2×2	112×112×64	-	-	-	-	-	-
Max pooling	3×3	2×2	56×56×64	-	-	-	-	-	-
Convolution	3×3	1×1	56×56×192	-	64	192	-	-	-
Max pooling	3×3	2×2	28×28×192	-	-	-	-	-	-
Inception (3a)	-	-	28×28×256	64	96	128	16	32	32
Inception (3b)	-	-	28×28×480	128	128	192	32	96	64
Max pooling	3×3	2×2	14×14×480	-	-	-	-	-	-
Inception (4a)	-	-	14×14×512	192	96	208	16	48	64
Inception (4b)	-	-	14×14×512	160	112	224	24	64	64
Inception(4c)	-	-	14×14×512	128	128	256	24	64	64
Inception (4d)	-	-	14×14×528	112	144	288	32	64	64
Inception (4e)	-	-	14×14×832	256	160	320	32	128	128
Max pooling	3×3	2×2	7×7×832	-	-	-	-	-	-
Inception (5a)	-	-	7×7×832	256	160	320	32	128	128
Inception (5b)	-	-	7×7×1,024	384	192	384	48	128	128
Average pool	7×7	1×1	1×1×1,024	-	-	-	-	-	-
Dropout	-	-	1×1×1,024	-	-	-	-	-	-
Dense	-	-	1×1×1,000	-	-	-	-	-	-
Softmax	-	-	1×1×1,000	-	-	-	-	-	-

7.2.3.2 Limitations

GoogLeNet does not have a disadvantage as such, but there is a proposed change in the form of Xception Network, which increases the limit of divergence of the inception module.

7.2.4 VGG

The Visual Geometric Group (VGG) architecture was introduced by K. Simonyan and A. Zisserman [4] from the University of Oxford in 2014. The main aim behind creating this architecture was to understand the effect of depth of networks on the accuracy of the model for large-scale image classification and recognition. The network architecture has various subtypes, viz., VGG11 to VGG19, each varying in the number of cascaded convolution layers, from 8 layers in VGG11 to 16 layers in VGG19. In this chapter, we will focus on the detailed architecture of VGG16.

7.2.4.1 Architecture Description

As the name suggests, VGG16 has a total of 16 layers, 13 convolution layers, and 3 fully connected layers. It takes a (224 × 224) image with three channels as an input and provides an output vector with dimension (1,000 × 1). The average pixel RGB value is calculated for the entire dataset, which is then subtracted from each pixel of each image. This is the only pre-processing done. The padding for all the layers is 'same', i.e., 1 pixel for a filter of size (3×3). This is done so as to preserve spatial resolution after the convolution is carried out. The specific details for each layer are as follows:

- The first two layers are convolution layers with 64 filters each, filter size of (3×3), and same padding with ReLu activation. This is followed by a maximum pooling layer with pool size (2×2) and strides (2×2).
- This same configuration is repeated once, with the exception of two convolution layers with 128 filters and the same maximum pooling layer.
- The next three layers are convolution layers with 256 filters each, filter size of (3×3), same padding, and ReLu activation. They are followed by a maximum pooling layer with pool size (2×2) and strides (2×2).
- This same configuration is repeated twice after the above set, with the exception of 512 filters each and a maximum pooling layer after a set of three convolution layers.
- Finally, we have a flattened and three dense, fully connected layers, with 4,096 parameters in the first two layers and 1,000 parameters in the last layer (Table 7.3).

TABLE 7.3

Structure of VGG16

Layers	Filter size	Strides	Number of Filters/Nodes	Output Size
Convolution	3×3	-	64	$224 \times 224 \times 64$
Convolution	3×3	-	64	$224 \times 224 \times 64$
Max pooling	2×2	2×2	-	$112 \times 112 \times 64$
Convolution	3×3	-	128	$112 \times 112 \times 128$
Convolution	3×3	-	128	$112 \times 112 \times 128$
Max pooling	2×2	2×2	-	$56 \times 56 \times 128$
Convolution	3×3	-	256	$56 \times 56 \times 256$
Convolution	3×3	-	256	$56 \times 56 \times 256$
Convolution	3×3	-	256	$56 \times 56 \times 256$
Max pooling	2×2	2×2	-	$28 \times 28 \times 256$
Convolution	3×3	-	512	$28 \times 28 \times 512$
Convolution	3×3	-	512	$28 \times 28 \times 512$
Convolution	3×3	-	512	$28 \times 28 \times 512$
Max pooling	2×2	2×2	-	$14 \times 14 \times 512$
Convolution	3×3	-	512	$14 \times 14 \times 512$
Convolution	3×3	-	512	$14 \times 14 \times 512$
Convolution	3×3	-	512	$14 \times 14 \times 512$
Max pooling	2×2	2×2	-	$7 \times 7 \times 512$
Dense	-	-	4,096	$1 \times 1 \times 4,096$
Dense	-	-	4,096	$1 \times 1 \times 4,096$
Dense (Softmax)	-	-	1,000	$1 \times 1 \times 1,000$

7.2.4.2 Limitations

The main problem faced by VGG was that it was very slow to train. Even on high-end hardware for the time, it took a couple of weeks to train. Another problem was that the model weights took a lot of disk space and bandwidth, making it inefficient.

7.2.5 ResNet

ResNet architecture was introduced by researchers from Microsoft [5] in 2015. The main aim behind the creation of the ResNet architecture was to resolve the vanishing gradient problem, which caused architecture with a large number of layers to have lower performance and accuracy. This was resolved by the introduction of a residual block in the ResNet structure.

7.2.5.1 Architecture Description

The residual block is one of the most important blocks of the architecture. For deep convolution networks, there is a problem of vanishing gradient, which is the lowering or saturation of the network accuracy as the network gets deeper. During the backpropagation phase, the loss is determined and the corresponding gradient values are calculated. These values are used to update the weights of the previous layers. The weight matrix for each layer is updated until the input layer is reached. It takes a longer time for the input layer to converge as the gradient decreases for each hidden layer present in the network. This results in ineffective learning for the initial layers. ResNet utilizes a method called skip connections, to solve this problem, which skips the training for a few layers as shown in Figure 7.9. This is done by applying an identity mapping to the input, which ensures that the output is the same as the input.

As we can see from the above diagram, there exist two paths during backpropagation: the identity mapping pathway and residual mapping pathway. During training, the gradient value will pass through the weight layers in the residual mapping path and new gradient values will be computed. This results in the initial layer gradient values being very small or vanishing entirely. To avoid this, the identity path is provided. This ensures that the initial layers receive the correct gradient values and learn the correct weights (Table 7.4).

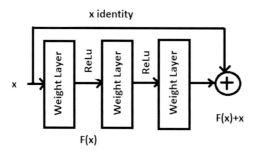

FIGURE 7.9
Skip connections in residual model.

TABLE 7.4

Structure of ResNet

Layer Name	Output Size	Filter Size
Conv1	112×112	7×7, 64 stride 2
Conv2_x	56×56	3×3 max pooling stride 2
		$\left.\begin{array}{l} 1 \times 1,\,64 \\ 3 \times 3,\,64 \\ 1 \times 1,\,256 \end{array}\right\} \times 3$
Conv3_x	28×28	$\left.\begin{array}{l} 1 \times 1,\,128 \\ 3 \times 3,\,128 \\ 1 \times 1,\,512 \end{array}\right\} \times 4$
Conv4_x	14×14	$\left.\begin{array}{l} 1 \times 1,\,256 \\ 3 \times 3,\,256 \\ 1 \times 1,\,1024 \end{array}\right\} \times 6$
Conv5_x	7×7	$\left.\begin{array}{l} 1 \times 1,\,512 \\ 3 \times 3,\,512 \\ 1 \times 1,\,2048 \end{array}\right\} \times 3$
	1×1	Average pool, 1000-d fc, softmax

7.2.5.2 Limitations

Even though ResNet has solved certain problems faced by its predecessors, one of the major drawbacks of this architecture is that due to the depth of the network it takes a long time to train the model.

7.2.6 R-CNN

Region-based CNN was developed by researchers at University of Berkeley [6] in 2014. This was used to locate objects inside images without using the brute force method. Objects in images can have various locations and aspect ratios, which can be very computationally expensive to find. This is one of the problems that the R-CNN architecture aims to solve.

7.2.6.1 Architecture Description

To solve the problem of the sliding window brute force object detection method, R-CNN used region proposals. These were a set of 2,000 regions in the image which were then searched for objects using a greedy algorithm called selective search. This algorithm combines smaller sectors to generate proposals on the image with the advantage that it has a limited number of proposals, limited by the number of regions in the image.

Although the architecture was innovative, it came with a few drawbacks. First, the training time was still considerably large as each image had a total of 2,000 regions which had to be classified. Second, the selective algorithm was fixed, which meant that no learning took place at that stage, which could result in incorrect proposals. Lastly, it could not be implemented in real time as it took around 47 s to detect objects in a single image.

7.2.6.1.1 Fast R-CNN

This was an improvement to R-CNN proposed by the author of the original R-CNN paper. He proposed that instead of generating the region proposals before feeding the image to the CNN, we let the CNN generate a feature map and then have region proposals based on these feature maps. These region proposals are then warped into squares using region of interest (RoI) pooling, which are then given to a fully connected layer and a softmax layer to predict its class and offset values for the bounding boxes. Even though region proposals were one of the defining features of R-CNN, during the testing of the fast R-CNN network, the region proposals seemed to slow down the algorithm as compared to when tested without region proposals. This limited the performance of the fast R-CNN architecture as a whole.

7.2.6.1.2 Faster R-CNN

Faster R-CNN was proposed by S. Ren et al. [7] in 2016. The major improvement that this architecture brought was that it removed the fixed selective search algorithm, instead of using an algorithm that lets the network learn the regions by itself. An image is given as an input to the network, which it feeds to a network and is used to predict the region proposals in the image. These region proposals are then reshaped using RoI pooling, and then, softmax is used to classify the proposed regions and provide values for bounding boxes. The structural flow that faster R-CNN follows is illustrated in Figure 7.10.

7.2.6.2 Limitations

As this architecture does not look at the image as a whole for object detection, this causes problems such as requiring multiple passes for feature extraction. Also, performance is not guaranteed as the object classification is heavily dependent on the proper functioning of the region proposal network.

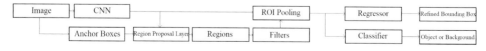

FIGURE 7.10
Structural flow of faster R-CNN.

7.2.7 You Only Look Once (YOLO)

YOLO architecture was introduced by J. Redmon [8] in 2015. It was created for the purpose of real-time image recognition, which was a problem faced by CNNs of the time. Its main advantage over other architectures is its speed, enabling it to process multiple images and predict objects within the image. Where some architectures took multiple seconds to process a single image, YOLO could process multiple images per second, which provided it an edge over other competing networks.

7.2.7.1 Architecture Description

7.2.7.1.1 YOLO v1

YOLO v1, as shown in Figure 7.11a, has a total of 30 layers, 24 convolution layers, 4 maximum pooling layers, and 2 fully connected layers. It takes an image as input and resizes it to have the same aspect ratio but dimensions of 448×448 by padding the image. It uses a 1×1 convolution followed by a 3×3 convolution for reducing the number of channels, leaky ReLu as an activation function, and batch normalization for regularization.

7.2.7.1.2 YOLO v2

YOLO v2 [20] provided improvements on YOLO v1 while maintaining the speed of its predictions. Its layer structure is described in Figure 7.11b. Some of the improvements were as follows:

- **Batch normalization:** The addition of batch normalization layer resulted in faster convergence of the model, which resulted in faster training. It also meant that the need for other normalization types such as dropout was eliminated without the risk of over-fitting. The addition of batch normalization also increases the mean average precision (mAP).

- **High-resolution classifier:** As compared to YOLO v1, which took an input image of dimensions (224×224), YOLO v2 takes an image of dimensions (448×448) as input. This results in the network getting more time to adjust filters for higher-resolution images. The increase in the input size results in a 4% increase in mAP.

- **Anchor boxes:** The fully connected layers in YOLO v1 are used to predict bounding boxes on objects in images unlike newer architectures like R-CNN, which

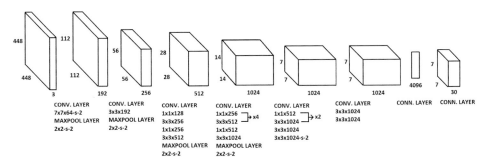

FIGURE 7.11a
Layers in YOLO v1. (Source: You Only Look Once [8].)

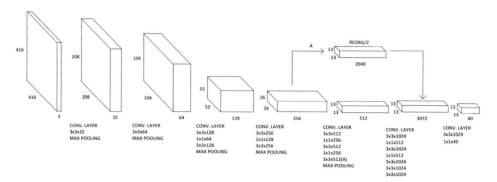

FIGURE 7.11b
Layers in YOLO v2. (Source: YOLO 9000 [20].)

predict coordinates of the object. In YOLO v2, the fully connected layers are removed and anchoring boxes are used instead of bounding boxes. In YOLO v2, the input image is divided into 13×13 cells and five anchor boxes are placed in each cell. The network predicts parameters of the bounding boxes such as center coordinates, height, and width, and then corrects the locations of the existing boxes.

7.2.7.1.3 YOLO v3

YOLO v3 is based on DarkNet, consisting of 106 convolution layers, the first 53 layers from DarkNet to train the network on the ImageNet dataset and the next 53 layers to perform detection. This causes the architecture to be slower as compared to its previous versions, but subsequently more accurate at detecting smaller clustered objects with more classes. YOLO v3 detects objects by applying a 1×1 filter on three different-sized feature maps at various points in the network (Table 7.5).

7.2.7.2 Limitations

YOLO uses localization to improve on the speed of detection of objects in images, which causes a few problems. Due to this technique, only a few objects can be detected per bounding box and can have only one class. It suffers from detecting multiple small objects that are clustered close to each other. It also struggles with images of unusual aspect ratios and uses coarse features of images to predict objects.

7.2.8 Generative Adversarial Networks (GANs)

GANs were first introduced by I. Goodfellow [9] and his colleagues in 2014. GANs served the purpose of generative modeling for unsupervised learning. The basic idea was to train two networks simultaneously in such a way that the gain of one network resulted in the loss of another network as shown in Figure 7.12a. The two networks are called generative and discriminative networks, respectively.

   Generator: It is a neural network that takes a random Gaussian noise as an input and is responsible to generate plausible data that attempts to map few new instances of the training data.

TABLE 7.5

Structure of YOLO v3

	Layer	Filters	Size	Output
	Convolution	32	3×3	256×256
	Convolution	64	$3 \times 3/2$	128×128
1×	Convolution	32	1×1	
	Convolution	64	3×3	
	Residual			128×128
	Convolution	128	$3 \times 3/2$	64×64
2×	Convolution	64	1×1	
	Convolution	128	3×3	
	Residual			64×64
	Convolution	256	$3 \times 3/2$	32×32
8×	Convolution	128	1×1	
	Convolution	256	3×3	
	Residual			32×32
	Convolution	512	$3 \times 3/2$	16×16
8×	Convolution	256	1×1	
	Convolution	512	3×3	
	Residual			16×16
	Convolution	1,024	$3 \times 3/2$	8×8
4×	Convolution	512	1×1	
	Convolution	1,024	3×3	
	Residual			8×8
	Average pool	Global		
	Connected	1,000		
	Softmax			

Discriminator: It is another neural network that takes the data generated by the generator model and tries to predict whether the data produced is real or fake.

The discriminator is trained separately from the generator on the real data. The two are then combined to form the complete GAN model. The discriminator consists of two loss functions: discriminator and generator loss. Discriminator loss is used to train the discriminator on the real data, and generator loss is used to train the generator.

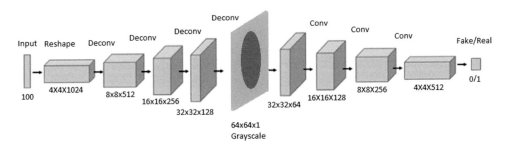

FIGURE 7.12a

Generator and discriminator networks combined to form GAN architecture.

The generator accepts random noise as an input to generate the data. The nature of the noise has little to no effect on the data generated, and thus, noise with a smaller dimension as compared to the output is chosen.

During the generator training phase, the generator produces data that is easily classified by the discriminator as fake, which results in a larger discriminator loss value. Through backpropagation, this value is then fed back to the generator to update its weights and biases. As this process repeats, the generator steadily improves its output, to a point where the discriminator is unable to tell the difference between the actual and generated data. Some of the types of GANs are as follows:

1. **Progressive GAN**: It is an extension to GAN that generates large high-quality images by starting from small low-quality images. This allows the generator network to be faster, stable, and versatile. This is achieved by starting with a small image and adding layers to the model till it achieves the required size and dimensions.

2. **Conditional GAN**: It is a type of GAN that is used to produce images of a certain class. It is trained on a dataset of labeled images and then generates images for the specified class. Its structure is detailed in Figure 7.12b.

7.2.8.1 Limitations

GANs require a large number of images to train and generate images from random Gaussian noise; for example, if we need GAN to generate plausible human faces, we must provide the generator and discriminator networks a large number of images of human faces. We need to maintain a balance between the training of generator and discriminator networks. If they are not in perfect synchronization, they might end up producing images similar to the given training dataset (Table 7.6).

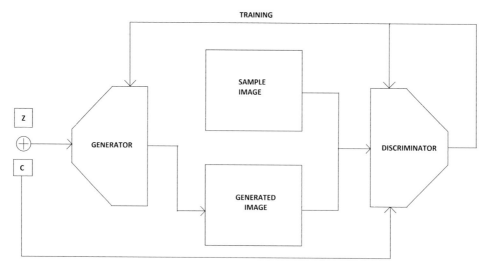

FIGURE 7.12b
Logical structure of conditional GAN.

TABLE 7.6

Comparison of the Parameters of Various CNN Architectures

Architecture Name	Year	Parameters	Error Rate	Depth	Category
LeNet	1998	0.060M	MNIST 0.8	5	Spatial exploitation
AlexNet	2012	60M	ImageNet 16.4	8	Spatial exploitation
VGG	2014	138M	ImageNet 7.3	19	Spatial exploitation
GoogLeNet	2015	4M	ImageNet 6.7	22	Spatial exploitation
ResNet	2016	25.6M	ImageNet 3.6	152	Depth and multi-path

7.3 Applications of Convolution Neural Networks in Industry 4.0

7.3.1 Healthcare Sector

Timely diagnosis of crucial diseases has proven to be very effective in treating and even preventing dangerous and life-threatening diseases in humans like diabetic retinopathy, breast cancer, and osteoarthritis. Delayed diagnosis of such diseases has led to serious consequences and even deaths of many patients in the past. CNNs with their incredible ability to process and analyze images better than humans will have a major impact on the healthcare sector. Using CNNs and image processing, we have the ability to detect and eventually cure diseases long before the patient starts showing any major symptoms of the disease.

For example, CNN-based image processing can be used to identify and treat breast cancer before it has a chance to spread. There are two types of tumors based on their characteristics and behavior: benign and malignant tumors. Benign tumors are noncancerous and cannot spread, whereas malignant tumors are cancerous and can spread to other parts of the body. The correct identification of these tumors is crucial to the eventual treatment of the patient. Here is where CNNs come to the rescue. Using CNNs, tumors can be detected and classified using ultrasonic images as shown in Figure 7.13a.

Another case in which CNNs are useful is about the detection of diabetic retinopathy [12], as shown in Figure 7.13b. Diabetic retinopathy is one of the major reasons for blindness among diabetic patients. The major issue with the detection of diabetic retinopathy is that it does not show any symptoms until it is too late and the patient already shows

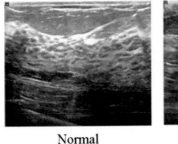
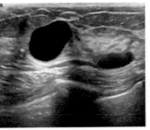
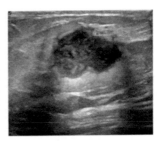

Normal   Benign   Malignant

FIGURE 7.13a

Comparison of the parameters of various architectures. (Source: Breast ultrasound image dataset [10].)

Normal   DR

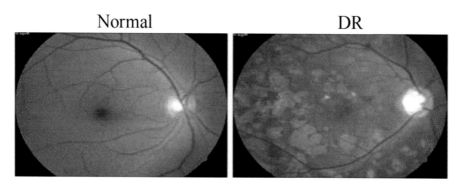

FIGURE 7.13b
Comparison of normal and diabetic retina. (Source: Ocutech retina comparison [11].)

symptoms like vision loss. Even with a highly trained doctor, the chance of detection of diabetic retinopathy is very low. As we know, CNNs are amazing at analyzing images and thus can be used for the early detection and prevention of the disease.

7.3.2 Automotive Sector

Cars and other vehicles have become more and more sophisticated as the years go by, gaining features such as automatic unlock and keyless ignition. These features provide a quality of life improvement over the previous models, but safety is still the major concern when it comes to motor vehicles. Features such as airbags provide a layer of safety to the driver in case of an accident but there has been a lot of work done recently to prevent the occurrence of accidents altogether. Most accidents occur due to a mistake on the part of the driver. This is where CNNs come in handy, providing faster and more accurate responses to situations where an accident could occur, subsequently reducing the risk of an accident or preventing them altogether. A number of methods of automation have been developed in the automotive sector to forward this goal of increased motor vehicle safety.

One of the major advancements in automotive safety has been in the form of self-driving cars [13]. Modern cars have implemented an array of cameras in their design so as to have a level of automation. This is done so as to reduce dependency of the vehicle to be maneuvered by humans, which reduces the chance for a mishap. This level of autonomy is achieved through algorithms such as object detection (shown in Figure 7.14a), lane detection, and depth estimation [15] (shown in Figure 7.14b), which are implemented using various CNN architectures.

7.3.3 Fault Detection

The current maintenance and quality checks are heavily dependent on human perception. Due to the sheer number of products manufactured on a day-to-day basis, some faults might be overlooked by humans during quality checks. The diagnostic software that is currently being used requires complex manufacturing data, which requires the technicians to physically probe the faulty component, thus requiring a longer time for diagnosing and correcting it. Humans are prone to negligence and making mistakes,

FIGURE 7.14a
Object detection using CNN. (Source: Collaborative vehicular vision [14].)

FIGURE 7.14b
Depth estimation using CNN. (Source: Depth prediction without sensors [15].)

which may result in a large number of faulty products being supplied to the end user. Such products might include faulty motors, bearings, or even power lines, to name a few. Fault detection using computer vision is a very crucial part of Industry 4.0. Fault detection and diagnosis consists of data collection, data processing, and fault classification. This can be efficiently done with the help of computer vision and CNNs.

For example, bearings might be considered a small part of a mechanical system, but play an extremely important role in the smooth functioning of the system as a whole. Faults in bearings occur on a very minute scale and can be very difficult to detect with the naked eye. The deformations in the bearings can be efficiently identified using computer vision and CNNs in conjunction [16] as shown in Figure 7.15a. Using this method will result in a reduction in the faults in the system and increase overall system efficiency.

Another field where fault detection is useful is in the case of power cables and various components on the electric poles. Uninterrupted power delivery to various sectors of the industry is of utmost importance. Considering this, detecting faults in power lines is an important venture for many industries. If a person is assigned to manually find faults, it might take them a lot of time depending on the length of power lines and might result in missing out on a few of the faults. CNNs, on the other hand, operate completely autonomously and have a very high accuracy rate. CNNs can also be trained in such a way that they can even predict the overall lifespan of the power cables being used currently as shown in Figure 7.15b. This ensures that continuous power will be delivered to the industries and faults in the cables can be found and replaced before they cause any major issue such as causing problems due to power failure in heavy and costly machinery or delayed production times.

The above image is an IR image taken from a drone to detect faults in the power cables. The image on the right shows the places on the cable where a fault is present.

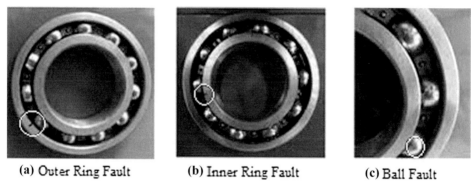

(a) Outer Ring Fault   (b) Inner Ring Fault   (c) Ball Fault

FIGURE 7.15a
Types of faults in bearings. (Source: A review of CNN in bearing fault detection, p. 2 [16].)

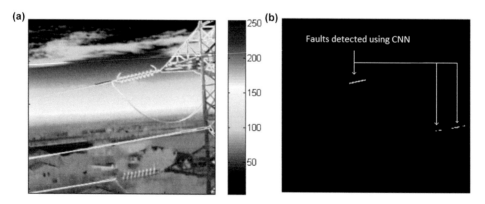

FIGURE 7.15b
Fault detection in power lines using IR images. (Source: Drones for fault detection [17].)

7.4 Conclusion

There has been a significant improvement in the field of CNN architectures, resulting in remarkable progress in the domains of image processing and computer vision. This chapter acts as a one stop solution for all major CNN architecture innovations. In this chapter, we have evaluated and analyzed the various architectural innovations that CNNs have undergone over the period of 1998–2016. We saw the increased use of block-based architectures as compared to layer-based structure of CNN. Our analysis points out that these architectural innovations in the field of CNN such as block-based structure, stacked convolution layers, various activation functions like sigmoid, ReLu and increase in overall depth of the network architecture have resulted in a significant rise of applications of CNN in healthcare and automotive sectors in Industry 4.0.

References

1. Y. Lecun, L. Bottou, Y. Bengio, and P. Haffner, (1998). Gradient-based learning applied to document recognition. *Proceedings of the IEEE*, vol. 86, no. 11, pp. 2278–2324. doi: 10.1109/5.726791.
2. A. Krizhevsky, I. Sutskever, and G. Hinton, (2012). ImageNet classification with deep convolutional neural networks. *Neural Information Processing Systems*, 25. doi: 10.1145/3065386.
3. C. Szegedy et al., (2015). Going deeper with convolutions, *2015 IEEE Conference on Computer Vision and Pattern Recognition (CVPR)*, Boston, MA, pp. 1–9. doi: 10.1109/CVPR.2015.7298594.
4. K. Simonyan, and A. Zisserman, (2014). Very deep convolutional networks for large-scale image recognition. arXiv 1409.1556.
5. K. He, X. Zhang, S. Ren, and J. Sun, (2016). Deep residual learning for image recognition, *2016 IEEE Conference on Computer Vision and Pattern Recognition (CVPR)*, Las Vegas, NV, pp. 770–778. doi: 10.1109/CVPR.2016.90.
6. R. Girshick, J. Donahue, T. Darrell, and J. Malik, (2014). Rich feature hierarchies for accurate object detection and semantic segmentation, *2014 IEEE Conference on Computer Vision and Pattern Recognition*, Columbus, OH, pp. 580–587. doi: 10.1109/CVPR.2014.81.
7. S. Ren, K. He, R. Girshick, and J. Sun, (2015). Faster R-CNN: Towards real-time object detection with region proposal networks. *IEEE Transactions on Pattern Analysis and Machine Intelligence*, 39. doi: 10.1109/TPAMI.2016.2577031.
8. J. Redmon, S. Divvala, R. Girshick, and A. Farhadi, (2016) You only look once: Unified, real-time object detection, *2016 IEEE Conference on Computer Vision and Pattern Recognition (CVPR)*, Las Vegas, NV, pp. 779–788. doi: 10.1109/CVPR.2016.91.
9. I. Goodfellow, J. Pouget-Abadie, M. Mirza, et al., (2014). Generative adversarial networks. *Advances in Neural Information Processing Systems*, 3. doi: 10.1145/3422622.
10. P. S. Rodrigues, (2017). Breast ultrasound image, *Mendeley Data*, V1. doi: 10.17632/wmy84gzngw.1
11. A. Gupta, and R. Chhikara, (2018). Diabetic retinopathy: Present and past, *Procedia Computer Science*, vol. 132, pp. 1432–1440. doi: 10.1016/j.procs.2018.05.074. http://www.sciencedirect.com/science/article/pii/S1877050918308068.

12. H. Pratt, F. Coenen, D. M. Broadbent, S. P. Harding, and Y. Zheng, (2016). Convolutional neural networks for diabetic retinopathy, *Procedia Computer Science*, vol. 90, pp. 200–205. doi: 10.1016/j. procs.2016.07.014. http://www.sciencedirect.com/science/article/pii/S1877050916311929.

13. H. Fujiyoshi, T. Hirakawa, and T. Yamashita, (2019). Deep learning-based image recognition for autonomous driving. *IATSS Research*, 43. doi: 10.1016/j.iatssr.2019.11.008.

14. S. Dey and T. Nguyen, Collaborative vehicular vision, UC San Diego, http://globalstip.ucsd. edu/research/collaborative- vehicular- vision.

15. V. Casser, S. Pirk, R. Mahjourian, and A. Angelova, (2019). Depth prediction without the sensors: Leveraging structure for unsupervised learning from monocular videos. *Proceedings of the AAAI Conference on Artificial Intelligence*, vol. 33, pp. 8001–8008. doi: 10.1609/aaai. v33i01.33018001.

16. N. Waziralilah, A. Abu, M. Lim, L. Quen, and A. Elfakharany, (2019). A review on convolutional neural network in bearing fault diagnosis. *MATEC Web of Conferences*, vol. 255, p. 06002. doi: 10.1051/matecconf/201925506002.

17. P. Smith, Drones for fault detection in power lines, DroneBelow, https://dronebelow.com/ 2019/07/15/drones-for-fault-detection-in-power-lines/.

18. C. Versloot, What are max pooling, average pooling, global max pooling and global average pooling? Machine Curve. https://www.machinecurve.com/index.php/2020/01/30/ what-are-max-pooling-average-pooling-global-max-pooling-and-global-average-pooling/.

19. U. Udofia, Basic overview of convolution neural network, medium. https://medium.com/ dataseries/basic-overview-of-convolutional-neural-network-cnn-fcc7dbb4f17.

20. J. Redmon and A. Farhadi, (2017). YOLO9000: Better, faster, stronger, *2017 IEEE Conference on Computer Vision and Pattern Recognition (CVPR)*, Honolulu, HI, pp. 6517–6525. doi: 10.1109/ CVPR.2017.690.

21. M. Lin, Q. Chen and S. Yan, (2013). *Network in Network*. https://arxiv.org/abs/1312.4400.

8

EMD-Based Triaging of Pulmonary Diseases Using Chest Radiographs (X-Rays)

Niranjan Chavan

Institute for Thermal Energy Technology and Safety (ITES) Karlsruher Institute for Technology

Priya Ranjan

SRM-AP University

Uday Kumar

Delhi State Cancer Institute

Kumar Dron Shrivastav

Amity University Uttar Pradesh

Rajiv Janardhanan

Amity University Uttar Pradesh

CONTENTS

Abbreviations

AI: Artificial intelligence

ANN: Artificial neural networks

COVID-19: Coronavirus disease of 2019

DOI: 10.1201/9781003143505-8

COPD: Chronic obstructive pulmonary disease
EMD: Earth mover's distance
LLMICs: Low- and lower middle-income countries
LMICs: Low- and middle-income countries
NCDs: Noncommunicable diseases
PHCs: Primary healthcare centres
SCCT: Smart Cyber Clinical Technology
SES: Socioeconomic status
WHO: World Health Organization

Symbols

X: First signature with n clusters
X_i: Cluster representative of X
W_{xi}:   Weight of cluster X
Y: Second signature with m cluster
Y_j: Representative of Y
W_{yj}:   Weight of cluster Y
D: Ground distance matrix
D_{ij}:   Ground distance between clusters X_i and Y_j
F: Flow matrix
F_{ij}:   Flow matrix between clusters X_i and Y_j

Chapter Organization

The chapter is divided into two main parts: the first part covers the introduction and methodology, whereas other parts focus on result and discussion.

The introductory part gives us a glimpse of the public health situation in India. It also shows us how technology advancements can be leveraged to help the healthcare professional in India diagnose and treat patients in an affordable and an accessible manner. In the last few years, conventional healthcare systems have seen drastic changes across the world. Here in this paper, we have shown a roadmap to use Industry 4.0 in the healthcare systems. Other sections such as earth mover's distance (EMD), dataset, and parameter settings brief about the work executed.

In another part, we have discussed the results and scope of work in future.

8.1 Introduction

Noncommunicable diseases (NCDs) disproportionately affect people living within low- and lower middle-income countries (LLMICs) [1–3], with almost three quarters of all NCD deaths and 82% of premature deaths occurring within LLMICs [4]. The relationship between NCDs, poverty and social and economic development has therefore received high-level recognition [5], with NCDs seen to pose a major challenge to development in the 21st century [6,7]. The poor may be more vulnerable to NCDs for many reasons, including material deprivation, psychosocial stress, higher levels of risk behaviour, unhealthy living conditions, limited access to high-quality health care, and reduced opportunity to prevent complications. Low socioeconomic status (SES) groups are more likely to use tobacco products and consume unhealthy foods, should be physically inactive, and be overweight or obese. In recent decades, the socioeconomic gradient of NCD risk factors has broadened from high-income countries to low- and middle-income countries (LMICs).

Among LMICs, India is considered a particularly important nation to study the emerging burden of NCDs as it indeed presents a unique landscape of disease burden. India is projected to experience more deaths from NCDs than any other country over the next decade, due to the size of the population and worsening risk factor profile, associated with recent dramatic economic growth. The country has deep and entrenched social and economic disparities, with affordable healthcare being beyond the reach of large sections of society. The epidemiological evidence on the SES-related patterning of NCDs remains limited in LMICs.

In the last two decades, a significant leap of thought has resulted in the translation of the fundamental concepts of precision medicine at a community level to understand the patterns and processes associated with the landscape of disease burden in the LMICs having fractious and fractionated healthcare ecosystems. India endowed with heterogeneous genetic base, wide variations in socio-cultural norms, along with divergent geological relief structures, provides a veritable landscape of disease burden resulting in prevalence of healthcare disparities. The need of the hour is to create fruganomic community-empowering solutions aimed at alleviating the healthcare disparities.

Although India has witnessed an epidemiological transition from communicable diseases to NCDs in the last two decades with cardiovascular diseases and cancer accounting for a significant proportion of morbidities and mortalities, the unfinished agenda of communicable diseases has led to emergence of recent pandemics such as COVID-19 [8]. The World Health Organization (WHO) has recognized the outbreak of COVID-19 to be a Public Health Emergency of International Concern on January 30, 2020, and declared it as a pandemic on March 11, 2020, resulting in enforced nationwide locked downs in several countries across the world, including India [9]. This has had a statutory impact on the economy of all the countries, including India.

COVID-19 infection primarily spreads through contact with an infected person and through respiratory droplets when people cough or sneeze. It shows the initial symptoms like viral pneumonia such as fever, cough, myalgia, fatigue and shortness of breath. In serious cases, it shows complications like acute respiratory distress syndrome and cardiac failure. About 5% can develop serious stages of pulmonary distress syndrome and can lead to death [10].

WHO report on NCDs and COVID-19 says that the new coronavirus can affect people in all age groups, but the risk is high in older age group and people with pre-existing NCDs [9]. These NCDs include cardiovascular diseases, chronic respiratory disease (COPD), diabetes, and cancer. Recent pieces of evidence from China and European countries show that people with a history of NCDs were susceptible to COVID-19-associated morbidities and mortalities [11]. According to WHO [6], four major NCDs, namely cardiovascular diseases, diabetes, cancer, and chronic pulmonary diseases, are the "leading killers" and "slow motion disaster", account for 71% (77 million) of deaths globally.

Global Health Estimates (2018) also reported, out of total deaths, about 85% of the NCD deaths occurred in the European region. Japan (82%), the United States (88%), and the UK (89%) also recorded incidences of deaths due to NCDs. The emergence of COVID-19 has worsened the situation in these countries. Old age and pre-existing NCD conditions can reduce the immunity level and will lead to worse outcomes. The global statistics on COVID-19 also shows that the death rate is 21.9% in the age group of above 80 years, 8% in the age group between 70 and 79 years, and further 3.6% in the age group of 60–69 years [12]. New York University Hospital reported that out of 400 COVID-19 patients admitted, 90% were from >45 years and 60% reported as >65 age group.

The most common comorbidity observed in COVID-19 patients who were admitted in the critical care units in most countries was due to hypertension, diabetes, and chronic respiratory problems. Chronic respiratory problems account for 8% of the mortalities [12]. India has 18% of the global population and an increasing burden of chronic respiratory diseases, including chronic obstructive pulmonary disease (COPD), asthma, pneumoconiosis, interstitial lung diseases, and pulmonary sarcoidosis [13]. The existence of environmental factors such as air pollution is known to significantly contribute to premature mortality and disease burden globally, with the highest impact in low-income and middle-income countries such as the Indian subcontinent endowed with resource-limited healthcare systems [14,15].

Recent pieces of evidences suggest that the potential of Industry 4.0 revolution could be harnessed as a contributor to the Sustainable Development Goals (SDGs) [33] that acts as a bridge between the industry and sustainability by finding a significant correlation between their components. As intended by the United Nations SDGs [33] for 2030, technological progress is driving the challenge of transition from traditional technology into intelligent machines without limiting the sustainability of the industrial economy. The combination of artificial intelligence (AI), robotics, and other advanced technologies applied across many sectors in the provisioning of healthcare will significantly contribute to alleviating the healthcare disparities that exist in resource-limited healthcare ecosystems prevalent in LMICs. The emergence of Industry 4.0 has enabled the automation of clinical chores with a view to leverage a greater efficiency of functions, e.g., reduction of resource consumption such as carbon foot printing. Therefore, Industry 4.0 might provide the most suitable platform to develop fruganomic digital sustainable operations allowing even LMICs to meet SDG.

Recent evidences suggest that the use of multimodal multisensor fusion technologies provides a unique big data-enabled platform, for strengthening resource-deprived healthcare systems prevalent in countries such as the Indian subcontinent. It is our belief that the transition from precision medicine to precision public health must be enabled by the state-of-art technology platforms with a view to create affordable and accessible healthcare solutions intrinsic to the needs of the Indian subcontinent. The various healthcare solutions in vogue so far have not had a major impact due to the neglect of system-level issues, which if addressed appropriately would enable seamless uptake of the solutions into the existing framework of the healthcare systems.

A major drawback of the existing interpretation algorithms based on artificial neural network (ANN) is its black-box nature. When you feed an image into a neural network and it matches it to another image, it is difficult to understand why and how it came up with a match [16–18]. Further, the nonlinear dynamical behaviour of deep neural networks is prone to chaotic nature and fundamental underlying unpredictability [19]. Adding to this issue is an extreme demand for computation in deep learning, which is forcing researchers to explore other techniques [20,21]. On the other hand, static and predictable algorithms like EMD perform image match by computing perceptual similarity and provide more meaningful and interpretable solutions to matching problems. To re-emphasize, ANNs have a long and very well-researched history of inherent instability, and its automated decisions can't be entrusted to make decisions critical to the survival of a patient afflicted with a severe case of a cardiac or a COVID-19 episode. India faces many challenges in delivering healthcare, especially in the rural domain due to shortage of resources and skilled personnel. The fact that chronic respiratory diseases account for 8% of mortalities in India clearly undermines the magnitude of the public health challenge it imposes. This is further compounded by the lack of suitable algorithms to map out the clinical resource allocation especially in rural underserved population where doctor/patient ratio is low and access to affordable and quality healthcare options is significantly marginalized.

Chest radiographs are still mostly relied upon modality for diagnosing lung disease conditions despite the limitations of extant algorithms in vogue to accurately triage the patients in the order of urgency. A more prudent step, therefore, will be to develop tools and applications that can be easily integrated into the current healthcare system. Additionally, the lack of well-structured databases for referencing and analysis hinders the progression of research from aiding and optimizing processes and clinical decision making with the help of AI.

India endowed with diverse genetic base and socio-cultural norms presents a unique landscape of disease burden necessitating the need for niche-specific databases for enhancing the accuracy of AI tools. Apart from the above-mentioned complexities, the development of these technologies comes with substantial technical, ethical, confidentiality and clinical challenges. Despite the aforementioned hindrances, the socioeconomic impact and benefits of AI-based automation of ECG traces, Cervigrams, abdominal CTs, as well as chest radiographs analysis for LMICs like India will be significant [34–44]. The potential healthcare application of AI-based platforms is vast, encompassing screening, disease detection, patient risk stratification along with niche-specific optimal intervention strategies. Since the penetration of the mobile platforms and Internet is extensive across the Indian subcontinent, the development of AI-enabled computational applications on these platforms would provide a valuable, precision public health tool for better management of lung disease epidemic alleviating a significant burden on the national exchequer.

While there is a discussion on how to organize modern medical/clinical infrastructure, Industry 4.0 naturally facilitates the use of automation and data exchange in clinical service provisioning. If we add smart clinical devices, smart software-aided diagnosis, and closed-loop clinical resource allocation along with a deck of principle technologies like smart software-aided diagnosis, we envision a Smart Cyber Clinical Technology

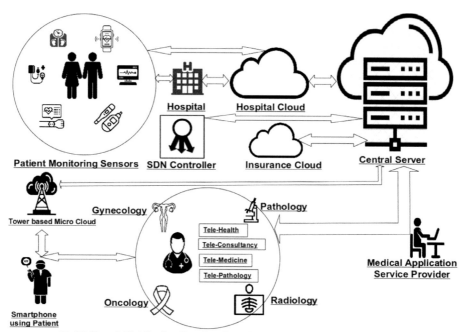

FIGURE 8.1
Industry 4.0 and Internet of Things integrated with Smart Cyber Clinical Technology and infrastructure.
Together, they boost clinical performance for the patients and save resources for society.

(SCCT), which can lead to this modern vision. These technologies can help us to build a "smart healthcare centre" where machines, systems, and humans collaborate with each other in order to facilitate and monitor progress along with the clinical and health service provisioning. Networked clinical devices over the nearest microcloud provide sensor/software-processed data and help the digital platforms in improving health services.

The fundamental impact of integrating smart clinical devices, IoT, and Industry 4.0 with clinical software and closed-loop resource allocation is the ability to rapidly deploy medical infrastructure in challenging places during natural calamities like flood, earthquake, drought, Tsunamis etc., while drastically reducing costs and reacting to demands in patients' preferences, pharma industry changes, the supply chain, and technology upgrades as shown in Figure 8.1.

8.1.1 Motivation for Building This Tool

This work is motivated by the following two objectives.

1. Development and validation of an Intelligent Decision Support System for segregating chest radiographs to detect lung diseases such as tuberculosis, pneumonia, COVID-19-associated lung diseases in both tertiary care settings and extended community along with tracking of patients through low-end mobile health applications.

2. Integration and validation of multimodal tool in clinical practice involving automated processing of anonymized chest radiographs along with conventional molecular biomarkers of tissue hypoxia in both angiogenic and fibrotic phases of the lung disease progression forming the rationale of effective triage methods for prioritizing the most urgent conditions to wait-listed ones.

The race- and sex-specific variations in the levels of conventional biomarkers such as angiogenesis/fibrosis indeed necessitate the validation and confirmation by a modality, which can crunch a large amount of data in an affordable and accessible manner. Our fruganomic data-intensive AI-enabled tool will facilitate the same by incorporating the clinical-epidemiological features of the subjects evaluated not only at tertiary care centres but also in the extended community. This computational modality, when integrated with digital signals from surrogate molecular markers, will form the rationale of a multimodal multifusion sensor technology [22–24], which will aim at not only resolving the dogma of missed and misdiagnosis of lung diseases such as tuberculosis or pneumonia at tertiary care centres and extended community but also individualizing the risk assessment of patients with suspected lung diseases or categorizing patients into low- or high-risk groups.

We are further looking at directions to integrate signals from computational fluid dynamics to understand the metrics of pulmonary hypertension and its putative role in modulating the clinical outcomes of lung diseases. This when integrated with clinical-epidemiological features of the subjects would significantly help in the identification and delineation of modifiable vascular risk factors (particularly hypertension and smoking) along with social and environmental factors, including reduced exposure to air pollution through provisioning of health advisories and literacy modules, thereby empowering the vulnerable population to risks of even COVID-19 pandemic, whose clinical outcomes are extremely poor in patients with cardiopulmonary diseases or risk factors of cardiopulmonary diseases.

In recent years, various computer-based tools have been developed, which can be reliably used for computational disease tagging purposes. Healthcare professionals with the help of such tools can accurately computationally tag different disease conditions within a short time.

In the past, people have developed deep learning models which use X-ray images to predict COVID-19 in the patients [25]. In this paper, we have explored the possibility to predict the lung ailment by applying EMD algorithm to the X-ray images of the patients. EMD mimics the human perception of texture similarity. EMD outperforms many other texture similarity measures when used for texture classification and segmentation [26]. All the previously developed models have chosen the process-based approach, which has inherent instabilities built in. In this work, we have used a programmatic approach to detect the diseases using chest radiographs (lung X-ray) of patients. Our programmatic approach has the advantage of being free from any dynamic instability as compared to tools like deep learning.

The lung X-ray images of patients suffering from pneumonia, TB, COVID-19, and healthy persons were pooled together from various datasets [27–29]. Two random images from each dataset are selected. The focus of our current study is restricted to understanding the patterns and processes associated with incremental burden of lung diseases in the Indian populace. The following sections of the current study discuss the preliminary observations in the study.

8.1.2 Earth Mover's Distance

The ground distance between two single perceptual features can often be found by psychophysical experiments. For example, perceptual colour spaces were devised in which the Euclidean distance between two single colours approximately matches human perception of their difference. This becomes more complicated when sets of features, rather than single colours, are being compared. This correspondence is key to a perceptually natural definition of the distances between sets of features. This observation led to distance measures based on bipartite graph matching [30,31], defined as the minimum cost of matching elements between the two histograms.

EMD is a method to calculate the disparity between two multidimensional distributions in some space where a distance magnitude between single ones (ground distance) is given. Suppose there are two distributions: one can be considered as the area with the mass of earth, and the other as a collection of holes in that same area. Then, the EMD is the measure of the least amount of work required to fill the holes with earth.

Here, the unit of work is the force needed in transporting unit earth by a unit of ground distance. Hence, it can also be defined as the minimum cost that must be provided to convert one histogram into other. Measurement of EMD is based on a solution of transportation problem. For finding mathematical representation, firstly we formalized it as the following linear programming problem:

Let X be the first signature with n clusters, X_i is the cluster representative, and W_{xi} is the weight of cluster.

Let Y be the second signature with m clusters.

Y_i is the cluster representative, and W_{yi} is the weight of cluster.

Let D be the ground distance matrix, and D_{ij} is the ground distance between clusters x_i and Y_j.

Let F be the flow matrix, and F_{ij} is the flow between X_i and Y_j.

Then,

$$X = (X_1, W_{X1}), (X_2, W_{X2}), (X_3, W_{X3})\ldots(X_n, W_{Xn}) \tag{8.1}$$

$$Y = (Y_1, W_{Y1}), (Y_2, W_{Y2}), (W_3, Y_{Y3})\ldots(Y_n, W_{yn}) \tag{8.2}$$

$$D = \begin{bmatrix} D_{ij} \end{bmatrix} \tag{8.3}$$

$$F = \begin{bmatrix} F_{ij} \end{bmatrix} \tag{8.4}$$

Now, the $\text{WORK}(X, Y, F) = \sum_{i=1}^{n} \sum_{j=1}^{m} f_{ij} D_{ij}$

Subject to constraints:

$$(i)\, f_{ij} \geq 0 \quad \text{where} \quad 0 \leq i \leq n, 0 \leq j \leq m$$

$$(ii)\, \sum_{i=1}^{m} f_{ij} \leq w_{ij} \quad \text{where} \quad 0 \leq i \leq n$$

$$(iii)\, \sum_{i=1}^{n} f_{ij} \leq w_{yj} \quad \text{where} \quad 0 \leq j \leq m$$

$$(iv)\, \sum_{i=1}^{m} \sum_{i=1}^{n} f_{ij} = \min \sum_{i=1}^{n} w_{xi}, \sum_{j=1}^{m} w_{yj}$$

The constraint (i) enables mass moving from X to Y. (ii) and (iii) restrict the amount of mass that can be sent by the clusters in X to their weights and the clusters in Y to receive no more mass than their weights. (iv) One forces to move the maximum amount of mass possible. It is also known as the total flow. Once we solve the transportation problem, we will get the optimal flow F. Now the EMD is defined as the work normalized by the total flow:

$$\mathrm{EMD}(X,Y) = \sum_{i=1}^{n} \sum_{j=1}^{m} F_{ij} D_{(ij)} \Big/ \sum_{i=1}^{n} \sum_{j=1}^{m} F_{ij}$$

8.1.3 Dataset

X-ray images of subjects of Indian origin were accessed from multiple sources. The data of the patients was anonymized before subjecting it to algorithms for image processing. Data and code are available on the link "https://github.com/niranjan917/EMD_XRAY" (Table 8.1).

TABLE 8.1

Lists of Images of Chest Radiograph (X-Ray) Typifying a Variety of Images along with Their Sources

Image No	Image Title	Patient Condition	Dataset Source
1	PN-F1.jpg	Normal	https://www.kaggle.com/parthachakraborty/Pneumonia-chest-x-ray
2	PN-F2.jpg	Normal	https://www.kaggle.com/parthachakraborty/Pneumonia-chest-x-ray
3	TB-F1.jpg	Normal	https://www.kaggle.com/raddar/Chest-X-rays-tuberculosis-from-India
4	TB-F2.jpg	Normal	https://www.kaggle.com/raddar/Chest-X-Rays-tuberculosis-from-India
5	PN-T1.jpg	Pneumonia diagnosed	https://www.kaggle.com/Parthachakraborty/pneumonia-chest-x-ray
6	PN-T2.jpg	Pneumonia diagnosed	https://www.kaggle.com/parthachakraborty/Pneumonia-chest-X-ray
7	TB-T1.jpg	Tuberculosis diagnosed	https://www.kaggle.com/raddar/Chest-X-rays-tuberculosis-from-India
8	TB-T2.jpg	Tuberculosis diagnosed	https://www.kaggle.com/raddar/Chest-X-rays-tuberculosis-from-India
9	CV-T1.jpg	COVID-19 diagnosed	Dr. Uday Shankar. Rajiv Gandhi super speciality Hospital, Tahirpur
10	CV-T2.jpg	COVID-19 diagnosed	Dr. Uday Shankar. Rajiv Gandhi super speciality Hospital, Tahirpur

TABLE 8.2

Depiction of EMD Values of Chest Radiographs (X-Rays) Different Types of Lung Diseases

	TB-F1	TB-F2	PN-01	PN-02	Sum
TB-F1	0	31.13487	56.29595	108.9537	196.3845
TB-F2	31.10189	0	30.50083	97.48288	159.0856
PN-01	56.29595	30.50819	0	84.08909	170.8932
PN-02	108.9537	98.91861	83.81279	0	291.6851
SUM	196.3515	160.5617	170.6096	290.5256	

8.1.4 Parameter Settings

Open-source software 'R' version 3.6.3 has been used as scientific programming environment as it is equipped with highly advanced image analysis tools and techniques with magnificent visualizations. For image processing and EMD calculations, imager and EMDIST libraries were installed, respectively, in the R studio. The computing machine used for the image analysis had the following features:

Processor: Intel® Core™ i7-5500U CPU @ 2.40GHz×4

OS Name: Ubuntu 20.04.1 LTS

OS Type: 64-bit

Graphics: NVIDIA corporation GM108M [GeForce 840M]

Memory: 7.7 GiB

To maintain the homogeneity, images were converted to the greyscale for uniform basis and scaled down to 50×50 pixel due to the limited computational resource availability. It should be noted that the bigger the pixel size, the larger the time required to process the images. EMD can be calculated using Euclidean or Manhattan approach. Here, we have used Euclidean approach for the analysis. The X-ray images of healthy patients are fed to the algorithm. The pairwise EMD comparison of healthy X-ray images showed that Image No 4 resembles all other normal images the most, and hence, it is suitable for the analysis as a normal representative image (Table 8.2). This representative image is compared with all other images. The pairwise comparison of all images using EMD has been tabulated in the LibreOffice calc, which gives us an interesting comparison about the relative impact of COVID-19, tuberculosis, and pneumonia (Figure 8.2).

8.2 Results and Discussion

The EMD comparison for each image gave us EMD values. Based upon the computational analysis of greyscale chest/lung X-ray images, EMD values of X-rays of patients afflicted with pneumonia were the maximum followed by EMD values of X-ray of patients affected with tuberculosis, while the least EMD values were observed in the

FIGURE 8.2
Computer screenshot shows ongoing image analysis on Rstudio.

X-ray images of lungs of COVID-19 patients. A unique and indigenous scale based upon EMD values categorizing the chest radiographs in the order of disease severity. The EMD values of chest radiographs of patients afflicted with pneunmonia is the highest followed by EMD values of chest radiographs of patients afflicted with tuberculosis and COVID-19 (Figure 8.3 and Table 8.3).

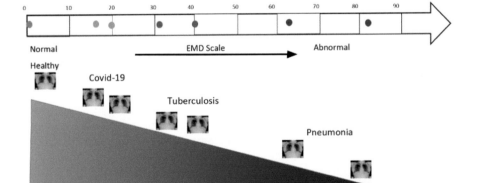

FIGURE 8.3
Severity scale of lung diseases. A novel triaging system of grading the chest radiographs (X-rays) based upon the EMD values. Clearly, it is visible that pneumonia damage is highest while COVID impact is least.

TABLE 8.3

Pairwise X-Ray Image Comparison between Reference Healthy Image and other Images

Reference Image No (Healthy)	Image No (Unhealthy)	EMD
4	4	0.0
4	9 (COVID-19 diagnosed)	16.79990
4	10 (COVID-19 diagnosed)	20.45182
4	8 (tuberculosis diagnosed)	31.82152
4	7 (tuberculosis diagnosed)	40.85552
4	5 (pneumonia diagnosed)	60.61876
4	6 (pneumonia diagnosed)	83.56622

8.2.1 Conclusion and Anticipated Outcomes

Data analytical processes often require a large amount of data for showing accurate analyses. Our preliminary analyses suggest that EMD can be used for tagging the chest radiographs and X-ray images with different diseases in a more efficient way. We are trying to acquire bigger datasets for studies so the accuracy level of our tool will increase, and the tool can also be used for tagging the untagged X-ray images.

8.2.2 Anticipated Outcomes

- The current study aims to form the rationale for the application of innovative concepts from divergent realms of science, engineering, and technology to improve health outcomes in an affordable and equitable manner within the overall context and framework of the existing healthcare system.

- These innovations will involve working with partners in order to understand the intricate nuances of genuine needs assessment, health systems integration, sociotechnical aspects, along with the identification and mitigation of barriers with an aim to validate and scale up prototypes to sustainably viably ventures through strong partnerships with the relevant stakeholders in the healthcare ecosystem.

In the face of the COVID-19 outbreak, LMICs are waking up to the limitations and inadequacies of their analogue healthcare system. It seems that the healthcare systems across the world need a rapid makeover although the healthcare system is still managing this crisis largely through risky brick-and-mortar visits.

As an analogue monolithic system, healthcare is ill-equipped to cope with this swiftly emerging epidemic. Although the healthcare industry is structured on the historically necessary model of in-person interactions between patients and their clinicians, clinical workflows and economic incentives have largely been developed to support and reinforce a face-to-face model of care, resulting in the congregation of patients in emergency departments and waiting areas during this crisis. This care structure contributes to the spread of the virus to uninfected patients who are seeking medical evaluation at the tertiary healthcare centres situated across the Indian subcontinent. Vulnerable populations

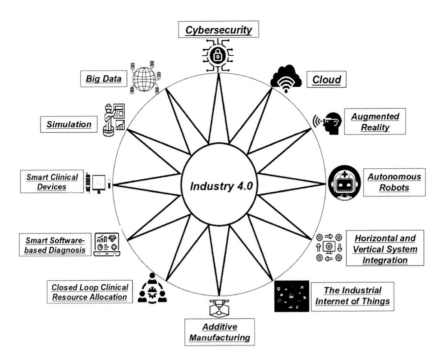

FIGURE 8.4
Applicability of tools of industry 4.0 towards the provision of affordable and accessible healthcare options.

such as patients with multiple chronic conditions or immunosuppression will face the difficult choice between risking iatrogenic COVID-19 exposure during a clinician visit and postponing needed care (Figure 8.4).

Although tools of Industry 4.0 such as AI, IoT, 5G, and other tools have made a significant impact in other industries, its deployment and acceptability in both the healthcare and pharmaceutical industries have been albeit a bit slow. However, professionals from these industries are quickly discovering that the right technology application could also revolutionize the delivery of healthcare to patients in a more affordable and accessible manner.

The era of digital transformation has surely created a novel platform to pave the way for exciting new opportunities for pharmaceutical and healthcare industry around the world. We believe that the use of tools of Industry 4.0 such as AI and machine learning could enable the clinicians/healthcare providers to learn from the enormous amounts of data in every industry, and we can unlock endless opportunities. For instance, AI can examine millions of pictures of medical imaging technology, helping doctors to rule out conditions and diseases more accurately. Additionally, using wearable devices coupled with AI and ML technology will make it much easier and economical to conduct large-scale clinical studies on a temporal scale. This could help the clinical fraternity in quickly unravelling the classical features of diseases that either go undetected or are difficult to diagnose. Although current estimates suggest that the healthcare industry's use of AI will be worth $200 billion by 2025, we believe that the full potential of AI enabled in healthcare and pharmaceutical industry is yet to be exploited to its complete capacity.

Use of wearables and sensors is changing the landscape of healthcare by enabling the clinical fraternity to record data in a seamless manner. For example, vital clinical attributes of the patient over a period could be collected and stored in cloud at ease in a secure and encrypted manner with the capability to be retrieved at ease even in remote areas. Using data and insights gathered through smart wearables appropriately analysed by AI and ML, healthcare providers can recognize changes in a patient's condition and quickly respond by scheduling a preventive check-up or sending out drug prescriptions. The leveraging of the cost-effective mobile health platforms will help in seamless connect of patients with their primary care physicians as has been evidenced recently during the peak of COVID-19 pandemic. Similarly, the application of robotics and robotic process automation is another area of opportunity for the healthcare sector. It is anticipated that collaborative robots, like the Da Vinci surgical robot, would assist healthcare professionals in making quicker and accurate decisions during complicated surgical procedures and significantly reduce the risk of mistakes.

Healthcare systems are able to adopt to this new world, and suddenly, the old ways of working which relied largely on an analogue world and the bricks and mortar of the 'office visit' are starting to look more outdated and out of step with the environment they are working within.

There is an interesting parallel here. Human coronavirus is a group of viruses that change small parts of their genetic code as part of their life cycle. Thus, they mutate as a matter of course. *The challenge is for our healthcare systems to also be able to "mutate" at pace and scale. We have never seen this happening before but happening it is, with digital transformation, the use of medical technological devices, the application of AI in the care of people becoming more commonplace.*

This is also being accompanied by other significant changes, particularly around the deployment of a whole gamut of new devices and products, including wearables which together with the dissemination of a 5G infrastructure and the very sudden increase in take up, are leading to a veritable explosion in the number of data points, which are going to become available to different healthcare systems globally.

There is also another and often forgotten dimension. The monitoring of existing NCDs, largely displaced as the central activity of health systems who are totally focused on managing the pandemic, will require new solutions, and the potential here for technology and digital solutions to enable better self-care is considerable.

It is therefore unlikely that this is a temporary phase. The post-COVID-19 world in the 2020s is going to be very different to what came before it.

Intrinsic Features of Affordable and Accessible Healthcare Technologies Developed in Post-COVID-19 Era

Obviously, they need to fulfil a tangible need and be "good enough" in terms of accuracy, reliability, safety and reproducibility to be deployed at scale. This is largely self-evident but there are five other aspects of how they are deployed, which may appear to be less obvious but in fact are equally important.

I. They need to fit within a governance structure so there is clarity around who is responsible for monitoring and action when required. This has in the past been all too often an afterthought. It is all too easy to get enticed by exciting technological advances and deploy them, because it is possible without spending the requisite time ensuring they fit within a clinical workflow and that the workforce implications around their deployment are managed with the same rigour as any other aspect.

II. They need to be built into existing clinical pathways and flows – the technologies that succeed will be the ones that assist in clinical decision support and preferably are "baked into" EMRs.

III. They need to be personalized. Ideally, they need to incorporate existing data to provide data, which is relevant to the individual and also relevant to the consultation. The age of metadevice is upon us now in this *era of precision public health and precision medicine*.

IV. They need to have interoperability built into them via open APIs. Unless one can easily and effortlessly integrate them within an EMR, it is unlikely they can form part of the system into which AI can work its magic and develop the insights we are desperate for to better manage subsequent waves.

V. They must be secure. Cybersecurity takes on an even greater importance and prominence in the age of pandemics.

Thus, the future for med tech is rosy. First movers and fast followers will reap the benefits of easier adoption and incorporation into the mainstream if they are true to these principles.

References

1. United Nations. Report of the director-general of the World Health Organization on the prevention and control of non-communicable diseases. Document A/R68/650, 2013. Accessed: 15 November 2020.
2. Dutta, U., Nagi, B., Garg, P.K., Sinha, S.K., Singh, K., Tandon, R.K. Patients with gallstones develop gallbladder cancer at an earlier age. *European Journal of Cancer Prevention* 14, 381–5, 2005. doi: 10.1097/00008469-200508000-00011.
3. Gajalakshmi, C.K., Shanta, V., Swaminathan, R., Sankaranarayanan, R., Black, R.J. A population-based survival study on female breast cancer in Madras. *British Journal of Cancer Research* 75, 771–5, 1997. doi: 10.1038/bjc.1997.137.
4. Mendis, S. Global status report on noncommunicable diseases 2014, 2014. Available: http://www.who.int/nmh/publications/ncd-status-report-2014/en/. Accessed: 15 November 2020.
5. United Nations. Outcome document of the high-level meeting of the General Assembly on the comprehensive review and assessment of the progress achieved in the prevention and control of non-communicable diseases. A/RES/68/300. 2014. Accessed: 15 November 2020.
6. World Health Organization. Novel coronavirus – China. Available from: http://www.who.int/csr/zxcvXDdon/12-january-2020-novel-coronavirus-china/en/, accessed on Accessed: 15 November 2020.
7. World Health Organization. Naming the coronavirus disease (COVID-19) and the virus that causes it. Available from: https://www.who.int/emergencies/diseases/novel-coronavirus2019/technical-guidance/naming-the-coronavirus-disease-(COVID-2019)-and-the-virus-that-causes-it, Accessed: 15 November 2020.

8. Huang, C., Wang, Y., Li, X., Ren, L., Zhao, J., Hu, Y., et al. Clinical features of patients infected with 2019 novel coronavirus in Wuhan, China. *Lancet* 395, 497–506, 2020.

9. World health organization (WHO), COVID 19 and NCDs, https://www.who.int/emergencies/diseases/novel-coronavirus-2019, Accessed: 15 November 2020.

10. Leung, C. Clinical features of deaths in the novel coronavirus epidemic in China. *Reviews in Medical Virology*, 30(3), e2103, 2020.

11. Chen, N., Zhou, M., Dong, X., Qu, J., Gong, F., Han, Y., et al. Epidemiological and clinical characteristics of 99 cases of 2019 novel coronavirus pneumonia in Wuhan, China: A descriptive study. *Lancet* 395, 507–13, 2020.

12. GBD, 2015 Chronic Respiratory Disease Collaborators. Global, regional, and national deaths, prevalence, disability-adjusted life years, and years lived with disability for chronic obstructive pulmonary disease and asthma, 1990–2015: A systematic analysis for the Global Burden of Disease Study 2015. *The Lancet Respiratory Medicine* 5, 691–706, 2017.

13. India State-Level Disease Burden Initiative CRD Collaborators. The burden of chronic respiratory diseases and their heterogeneity across the states of India: The Global Burden of Disease Study 1990-2016. *Lancet Global Health*, 6(12), e1363–e1374, 2018. doi: 10.1016/S2214-109X(18)30409-1.

14. Cohen, A.J., Brauer, M., Burnett, R., et al. Estimates and 25-year trends of the global burden of disease attributable to ambient air pollution: An analysis of data from the Global Burden of Diseases Study 2015. *Lancet*, 389, 1907–1918, 2017.

15. Landrigan, P.J., Fuller, R., Acosta, N.J.R., et al. The lancet commission on pollution and health. *Lancet*, 391, 462–512, 2018.

16. Yampolskiy, R.V. Unpredictability of AI, Arxiv-1905.13053, 2019.

17. Yampolskiy, R.V. Unexplainability and incomprehensibility of artificial intelligence, Arxiv-1907.03869, 2019.

18. Sohl-Dickstein, J., Weiss, E.A., Maheswaranathan, N., Ganguli, S. Deep unsupervised learning using non-equilibrium thermodynamics, Arxiv-1503.03585, 2018.

19. Li, H. Analysis on the nonlinear dynamics of deep neural networks: Topo-logical entropy and chaos, Arxiv-1804.03987, 2018.

20. Thompson, N.C., Greenewald, K., Keeheon, L., and Manso, G.F. The computational limits of deep learning. Arxiv-2007.05558, 2020.

21. https://venturebeat.com/2020/07/15/mit-researchers-warn-that-deep-learning-is-approaching-computational-limits/, Published on July 15, 2020. Accessed on 14 November 2020.

22. Chen, R.J., Lu, M.Y., Wang, J., Williamson, D.F.K., Rodig, S.J., Lindeman, N.I., and Mahmood, F., Pathomic fusion: An integrated framework for fusing histopathology and genomic features for cancer diagnosis and Prognosisi, https://arxiv.org/abs/1912.08937, 2019.

23. Carmichael, I., Calhoun, B.C., Hoadley, K.A., Troester, M.A., Geradts, J., Heather, D.C., Olsson, L., Perou, C.M., Niethammer, M., Hannig, J., and Marron, J.S. Joint and individual analysis of breast cancer histologic images and genomic covariates, 2019. https://arxiv.org/abs/1912.00434.

24. Guo, A., Chen, Z., Li, F., Li, W., and Luo, Q. Towards more reliable unsupervised tissue segmentation via integrating mass spectrometry imaging and hematoxylin-erosin stained histopathological image, 2020. https://www.biorxiv.org/content/10.1101/2020.07.17.208025v1.

25. Ozturk, T., Talo, M., Yildirim, E.A., Baloglu, U.B., Yildirim, O., and Acharya, U. Automated detection of COVID-19 cases using deep neural networks with. *Computers in Biology and Medicine*, 1, 11, 2020.

26. Bhaskar, P. and Bandyopadhyay, T. Beyond job security and money: Driving factors of motivation for government doctors in India. *Human Resources for Health*, 12, 12, 2014.

27. Partha, C. Dataset, 2018. Retrieved from Kaggle: https://www.kaggle.com/parthachakraborty/pneumonia-chest-x-ray.

28. Puzicha, J., Rubner, Y., Tomasi, C., and Buhmann, J. Empirical evaluation of dissimilarity measures for color and texture. *Proceedings of the IEEE International Conference on Computer Vision*, pp. 1165–1173. Corfu, Greece, September 1999.

29. Raddar. Dataset, 2020. Retrieved from Kaggle: https://www.kaggle.com/raddar/chest-xrays-tuberculosis-from-india.

30. Peleg, S., Werman, M., and Rom, H. *A Unified Approach to the R.O. Duda, P.E. Hart, D.G. Stork, Pattern Classi1cation*, 2nd Edition. Wiley, New York, 2000.

31. Zikan, K. The theory and applications of algebraic metric-spaces. Ph.D. Thesis, 1990, Stanford University.

32. Duda, R.O., Hart, P.E., Stork, D.G., *Pattern Classification*, 2nd Edition, Wiley, New York, 2000.

33. United Nations. A/RES/70/1 transforming our world: The 2030 Agenda for sustainable development. Available online: https://www.un.org/en/development/desa/population/migration/generalassembly/docs/globalcompact/A_RES_70_1_E.pdf (accessed on 09 February 2021).

34. Ranjan, Priya, Kumar Dron Shrivastav, Satya Vadlamani, and Rajiv Janardhanan. "HRIDAI: A Tale of Two Categories of ECGs." In International Symposium on Signal Processing and Intelligent Recognition Systems, pp. 243–263. Springer, Singapore, 2020.

35. Shrivastav, Kumar Dron, Priyadarshini Arambam, Ankan Mukherjee Das, Shazina Saeed, Upendra Kaul, Priya Ranjan, and Rajiv Janardhanan. "Earth Mover's Distance-Based Automated Geometric Visualization/Classification of Electrocardiogram Signals." Trends in Communication, Cloud, and Big Data (2020): 75–85.

36. Arti Taneja, Amit Ujlayan, Rajiv Janardhanan, Priya Ranjan, "Pancreatic Cancer Detection by an Integrated Level Set-Based Deep Learning Model," Big Data and Artificial Intelligence for Healthcare Applications, 2021, DOI: 10.1201/9781003093770-8

37. Vandana Bhatia, Priya Ranjan, Neha Taneja, Harpreet Singh, Rajiv Janardhanan,"Early and Precision-Oriented Detection of Cervical Cancer," Big Data and Artificial Intelligence for Healthcare Applications, 2021, DOI: 10.1201/9781003093770-9

38. Khatri, Archit, Rishabh Jain, Hariom Vashista, Nityam Mittal, Priya Ranjan, and Rajiv Janardhanan. "Pneumonia identification in chest X-ray images using EMD." Trends in Communication, Cloud, and Big Data (2020): 87–98.

39. Goyal, Ayush, Sunayana Tirumalasetty, Disha Bathla, Manish K. Arya, Rajeev Agrawal, Priya Ranjan, Gahangir Hossain, and Rajab Challoo. "A computational segmentation tool for processing patient brain MRI image data to automatically extract gray and white matter regions." In Emerging Research in Computing, Information, Communication and Applications, pp. 1–16. Springer, Singapore, 2019.

40. Shrivastav, Kumar Dron, Ankan Mukherjee Das, Harpreet Singh, Priya Ranjan, and Rajiv Janardhanan. "Classification of Colposcopic Cervigrams Using EMD in R." In International Symposium on Signal Processing and Intelligent Recognition Systems, pp. 298-308. Springer, Singapore, 2018.

41. Taneja, Arti, Priya Ranjan, and Amit Ujlayan. "Multi-cell nuclei segmentation in cervical cancer images by integrated feature vectors." Multimedia Tools and Applications 77, no. 8 (2018): 9271–9290.

42. Taneja, Arti, Priya Ranjan, and Amit Ujlayan. "An efficient SOM and EM-based intravascular ultrasound blood vessel image segmentation approach." International Journal of System Assurance Engineering and Management 7, no. 4 (2016): 442–449.

43. Taneja, Arti, Priya Ranjan, and Amit Ujlayan. "Novel Texture Pattern Based Multi-level set Segmentation in Cervical Cancer Image Analysis." In Proceedings of the International Conference on Image Processing, Computer Vision, and Pattern Recognition (IPCV), p. 76. The Steering Committee of The World Congress in Computer Science, Computer Engineering and Applied Computing (WorldComp), 2016.

9

Adaptive Neuro Fuzzy Inference System to Predict Material Removal Rate during Cryo-Treated Electric Discharge Machining

Vaibhav S. Gaikwad
K.K. Wagh Institute of Engineering Education and Research

Vijaykumar S. Jatti
D.Y. Patil College of Engineering

Satish S. Chinchanikar
Vishwakarma Institute of Information Technology

Keshav N. Nandurkar
K.K. Wagh Institute of Engineering Education and Research

CONTENTS

Abbreviations

EDM: Electric Discharge Machining
MRR: Material Removal Rate
TWR: Tool Wear Rate
ANFIS: Adaptive Neuro Fuzzy Inference System
NiTi: Nickel Titanium
AISI: American Iron and Steel Institute
MF: Membership Function
ANN: Artificial Neural Network

DOI: 10.1201/9781003143505-9

9.1 Introduction

In today's competitive environment machining of difficult to cut material with accuracy is a major challenge in manufacturing sector. NiTi alloy is one of such advanced material, which is known for its high strength, biocompatibility, and high wear resistance, and hence, it finds its applications in the field of aerospace, defence, medical etc. Many attempts have been made by the researchers to machine this alloy using electric discharge machining (EDM) as this material has machinability issues when cut by conventional processes. Moreover, the application field of NiTi alloy demands for high degree of machining accuracy. EDM is the nontraditional process wherein there is no direct contact between the workpiece and the tool, and hence, the machining process is comparatively less dependent on the mechanical properties of work materials. Attempts have also been made by the researchers using cryogenic cooling during machining. This chapter discusses the available open literature about the EDM processes, methodology used for experimentation, discussion about the obtained results and conclusion.

Kaynak et al. (2013) studied tool wear during machining of NiTi shape memory alloy using cryogenic cooling. Experimental results of their study indicated that cryogenic cooling significantly reduces progressive tool wear and tool wear rate (TWR). Liu et al. (2019) during their study on micro-EDM process developed a model to investigate the effect of surface layer on material removal rate (MRR) when using AISI 304 as a work material. Their experimental observations concluded that MRR increases with an increase in surface free energy. Further, they observed lower values of MRR in case of workpieces without oxidation treatment as compared to workpieces with the oxidation treatment. Sharma and Gupta (2019) optimized the machining performance using genetic algorithm in terms of cutting rate during wire EDM of NiTi alloy. Experiments were designed by them using Taguchi's L16 orthogonal array. Sivaprakasam et al. (2019) investigated the powder mixed wire EDM of inconel-718 alloy. They used suspended particles of graphite powder in dielectric medium during machining. From their study, it is seen that mixing of powder improves the surface topography, roughness and considerable improvement in MRR with an increase in concentration of powder in dielectric medium within the domain of their experimental work. Bellotti et al. (2019) investigated breakthrough phenomena in drilling microholes by EDM. They observed an abrupt increase in arc when tool first breaks the duration of interval of time between short circuits. On the other hand, they observed a decreasing trend of arc during open circuit even after breakthrough. Kaynak et al. (2014) investigated the surface integrity characteristics of NiTi shape memory alloys during dry and cryogenic machining. Their experimental study concluded that cryogenic machining gives better surface quality as compared with the dry machining. Zailani and Mativenga (2016) investigated the machining of NiTi alloy using chilled air. From this study, it was noted that the TWR reduces due to the application of chilled air and lubrication. Moreover, they observed that the formation of burr during the process when using chilled air was uniformly characterized as smaller in height. Kumar and Goyal (2020) investigated the wire EDM of $^{20}MnCr_5$ alloy steel. In this study, optimization of MRR and surface roughness was performed using Taguchi's method. Moreover, it was observed from the analysis of variance that servo voltage was the significant process parameter for MRR and surface roughness. Sharma et al. (2019) investigated the effect of cryogenically treated wire on EDM of tool steel (AISI D3). MRR and surface roughness were evaluated during the study. It was noted from the study that better surface finish was obtained by cryogenically treated wire; also, pulse width was a significant parameter that affected the MRR. Jiang and Xiao (2021) developed the constitutive model for grain size and strain rate using finite element analysis. From this modelling, it was concluded that the grain size with 68 and 90 nm gives strong mechanical coupling. Singh et al., (2020) performed the predictive analysis of surface roughness using ANN and adaptive neuro fuzzy inference system (ANFIS). From this study, it was concluded that the model developed using ANFIS gives more accurate results as compared with ANN. From the literature, it has been observed that various attempts have been made by researchers using machining of NiTi shape memory alloys. However, very few attempts have been made by the researchers in modelling of cryo-treated EDM process when using NiTi shape memory alloys. With this view, in this chapter, a model to predict MRR during EDM of cryo-treated NiTi alloy as a workpiece material and electrolytic copper as tool material considering the effect of gap current, pulse-on time, pulse-off time, electrical conductivity of workpiece and tool materials during EDM process using ANFIS.

9.2 Materials and Experimental Set-up

In this study, NiTi alloys are being used as a workpiece material and electrolytic copper as a tool material. Samples of NiTi alloy are cut in size of diameter of 20 mm, and meso-scale holes of square size 3 mm×3 mm are drilled into the workpiece, and MRR is measured. Figure 9.1 shows the experimental set-up of EDM. In this study, cryogenic treatment for workpiece and tool material is performed at −185°C.

Table 9.1 gives the levels of EDM process parameters during cryo-treated EDM process drilling of NiTi alloy as a workpiece material and electrolytic copper as a tool electrode.

FIGURE 9.1
Experimental set-up.

TABLE 9.1

Various EDM Processes Parameters

	Levels				
Factors	1	2	3	4	5
Workpiece electrical conductivity (S/m)	3,267	4,219	-	-	-
Tool electrical conductivity (S/m)	10,000	26,316	-	-	-
Gap current (A)	8	10	12	14	16
Pulse-on time (µs)	13	26	38	51	63
Pulse-off time (µs)	5	6	7	8	9

Electrical conductivity of workpiece material (A), electrical conductivity of tool material (B) are varied at two levels, whereas gap current (C), pulse-on time (D), and pulse-off time (E) were varied at three different levels. Parameters for the study were selected based on the pilot experimentation performed. ANFIS model is developed using MATLAB tool. Based on the experiments performed, 23 data sets were chosen for testing data and 50 data sets for training data. These data sets are chosen in such a way that it covers the entire range of each process parameter. The flow chart for ANFIS is shown in Figure 9.2.

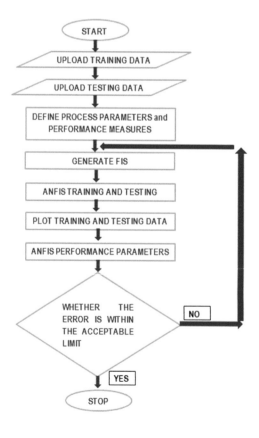

FIGURE 9.2
Flowchart ANFIS model.

9.3 Results and Discussion

This section discusses the experimental results obtained during cryo-treated EDM of drilling of *NiTi* alloy using electrolytic copper tool by varying the input parameters as shown in Table 9.1. Experimental results of EDM carried out on die sink-type electric discharge machine showed an increase in MRR with an increase in pulse-on time and gap current. Further, a significant improvement in MRR has been observed when using cryo-treated workpiece and tool materials as against non-cryo-treated workpiece and tool materials during drilling of *NiTi* shape memory alloy. Further, experimental results were used to develop the prediction model using ANFIS. Table 9.2 shows the experimental matrix and predicted results of MRR by ANFIS when using triangular and trapezoidal membership functions (MFs), respectively. ANFIS is a technique developed in the early 1990s that integrates principles of both neural networks and fuzzy logic. MFs, which are the building blocks of fuzzy set theory, determine fuzziness in a fuzzy set. However, accuracy of the fuzzy inference system primarily depends upon the shape of the MFs. Among the various MF shapes, triangular and trapezoidal are more often used as they are simple and flexible, and calculations with these MFs are easy. In this study, a model to predict MRR during cryo-treated *NiTi* alloy as a workpiece material and electrolytic copper as tool material considering the effect of gap current, pulse-on time, pulse-off time, electrical conductivity of workpiece and tool materials during EDM process using ANFIS tool in MATLAB. The percentage error between the predicted results of MRR using ANFIS when using both the MFs, respectively, and experimental values of MRR for each experimental condition was calculated by Equation 1. The average of all such errors gives the average % error of ANFIS model when using triangular and trapezoidal MFs, respectively.

$$\frac{|\text{Experimental value} - \text{predicted value}| \times 100}{\text{Experimental value}} \tag{9.1}$$

TABLE 9.2

Experimental Matrix and Predicted Results of MRR by ANFIS

Process Parameters (Factors) as Designated Mentioned in Table 9.1						Predicted Values of MRR Using Membership Function	
A	B	C	D	E	Expt. MRR (mm³/min)	Triangular	Trapezoidal
3,267	10,000	16	63	9	5.56	5.56	5.56
3,267	10,000	8	13	5	1.08	1.08	1.08
3,267	10,000	12	38	7	4.57	3.76	3.76
3,267	10,000	16	63	9	5.57	5.56	5.56
3,267	10,000	8	13	7	1.06	1.06	1.06
3,267	10,000	12	38	9	4.51	3.75	3.75
3,267	26,316	8	38	9	1.97	1.97	1.97
3,267	26,316	12	63	5	3.62	3.62	3.62
3,267	26,316	16	13	7	5.07	5.07	5.07

(Continued)

TABLE 9.2 (*Continued*)

Experimental Matrix and Predicted Results of MRR by ANFIS

Process Parameters (Factors) as Designated Mentioned in Table 9.1						Predicted Values of MRR Using Membership Function	
A	B	C	D	E	Expt. MRR (mm³/min)	Triangular	Trapezoidal
4,219	10,000	8	38	5	3.91	3.91	3.91
4,219	10,000	12	63	7	4.23	3.55	3.55
4,219	10,000	16	13	9	6.4	6.4	6.4
4,219	10,000	8	38	7	3.68	2.67	2.67
4,219	10,000	12	63	9	4.14	4.14	4.14
4,219	10,000	16	13	5	6.82	6.82	6.82
4,219	26,316	8	63	9	2.29	2.29	2.29
4,219	26,316	12	13	5	5.84	5.84	5.84
4,219	26,316	16	38	7	6.34	4.42	4.42
4,219	26316	8	63	5	2.98	2.98	2.98
4,219	26,316	12	13	7	5.78	3.13	3.14
4,219	26,316	16	38	9	6.31	6.31	6.31
3,267	10,000	8	38	7	1.465	1.51	1.51
3,267	10,000	10	38	7	2.721	2.63	2.63
3,267	10,000	12	38	7	2.741	3.76	3.76
3,267	10,000	14	38	7	3.523	3.52	3.52
3,267	26,316	10	38	7	2.049	2.05	2.05
3,267	26,316	12	38	7	3.005	3.01	3.01
3,267	26,316	14	38	7	3.419	3.41	3.41
3,267	26,316	16	38	7	3.818	3.82	3.82
4,219	10,000	8	38	7	1.872	2.67	2.67
4,219	10,000	10	38	7	2.44	2.85	2.85
4,219	10,000	12	38	7	2.56	3.03	3.03
4,219	10,000	14	38	7	3.424	3.43	3.42
4,219	26,316	10	38	7	2.259	2.26	2.26
4,219	26,316	12	38	7	2.499	2.45	2.45
4,219	26,316	14	38	7	2.615	3.43	3.43
4,219	26,316	16	38	7	2.908	4.42	4.42
3,267	10,000	12	13	7	2.315	2.44	2.42
3,267	10,000	12	26	7	3.381	3.12	3.16
3,267	10,000	12	38	7	2.391	3.76	3.76
3,267	10,000	12	51	7	2.305	2.84	2.8
3,267	10,000	12	63	7	2.277	2	2.01
4,219	26,316	12	13	7	2.066	3.13	3.14
4,219	26,316	12	26	7	2.876	2.78	2.76
4,219	26,316	12	38	7	2.542	2.45	2.45
4,219	26,316	12	51	7	2.645	2.33	2.33
4,219	10,000	12	26	7	3.602	3.6	3.6
4,219	10,000	12	38	7	3.488	3.03	3.03
4,219	10,000	12	51	7	3.154	3.3	3.32
4,219	10,000	12	63	7	2.957	3.55	3.55

(*Continued*)

TABLE 9.2 (*Continued*)

Experimental Matrix and Predicted Results of MRR by ANFIS

Process Parameters (Factors) as Designated Mentioned in Table 9.1						Predicted Values of MRR Using Membership Function	
A	B	C	D	E	Expt. MRR (mm³/min)	Triangular	Trapezoidal
4,219	26,316	12	13	7	1.599	3.13	3.14
4,219	26,316	12	26	7	2.592	2.78	2.76
4,219	26,316	12	38	7	1.963	2.45	2.45
4,219	26,316	12	51	7	2.018	2.33	2.33
3,267	10,000	12	38	6	4.618	4.27	4.27
3,267	10,000	12	38	7	4.596	3.76	3.76
3,267	10,000	12	38	8	4.372	3.75	3.76
3,267	10,000	12	38	9	3.171	3.75	3.75
3,267	10,000	12	38	5	4.711	4.79	4.79
3,267	10,000	12	38	6	4.072	4.27	4.27
3,267	10,000	12	38	7	4.093	3.76	3.76
3,267	10,000	12	38	9	3.171	3.75	3.75
4,219	10,000	12	38	5	5.517	5.52	5.52
4,219	10,000	12	38	8	4.134	3.55	3.55
4,219	10,000	12	38	9	3.776	4.07	4.07
4,219	26,316	12	38	5	3.773	3.85	3.85
4,219	26,316	12	38	6	3.29	3.14	3.15
4,219	26,316	12	38	7	3.019	2.45	2.45
4,219	26,316	12	38	8	3.008	2.73	2.73
4,219	26,316	12	38	9	2.879	3.02	3.02
Average % Error						**3.567**	**3.563**

From the predicted results of MRR with triangular and trapezoidal MFs in ANFIS can be seen as in better agreement with experimental values of MRR during cryo-treated EDM of NiTi alloy with electrolytic copper as tool material in the domain of the process parameters selected in the present study. It has been observed that results predicted by trapezoidal as well as triangular MFs are almost similar with the accuracy of 96.45% as average percentage error between the predicted results of MRR using ANFIS when using both the MFs, respectively, and experimental values of MRR can be seen around 3.56%. It can be seen that the developed model could be effectively used to predict MRR during cryo-treated EDM within the domain of the parameters selected in the present study using either of the trapezoidal or of the triangular MF.

9.4 Conclusions

In this paper, an attempt has been made to develop a model using ANFIS to predict MRR during EDM of cryo-treated NiTi alloy with electrolytic copper as a tool material. Experimental results of EDM carried out on die sink-type electric discharge machine showed an increase in MRR with an increase in pulse-on time and gap current. A model to predict MRR was developed using MFs, namely triangular and trapezoidal MFs, respectively. It has been observed that results predicted by trapezoidal as well as triangular MFs are almost similar with the accuracy of 96.45%, and the developed model using either of the trapezoidal or of the triangular MF could be effectively used to predict MRR during cryo-treated EDM within the domain of the parameters selected in this study.

References

Bellottia, M., Qiana, J., Reynaerts, D. 2019. Breakthrough phenomena in drilling micro holes by EDM. *International Journal of Machine Tools and Manufacture* 146, 103436. doi: 10.1016/j.ijmachtools.2019.103436.

Jiang, D., Xiao, Y. 2021. Modelling on grain size dependent thermomechanical response of superelastic NiTi shape memory alloy. *International Journal of Solids and Structures*, 210–211, 170–182. doi: 10.1016/j.ijsolstr.2020.11.036.

Kaynak, Y., Karaca, H.E, Noebe, R.D, Jawahir, I.S. 2013. Tool-wear analysis in cryogenic machining of NiTi shape memory alloys: A comparison of tool-wear performance with dry and MQL machining. *Wear* 306, 51–63. doi: 10.1016/j.wear.2013.05.011.

Kaynak, Y., Karacab, H.E., Jawahir, I.S. 2014. Surface integrity characteristics of *NiTi* shape memory alloys resulting from dry and cryogenic machining. *Procedia CIRP* 13, 393–398. doi: 10.1016/j.procir.2014.04.067.

Kumar, N., Goyal, K. 2020. Multi-objective optimization of wire electrical discharge machining of $^{20}MnCr_5$ alloy steel. *World Journal of Engineering* 17, 325–333. doi: 10.1108/WJE-09-2017-0304.

Liu, Q., Zhang, Q., Zhang, M., Yanga, F., Rajurkard, K.P. 2019. Effects of surface layer of AISI 304 on micro EDM performance. *Precision Engineering* 57, 195–202. doi: 10.1016/j.precisioneng.2019.04.006.

Sharma, N., Gupta, K. 2019. Wire spark erosion of Ni rich NiTi shape memory alloy for bio-medical applications. *Procedia Manufacturing* 35, 401–406. doi: 10.1016/j.promfg.2019.05.059.

Sharma, H., Goyal, K., Kumar, S. 2019. Performance evaluation of cryogenically treated wires during wire electric discharge machining of AISI D3 die tool steel under different cutting environments. *Multidiscipline Modeling in Materials and Structures* 15, 1318–1336. doi: 10.1108/MMMS-04-2019-0078.

Singh, N.K., Singh, Y., Kumar, S., Sharma, A. 2020. Predictive analysis of surface roughness in EDM using semi-empirical, ANN and ANFIS techniques: A comparative study. *Materials Today: Proceedings*, 25, 735–741. doi: 10.1016/j.matpr.2019.08.234.

Sivaprakasam, P., Hariharan, P., Gowri, S. 2019. Experimental investigations on nano powder mixed micro-wire EDM process of inconel-718 alloy. *Measurement* 147, 106844. doi: 10.1016/j.measurement.2019.07.072.

Zailani, Z.A., Mativenga, P.T. 2016. Effects of chilled air on machinability of *NiTi* shape memory alloy. *Procedia CIRP* 45, 207–210. doi: 10.1016/j.procir.2016.02.156.

10

A Metaheuristic Optimization Algorithm-Based Speed Controller for Brushless DC Motor: Industrial Case Study

K. Vanchinathan

Velalar College of Engineering and Technology

P. Sathiskumar

Applied Materials

N. Selvaganesan

Indian Institute of Space Science and Technology

CONTENTS

DOI: 10.1201/9781003143505-10

10.1 Introduction

Permanent magnet brushless direct current (BLDC) motors play a significant role in the modernized control engineering applications (Luo et al., 2017). Most of the world's manufacturing units depend on BLDC motors to achieve high efficiency with better controllability. Interestingly, it is noted that 60% of electrical energy is consumed by electrical motors. In future, due to the automation of the modern world, it may be increased to above 80%. The future world will be automated for every process without the help of human physical presence. In this situation, even a fraction of a second is equally important to accomplish the desired speed control of BLDC motor with the help of metaheuristic optimization algorithm (Dokeroglu et al., 2019). In addition, the presence of variable speed control (VSC) drive system produces excellent accuracy and quick recovery under various operating conditions (De la Guerra and Alvarez, 2020).

The BLDC motor used for many applications with constant loads includes single spindle drives, BLDC ceiling fans, quad copters, electric wheelchairs, laptop and desktop cooling fans, compact disk drive motors, medical analysers, medical pumps, diagnostic equipment pieces, dispensers, etc. Applications with varying loads include robotic arm controls, aircraft landing gear, aircraft on-board instrumentations, winding machines, spinning machines, room air conditioners, washing machines, air coolers, etc, and the positioning applications include computer numeric-controlled machines, etc. (Xia, 2012). Figure 10.1 shows the BLDC motors' market size for 2014–2025.

In general, permanent magnet BLDC motors are powered by a three-phase inverter that is operated using sensor signal, and the location of the rotor can be sensed using the Hall effect sensors. It is mainly used for determining the permanent magnet rotor

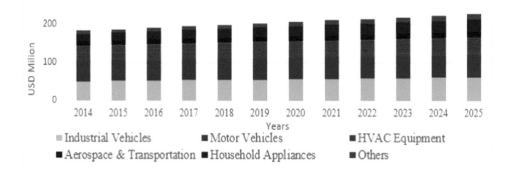

FIGURE 10.1
BLDC motors market size for 2014–2025 (USD million). (Source: Grand viewer search).

position, based on the rotor position; correspondingly electromagnetic stator windings are energized. These sensors increase the cost, poor reliability, bulky size and complication of speed control by reducing the controllability and efficiency of these BLDC motor drives (Krishnan, 2017). The sensorless techniques are developed for BLDC motor to reduce the high cost of the Hall effect sensors, sensor failures and complex in controller design. For robust control of the speed of BLDC motor drives, several works have reported on sensorless commutation processes. The sensorless BLDC motor drive provides an effective control of speed and position in a closed-loop feedback system without a shaft-mounted position sensor (Damodharan and Vasudevan, 2010). In addition, it also offers less weight, high starting torque and compact size. This makes BLDC motor as the perfect choice for an industrial application. Further, proportional integral derivative (PID) and fractional order PID (FOPID) controllers provide better speed control and time-domain characteristics in the sensorless speed control of the BLDC motor. Hence, the control design engineers and manufacturers are looking for better speed control techniques to improve the motor's speed control and its performance (Kim et al., 2007).

The speed control of BLDC motor is established by deploying controller for specific speed control applications. However, traditionally used speed controllers may be used often by industries as they possess a simple control structure and easy implementation. However, controllers have certain inherent difficulties to tackle the complexities like nonlinearity, load disturbances and parametric variations. Moreover, conventional controller has certain disadvantages like high settling time, maximum overshoot, sensitivity to controller gains and sluggish response because of sudden disturbances (Shanmugasundram et al. 2014). Another major disadvantage is the fact that conventional controllers need precise linear mathematical models. Since BLDC machine has inherent nonlinearity, the use of linear conventional controller is unsuitable (Valle et al., 2017).

Overcoming the limitations mentioned earlier for conventional controllers, this analysis considers various optimization techniques based on optimal tuning of PID/FOPID controllers for speed control of BLDC drive to get the minimum time-domain characteristic (Varshney et al., 2017). The metaheuristics algorithm-optimized PID/FOPID controllers are simulated in MATLAB-R2019a, M-file under Windows 10® on a PC Intel Core i5-4200U CPU, 2.30 GHz speed system. The major contributions of these industrial case studies are as follows:

- Focuses on various aspects of problem formulations for metaheuristic optimization algorithm-based optimal tuning of controller parameters.
- Literature review on the classification of optimization algorithm and tuning methods for controllers of BLDC motor is presented. Also, the functional blocks of PID/FOPID controller-based sensorless speed control of BLDC motor are detailed.
- Analysis of various optimization algorithms with MATLAB/Simulink 2019a implementation in order to measure time-domain parameters such as rise time, peak time, settling time, and steady-state error.
- Simulation is carried out to suggest the best controller by evaluating the various metaheuristic optimization algorithms, and the same is also recommended for the sensorless speed control of BLDC motor drive in industry.

The rest of this chapter is organized as follows: Section 10.2 covers the mathematical model and the speed control scheme for sensorless BLDC motor. Metaheuristics optimization algorithm-based PID/FOPID controllers analysed by various researchers are extensively discussed in Section 10.3. Metaheuristics optimization-based PID/FOPID controllers tuned for sensorless BLDC motor are described in Section 10.4. Section 10.5 presents the simulation results and discussions under various operating conditions. Section 10.6 concludes this chapter.

10.2 Speed Control of Sensorless BLDC Motor Drives

This section comprises the description of the modelling and sensorless speed control of BLDC motors. The mathematical modelling of the BLDC motor is presented, which is used to design PID/FOPID controllers using metaheuristic algorithm for sensorless speed control of BLDC motor.

10.2.1 Mathematical Model of BLDC Motors

The differential equation model of BLDC motor is built for a three-phase stator winding of BLDC motor. The stator winding is star-connected, the permanent magnet rotor has a non-salient pole construction, and the three Hall effect sensors are placed at 120° displacement for position measurement. The following assumptions are made to drive the differential equation of the BLDC motor: (i) the core saturation, eddy current losses and hysteresis losses are ignored; (ii) the armature reaction and the distribution of air-gap magnetic field that are considered as a trapezoidal wave with a flat-top width of 120° electrical angle are ignored; (iii) the cogging effect, suppose the conductors that are distributed continuously and evenly on the surface of the armature are ignored; and (iv) power switches and flywheel diodes of the inverter circuit have ideal switch features. Hence, the simplified schematic diagram of the BLDC motor is shown in Figure 10.2.

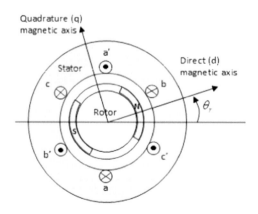

FIGURE 10.2
Basic structure of BLDC motor [4].

The phase voltage of each winding, which includes the resistance, voltage drop and the induced EMF, is expressed as

$$u_x = R_x i_x + e_{\Psi x} \tag{10.1}$$

The winding-induced EMF is equal to the change of rate of the flux, and hence, it can be written as

$$\text{Winding-induced EMF } e_{\Psi x} = \frac{d\Psi_x}{dt} \tag{10.2}$$

Taking phase A for example, the flux is given as

$$\Psi_A = L_A i_A + M_{AB} i_B + M_{AC} i_C + \Psi_{pm}(\theta) \tag{10.3}$$

The magnitude of $\Psi_{pm}(\theta)$ depends on the magnetic field distribution of the permanent magnet in the air gap. The radial component of permanent magnet air-gap magnetic field is distributed as a trapezoidal profile along the inner surface of the stator. When the rotor rotates anticlockwise, the effective flux of phase A will change with regard to the rotor position. When the rotor position is α, the PM flux of phase A is

$$\Psi_{pm}(\alpha) = N\varnothing_{pm}(\alpha) \tag{10.4}$$

$$\varnothing_{pm}(\alpha) = \int_{\frac{-\pi}{2}+\alpha}^{\frac{\pi}{2}+\alpha} B(\theta) S d\theta \tag{10.5}$$

Substituting (10.2)–(10.5) into (10.1), we get

$$u_A = Ri_A + \frac{d}{dt}\left(L_A i_A + M_{AB} i_B + M_{AC} i_C + \Psi_{pm}\right)$$

$$= Ri_A + \frac{d}{dt}(L_A i_A + M_{AB} i_B + M_{AC} i_C) + \frac{d}{dt}\left[NS \int_{-\frac{\pi}{2}+\theta}^{\frac{\pi}{2}+\theta} B(x) dx\right]$$

$$= Ri_A + \frac{d}{dt}(L_A i_A + M_{AB} i_B + M_{AC} i_C) + e_A \tag{10.6}$$

Equation (10.6) includes a derivative of the product of inductance and current, where the self-inductance and the mutual inductance of the winding are proportional to N^2 and the permeance of the corresponding magnetic circuit, that is

$$L_A = N^2 \wedge_A \tag{10.7}$$

$$M_{AB} = N^2 \wedge_{AB} \tag{10.8}$$

In general, the surface-mounted salient-pole rotor is used for BLDC motors, and hence, the winding inductance does not vary with respect to time. Further, as the three-phase stator windings are symmetrical, the self-inductances are equal and so as the mutual inductance, i.e. $L_A = L_B = L_C = L$, $M_{AB} = M_{BA} = M_{BC} = M_{CB} = M_{AC} = M_{CA} = M$. Substituting them into (10.6), we get,

$$u_A = Ri_A + L\frac{di_A}{dt} + M\frac{di_B}{dt} + M\frac{di_C}{dt} + e_A \tag{10.9}$$

where

$$e_A = \frac{d}{dt}\left[NS \int_{-\frac{\pi}{2}+\theta}^{\frac{\pi}{2}+\theta} B(x)dx \right]$$

$$= NS\omega\left[B\left(\frac{\pi}{2}+\theta\right) - B\left(-\frac{\pi}{2}+\theta\right) \right] \tag{10.10}$$

According to the distribution of magnetic density in the air gap with $B(\theta)$ having a period of 2π and $B(\theta+\pi) = -B(\theta)$, we get

$$e_A = NS\omega\left[B\left(\frac{\pi}{2}+\theta\right) - B\left(\frac{\pi}{2}+\theta+\pi-2\pi\right) \right]$$

$$= NS\omega B\left(\frac{\pi}{2}+\theta\right) \tag{10.11}$$

Here, θ-dependent back-EMF wave of phase A is $\pi/2$ ahead of the distribution of the magnetic density in the air gap, and hence, e_A can be expressed as

$$e_A = 2NS\omega B_m f_A(\theta) = \omega\Psi_m f_A(\theta) \tag{10.12}$$

Note that the trapezoidal distribution with the rotor position and its maximum and minimum values are 1 and –1, respectively. Similarly, phase B and C EMFs are also derived. For the three-phase symmetrical windings, there also exist $f_B(\theta) = f_A(\theta - 2\pi/3)$ and $f_C(\theta) = f_A(\theta + 2\pi/3)$. It can be seen from Equation (10.10) that e_A is a rotating back-EMF that is produced by the winding flux linkage caused by the rotating rotor. The currents of the three phases satisfy the following condition:

$$i_A + i_B + i_C = 0 \tag{10.13}$$

Equation (10.9) can be further simplified as

$$u_x = Ri_A + (L-M)\frac{di_A}{dt} + e_A \tag{10.14}$$

Then, the matrix form of phase voltage equation of the BLDC motor can be expressed as:

$$
\begin{bmatrix} u_A \\ u_B \\ u_C \end{bmatrix} = \begin{bmatrix} R & 0 & 0 \\ 0 & R & 0 \\ 0 & 0 & R \end{bmatrix} \begin{bmatrix} i_A \\ i_B \\ i_C \end{bmatrix} + \begin{bmatrix} L-M & 0 & 0 \\ 0 & L-M & 0 \\ 0 & 0 & L-M \end{bmatrix}
$$

$$
\times \frac{d}{dt} \begin{bmatrix} i_A \\ i_B \\ i_C \end{bmatrix} + \begin{bmatrix} e_A \\ e_B \\ e_C \end{bmatrix} \tag{10.15}
$$

Based on (10.15), the equivalent circuit of the BLDC motor is shown in Figure 10.3. The mathematical model based on line voltage is more suited to the practical system and can be obtained through subtraction calculation of the phase-voltage equation (Pillay and Krishnan, 1988) as

$$
\begin{bmatrix} u_{AB} \\ u_{BC} \\ u_{CA} \end{bmatrix} = \begin{bmatrix} R & -R & 0 \\ 0 & R & -R \\ -R & 0 & R \end{bmatrix} \begin{bmatrix} i_A \\ i_B \\ i_C \end{bmatrix} + \begin{bmatrix} L-M & M-L & 0 \\ 0 & L-M & M-L \\ M-L & 0 & L-M \end{bmatrix}
$$

$$
\times \frac{d}{dt} \begin{bmatrix} i_A \\ i_B \\ i_C \end{bmatrix} + \begin{bmatrix} e_A - e_B \\ e_B - e_C \\ e_C - e_A \end{bmatrix} \tag{10.16}
$$

The power transferred to the rotor, which is called the electromagnetic power, equals the sum of the product of current and back-EMF of the three phases. This is given by

$$
P_e = e_A i_A + e_B i_B + e_C i_C \tag{10.17}
$$

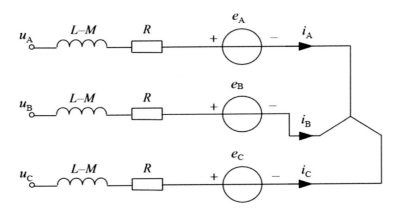

FIGURE 10.3
Equivalent circuits of the BLDC motor.

By ignoring the mechanical loss and stray loss, the electromagnetic power is totally converted into kinetic energy, so

$$P_e = T_e \Omega \tag{10.18}$$

Hence, from (10.17) and (10.18), we get

$$T_e = \frac{e_A\, i_A + e_B\, i_B + e_C\, i_C}{\Omega} \tag{10.19}$$

The torque equation can be obtained and presented as follows by substituting (10.12) into (10.19):

$$T_e = p\Big[\, \Psi_m f_A\,(\theta) i_A + \Psi_m f_B\,(\theta) i_B + \Psi_m f_C\,(\theta) i_C\,\Big] \tag{10.20}$$

Note that the symbols of $f(\theta)$ at the flat-top position are opposite to each other for different windings, and hence, (10.20) can be further simplified as

$$T_e = 2p\psi_m\, i_A = K_T i \tag{10.21}$$

In order to build a complete mathematical model of the electromechanical system, the motion equation has to be included and is as follows:

$$T_e - T_L = J\frac{d\Omega}{dt} + B_v \Omega \tag{10.22}$$

Thus, (10.15), (10.19) and (10.22) constitute the differential equation for the mathematical model of the BLDC motor.

10.2.2 Sensorless Speed Control Scheme of BLDC Motors

Many industries need to control the speed of BLDC motors by sensorless techniques for various applications (Lee et al., 2017). The conventional speed control strategies provide a less precise and ineffective performance. Hence, an optimal speed controller is designed for the sensorless BLDC motor drives using intelligent tuning techniques for obtaining the parameters of the PID/FOPID controllers (Yao et al., 2019).

10.2.2.1 Principle of Sensorless Position Detection

Generally, the PM rotor position is measured by position sensors, i.e. Hall effect sensors, optical encoders, variable reluctance sensors and accelerometers. These position sensors increase the system cost, complex electrical wiring arrangements and lack of system reliability. In order to decrease the system cost and make it highly reliable, sensorless position detection techniques are used (Jia et al., 2020). It depends on the estimation of the back-EMF that is induced by the rotor motion. The back-EMF-based method, flux linkage-based method, inductance-based method and artificial intelligence-based method are used for detecting the rotor position of BLDC motor drives by sensorless scheme (Hemalatha and Nageswari, 2020).

10.2.2.2 Sensorless Speed Control of BLDC Motors

The BLDC motor drives have poor performance because of the complex electric wiring arrangements and expensive rotor position sensors. In order to overcome these drawbacks, it is recommended to use sensorless drive systems for improving the performance at transient and steady-state conditions (Verma et al., 2019). The functional blocks of sensorless BLDC motor are shown in Figure 10.4.

It consists of the power converter, BLDC motor, electrical parameter measurements and speed estimation. The rotor speed is proportional to motor current and terminal voltage, which is varied by using power semiconductor devices with the PWM techniques. The sensorless drive is operated on the trapezoidal shape back-EMF induced by the movement of a PM rotor. The speed of the BLDC motor is estimated by using electrical parameters such as the terminal voltage and current (Gan et al., 2018).

10.2.2.3 Sensorless Control Strategy

The sensorless control techniques are normally used in BLDC motor for estimating the electrical parameters like speed, flux and torque. The following methods are used for estimating the speed of the BLDC motor, such as Kalman filter, disturbance observer, sliding mode observer, model reference adaptive system, etc (Zhou et al., 2016). In this research, the Kalman filter algorithm is used for estimating the speed of the motor. It is an ideal estimation algorithm for the linear system. The following factors are considered for estimating the speed of the motor:

a. Dynamic state model for a linear machine
b. Selection of machine model
c. Calculation of system gradient and measurement matrices.

The linear speed control of BLDC motor is applied with the Kalman filter algorithm for estimating the rotor position and speed of the motor with the help of the stator winding line voltage and current (Zaky et al., 2018).

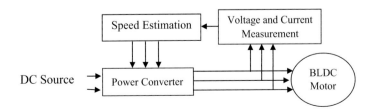

FIGURE 10.4
Functional blocks of sensorless BLDC motor.

10.3 Analysis of Metaheuristics Optimization Algorithm-Based Controller

This section explains the analysis of the speed control of BLDC motor by various controllers and the time-domain characteristics considered for design aspects. The time-domain characteristics are analysed, and a suitable controller with optimization algorithm is suggested for the speed control of BLDC motor. The various optimization techniques based on the controllers are shown in Figure 10.5.

10.3.1 Methods of Optimal Tuning of Controller

Optimal tuning of controllers is always a challenging task because the controller has three to five DOF gain parameters (K_p, K_i, K_d, λ and μ) to tune for achieving the required specifications. The optimal tuning techniques are classified as follows: (i) rule-based techniques, (ii) analytical techniques and (iii) numerical techniques. In addition to these techniques, the self-tuning regulator and model reference adaptive controller are also used for optimal tuning of controller parameters to meet the required closed-loop specification under large disturbances and parameter uncertainties.

10.3.1.1 Analysis of Optimization Techniques Based on PID Controller

The performance of various optimization techniques is analysed for PID controller in order to minimize time-domain parameters in speed control of BLDC motor drives. The optimization techniques are mainly used for tuning the PID controller parameters (Kp, Ki, Kd) for minimizing settling time and steady-state error for BLDC motor.

Table 10.1 shows an analysis of various optimization techniques with their capability based on optimal tuning of PID controller parameters. The method involves observing the system's closed-loop response by changing one of the controller tuning parameters. Based on the closed-loop performance requirements, a suitable value of the tuning parameters is selected.

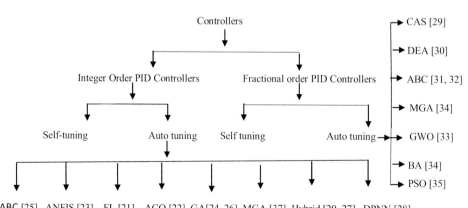

FIGURE 10.5
Various optimization techniques based on controllers.

TABLE 10.1

Analysis of Various Optimization Techniques Based on PID Controller

Reference Paper & Year	Conventional Methods	Objective	Steps Involved	Inference
Kiree et al. (2016)	Particle swarm optimization	PSO-optimized PID controller for a BLDC motor	• Fitness function • Generate new velocity and position values. • Update of inertia weight.	Minimized steady-state error and enhanced motor dynamics characteristics
Dasari et al. (2019)	Adaptive neuro-fuzzy inference system	ANFIS-based speed controller of BLDC motor	• Fuzzy if-then rules • Neural network • Fuzzy logic	Time-domain parameters are measured and analysed
Jigang et al. (2019)	PI and fuzzy logic controller	PI and fuzzy logic controller-based speed control of a PM BLDC motor	• Ruleset contained in the fuzzy rule • Membership function	It is seen that a peak overshoot occurs and poor steady-state error
Vanchinathan et al. (2015)	Ant colony optimization-based PID controller	Minimized setting time and steady-state errors	• Run the process model by random initialization • Determine the fitness function • Update the pheromone and the possibility	Better settling time over the GA and FL soft computing techniques. Reduced steady-state error
Ibrahim et al. (2019)	Conventional genetic algorithm	To analysis the performance of the proposed system	• Population size • Calculation of fitness function • Reproduction • Crossover and Mutation	Poor time-domain parameters due to the nonlinear system
El-Wakeel et al. (2015)	Hybrid BF–PSO	To improve the step response time-domain characteristics	• Chemotaxis • Swarming • Reproduction • Elimination • Dispersal	Experimental results are tested at various working conditions.
Gobinath et al. (2019)	Deep perceptron neural network	To study stability analysis and performance characteristics of BLDC motor	• Learning rate • No. of neurons • Learning rule • Membership function • Fuzzy inference system	It was proven to be more stable and effectiveness has been analysed at various working conditions

10.3.1.2 *Analysis of Optimization Techniques Based on FOPID Controller*

The FOPID controllers are generally used in control systems that need to be tuned with the help of various optimization techniques. The controller has five control parameters (K_p, K_i, K_d, λ and μ), and hence, optimal tuning of controller parameters is difficult. On the other hand, it produces precision output response, minimizes the steady-state error and reduces settling time for speed control of BLDC motor drives. Table 10.2 shows the comparative analysis of various optimization techniques used for tuning of FOPID controller parameters.

TABLE 10.2

Analysis of Various Optimization Techniques Based on FOPID Controller

Reference Paper & Year	Conventional Methods	Objective	Steps Involved	Inference
Zheng et al. (2016)	Differential evolutionary algorithm	In order to optimize FOPID controller parameters gain values by EOA	• Fractional order calculus, initial population • Objective function evaluation, fitness parent choice • Crossover mutation • Offspring evaluation • Choice of the best individual	Provides flexibility and robust reliability with traditional PID controllers compared to the same device
Arpaci et al. (2017)	Adaptive network-based fuzzy inference system	Determination of optimum control parameters and better robustness	• Neural network and fuzzy logic • Inter-nodal weights • Nodes • Characteristic activation function	Analysis of the transient, steady state and stability of the system
Verma et al. (2017)	Grey wolf optimizer	To improve the robustness of GWO-optimized FOPID controller	• It has been taken as an objective • Fitness function.	Robustness analysis of GWO/FOPID approach has been carried out with the minimization of time-domain parameters
Hekimoğlu (2019)	Chaotic atom search optimization	Determine the optimal parameters of the FOPID controller	• Benchmark functions are used	Best transient response profile and a good frequency response
Rahideh et al. (2010)	Modified genetic algorithm	To determine the optimal FOPID controller parameters	• An initialization of parameters • Crossover • Reproduction • Mutation • Evaluation function.	The validated enhanced performance of the MGA-optimized FOPID controller to minimize time-domain parameters
Vanchinathan et al. (2018)	Artificial bee colony	To find an optimal solution for the numerical optimization issues	• Employed bees • Unemployed bees – it contains onlooker bees and scouts.	The ABC-optimized FOPID controller clearly outperforms the various optimized FOPID controllers at all considered working conditions
Vanchinathan et al. (2018)	Bat algorithm	An optimal value is generated to attain the desired speed of the BLDC motor	• Velocity • Position vectors • Variations of loudness • Pulse emission	This controller performs more superior than other considered controllers by attained the desired speed and obtained good time-domain specifications under all working conditions

From the literature, it is noted that the PID controller has provided poor time-domain response due to three DOF parameters and is also not suitable for system containing inherent nonlinearities/nonlinear characteristics. On the other hand, the metaheuristics algorithm-optimized FOPID controllers are used for better time-domain parameters such as peak time, rise time, settling time and steady-state error.

10.4 Metaheuristic Optimization Algorithm-Based Controller Tuning

This section shows the outline of the various types of controller tuning for sensorless speed control of BLDC motor using different optimization techniques. The main reason for the success of these sensorless techniques over their sensor techniques is that the sensor failures yield poor reliability and controllability. Hence, sensorless techniques based on speed control of BLDC motor are very essential for many applications to attain the desired set speed of the motor.

10.4.1 Controller Design

The controller is generally used for various modern industrial applications, automation, robotics, electric vehicle, etc. Several control strategies are developed for VSC drives, and their algorithms are applied to practical control problems. Figure 10.6 shows the schematic diagram of the controller.

The input of the controller is as follows:

- Load torque (T_l), flux (φ), PM rotor speed (N) and Hall effect sensors signals
- The unpredictable load variations
- The measurements of load torque (T_l), flux (φ), PM rotor speed (N) and Hall effect sensors output for the feedback signal
- Maintain the rated currents (I), rated voltage (V), rated speed (N) and rated load torque (T_l)
- The gain constants of proportional gain (K_p), integral gain (K_i), differential gains (K_d), integral order (λ) and derivative order (μ).

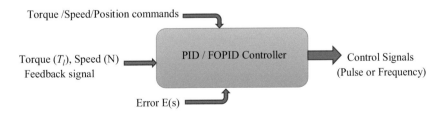

FIGURE 10.6
Basic structure of controller.

The speed of the BLDC motor is controlled by tuning PID/FOPID controllers with the help of various optimization techniques. There are several optimal tuning techniques that are used to attain the desired speed control characteristics of BLDC motor. More than 50 general tuning algorithms of PID/FOPID controllers are developed over the years.

10.4.2 Basic Structure of Optimal Tuning of Controller

Generally, the optimal tuning is done based on the frequency-domain and time-domain specification. The controller is tuned by intelligent methods to minimize the time-domain characteristics. This analysis focuses on the optimal tuning of the PID/FOPID controller using intelligent techniques for generating the optimal PID/FOPID controller parameters (Pritesh and Agashe, 2016). Figure 10.7 shows the basic structure of optimal tuning of PID/FOPID controller.

10.4.3 Optimization Techniques Based on Controller Tuning for Brushless DC Motor

The controllers are mathematical function, which for a given input value provides a required characteristics output value. In the controller, proportional gain (Kp) function is a multiplier, integral gain (Ki) function gives the cumulative value of the input, and derivative gain (Kd) function determines the rate of change input. The error detector receives two inputs: (i) estimated speed, which is the original speed by sensorless techniques, and (ii) set speed, in which the value is manually set. The error detector generates discrepancy values that are fed to the controller as error values. The PID controller output signal:

$$u(t) = K_p e(t) + K_i \int e(t)dt + K_p \frac{de}{dt} \tag{10.23}$$

The PID controller transfer function in Laplace transform form is

$$c(s) = K_P + \frac{K_i}{S} + K_d s \tag{10.24}$$

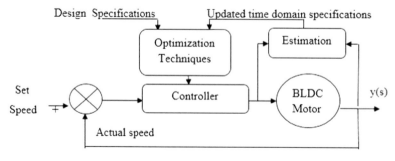

FIGURE 10.7
Basic structure of optimal tuning of PID/FOPID controller.

The optimization techniques are mainly used for fine-tuning of the PID controller in order to achieve the desired set speed of the BLDC machine. It generates optimal values for fine-tuning of PID controller parameters in order to minimize the rise time, peak time, settling time and steady-state error. However, the PID controller is not suitable for a nonlinear and sudden load variation because the controller has three control parameters that are only used. Hence, FOPID controllers are used for such operating conditions in order to achieve better transient and steady-state characteristics. They have five DOF parameters, namely proportional gain (K_p), integral gain (K_i), derivative gain (K_d), integral order (λ) and derivative order (μ). As a consequence, the error detector is continuously associated with the required set speed and the process variable $Y(s)$ is calculated. It provides greater flexibility and enhances the time-domain characteristics when compared to the PID controller.

The transfer function of FOPID is

$$G_c(s) = \frac{U(s)}{E(s)} = K_P + K_I S^{-\lambda} + K_D S^{\mu} \ (\lambda, \mu > 0) \tag{10.25}$$

The time domain of FOPID controller is

$$U(t) = K_P e(t) + K_I D^{-\lambda} e(t) + K_D D^{\mu} e(t) \tag{10.26}$$

Figure 10.8 shows the FOPID controller-based sensorless speed control of a BLDC motor. An inverter is a static equipment that produces AC voltage from DC voltage by triggering the power devices. The switching time is determined by the given switching logic. Voltage and current values are calculated in the sensorless technique, and the real rotor speed is determined by the Kalman filter algorithm. The error detector receives the estimated speed (N_e) in rpm and manually sets the reference speed, i.e. set speed (N_s) to 1,500 rpm. The error detector produces an error value that is fed to the FOPID controller.

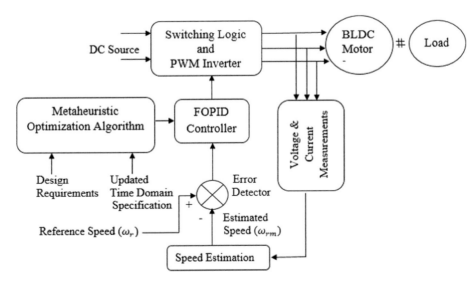

FIGURE 10.8
FOPID controller-based sensorless speed control of a BLDC motor.

The FOPID controller supplies alternating logic circuits with the control signal $u(t)$. The switching logic circuits produce a series of switching pulses to energize the winding of the stator. The PWM signal is produced by a pulse generator with the support of a varying firing angle to regulate the motor input voltage, and hence, the motor speed is controlled. Based on the optimization techniques, the FOPID controller produces optimal five DOF parameter values K_p, K_i, K_d, λ, and μ. Finally, enhanced performance for the sensorless BLDC motor speed is accomplished by changing the duty ratio (α) and evaluating the time-domain characteristics. These methods mainly focus on minimizing time-domain characteristics of sensorless BLDC motor under various operating conditions such as constant load, varying load and switching loads.

10.4.4 Effect of Controller Parameters

The PID/FOPID controller parameters are tuned with different optimization techniques in the closed-loop control for BLDC motor drive. The primary objective of controller is to produce very fast rise time, minimum settling time and zero values of maximum overshoot to sustain the system stability under various operating conditions. Based on objective functions, controller parameters are varied. The effect of PID/FOPID controller parameters is shown in Table 10.3.

10.4.5 Problem Formulation

An objective function is utilized to acquire the optimal tuning parameters of the PID/FOPID controller under varying load and set speed (N_s) conditions of the BLDC motor. The recommended controller is demonstrated in a sensorless speed control system by minimizing the objective function (f_{obj})

$$\text{Objective function } f_{obj} = \sqrt{\sum_{i=0}^{T} \frac{\left(N(t)_{set_i} - N(t)_{est_i}\right)^2}{T}} \qquad (10.27)$$

where $N(t)_{set_i}$ is the set speed in rad/s, $N(t)_{est_i}$ is the estimated speed of the BLDC motor by Kalman filter algorithm at each sample, and (T) is the total simulation time for the optimization of FOPID controller parameters. An optimization algorithm approach is used for the following two purposes:

TABLE 10.3

Effect of Controller Parameters

Controller Parameters	Rise Time (t_r)	Peak Time (t_p)	Settling time (t_s)	Overshoot Time (t_{oh})	Steady-State Error (e_{ss})
K_p	Decrease	Increase	Small change	Increase	Decrease
K_i	Decrease	Increase	Increase	Increase	Eliminate
K_d	Small change	Decrease	Decrease	Decrease	No change
λ	-	-	Decrease	-	Increase
M	-	-	Increase	-	Decrease

- To reduce the error value $e(s)$ between given set speed (N_s) and obtained estimated speed (N_e) from system models.
- To determine the optimized value of PID/FOPID controller parameters, namely K_p, K_i, K_d, λ and μ

The ranges for the optimal tuning parameter of the PID/FOPID controllers are given in Equations (10.22) to (10.26) as follows:

- $K_{p\text{-min}} \leq K_p \leq K_{p\text{-max}}$; $K_{p\text{-min}}=0$ and $K_{p\text{-max}}=3$
- $K_{i\text{-min}} \leq K_i \leq K_{i\text{-max}}$; $K_{i\text{-min}}=0$ and $K_{i\text{-max}}=2$
- $K_{d\text{-min}} \leq K_d \leq K_{d\text{-max}}$; $K_{d\text{-min}}=0$ and $K_{d\text{-max}}=3$
- $\lambda_{\min} \leq \lambda \leq \lambda_{\max}$; $\lambda_{\min} = 0.01$ and $\lambda_{\max} = 0.99$
- $\mu_{\min} \leq \mu \leq \mu_{\max}$; $\mu_{\min} = 0.01$ and $\mu_{\max} = 0.99$

The following constraints are used for the design of controller during optimization:

- Settling time ≤ 3 s
- Steady-state error $\leq 3\%$

10.5 Results and Discussions

In this section, the various optimization techniques based on PID/FOPID controllers are executed through a MATLAB 2019a M file in Windows 10® on a PC Intel (R) Core i5 CPU @ 2.30 GHz. The objective functions take more than ten iterations for determining the optimal three/five DOF parameters for the fine-tuning of the PID/FOPID controllers. The corresponding time-domain characteristics are also measured. A unit step input is applied to the closed-loop system for evaluating the transient and steady-state characteristics under varying load, varying set speed and combined conditions. To demonstrate the given case study, only MGA, ABC and BA are considered. The selection parameters of MGA, ABC and BA are shown in Table 10.4.

TABLE 10.4

Parameters Selection of MGA, ABC and BA

MGA	ABC	BA
Generation: 10	Generation: 10	Population size (X_i): 10
Lower bound [λ, μ]: [0.1, 0.1]	No. of bees: 20	No of dimensions: 3
Upper bound [λ, μ]: [2, 2]	No. of food sources: 2	Number of iteration: 10
Population size: 30	Population size: 30	Loudness (A): 90
Trial: 30	Trial: 30	Pulse emission rate: [1, 10] Changing frequency
No. of bits/gene: 30		($Q_{\min}$ & $Q_{\max}$): 1–90 KHz
Crossover function: 0.01		8 Echolocations: 8–10 ms
Mutation function: 0.03		

10.5.1 Simulink Model

The Simulink model for the PID/FOPID controller along with the plant has been designed. The model consists of a three-phase BLDC motor, three-phase voltage source PWM inverter, PID/FOPID controller, PWM generator, switching logic circuits and blocks for the measurement of input voltage and current. The Simulink model of PID/FOPID controller for sensorless speed control of BLDC motor is shown in Figure 10.9.

The Simulink model has a PWM generator, a comparator, a triangle wave generator, three AND gates and three NOT gates. In addition, a Simulink model is created for measuring rotor speed (by sensorless technique), rotor position, electromagnetic torque, back-EMF and stator current. The specifications of the BLDC motor are shown in Table 10.5.

The metaheuristics algorithm-optimized controllers have been proposed for minimizing the time-domain parameters and reducing the steady-state error for sensorless BLDC motor under constant and varying load conditions (20% load, 40% load, 60% load, 80% load and 100% load). For changing the operating speed, the following experiments (Exp.) are selected for simulation:

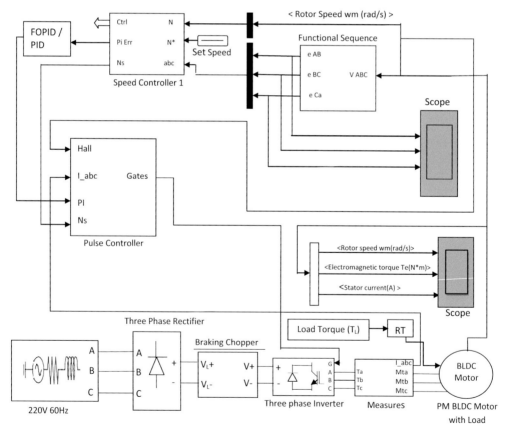

FIGURE 10.9
Simulink model of PID/FOPID controller for sensorless speed control of BLDC motor.

TABLE 10.5

Specifications of the BLDC Motor and Approximation for FOPID Controller

S.No	Specifications of BLDC Motor	Approximation of FOPID Controller
1	Rated voltage: 300 V	Order (N): 5
2	Rated current: 5 A	Frequency minimum : 0.001 rad/s
3	Rated power: 1.1 HP	Frequency maximum : 1,000 rad/s
4	Rated speed: 2,000 rpm	Algorithm: Oustaloup recursive approximation
5	Load torque (T_L): 2.5 Nm	-

- **Exp. 1**: Operating speed from 500 to 1,500 rpm
- **Exp. 2**: Operating speed from 500 to 1,000 rpm
- **Exp. 3**: Operating speed from 1,000 to 1,500 rpm
- **Exp. 4**: Operating speed from 1,500 to 1,000 rpm
- **Exp. 5**: Operating speed from 1,000 to 500 rpm

10.5.2 Speed Response under Constant Load Condition

The time-domain characteristics like rise time, peak time, settling time and steady-state error are evaluated under constant-load and full-load conditions of BLDC motor simulated with the tuned PID/FOPID controller parameters obtained using MGA, ABC and BA algorithms, which are shown in Table 10.6. From the table, it is observed that FOPID controller outperforms the conventional PID controllers in all the time-domain characteristics for BLDC motor under no-load and full-load conditions. It is also noted that BA has produced better results in comparison with MGA and ABC.

TABLE 10.6

Time-Domain Parameters for Constant Load Conditions

Reference Paper & Year	Controller	No-Load Conditions				Full-Load Conditions			
		Rise Time (s)	Peak Time (s)	Settling Time (s)	Steady-State Error (%)	Rise Time (s)	Peak Time (s)	Settling Time (s)	Steady-State Error (%)
Bagis (2007)	MGA-PID	1.021	1.152	1.135	21	1.954	1.992	1.193	16
Rajasekhar et al. (2014)	ABC-PID	0.954	1.116	0.984	20	1.201	1.198	1.065	15
Premkumar et al. (2016)	BA-PID	0.684	0.881	0.896	16	0.824	0.924	0.965	12
Chang et al. (2008)	MGA-FOPID	0.542	0.721	0.840	14	0.667	0.783	0.962	12
Rajasekhar et al. (2014)	ABC-FOPID	0.412	0.612	0.631	10	0.635	0.681	0.768	8
Vanchinathan et al. (2018)	BA-FOPID	0.034	0.061	0.082	4	0.060	0.081	0.096	2

10.5.3 Speed Response under Varying Load Conditions

The time-domain characteristics like rise time, peak time, settling time and steady-state error along with error indices like IAE, ITAE and ISE are evaluated under varying load conditions (20%, 40%, 60%, 80% and 100%) of BLDC motor simulated with the tuned PID/FOPID controllers obtained using MGA, ABC and BA algorithms, which are shown in Table 10.7. The following expressions are used for computing these performance indices:

$$IAE = \int |s|dt = \int |N(t)_{set_i} - N(t)_{est_i}|dt \qquad (10.28)$$

$$ITAE = \int t|s|dt = \int t * (|N(t)_{set_i} - N(t)_{est_i}|)dt \qquad (10.29)$$

TABLE 10.7

Time-Domain Response for Varying Loading Conditions

Reference Paper & Year	Various Controllers	Varying Load Conditions	Rise Time (s)	Peak Time (s)	Settling Time (s)	Steady-State Error (%)	IAE	ITAE	ISE
Bagis (2007)	MGA-PID	20% load	1.124	1.356	1.621	22	0.098	0.006	0.052
		40% load	1.148	1.384	1.635	21	0.017	0.003	0.013
		60% load	1.168	1.399	1.601	21	0.014	0.003	0.011
		80% load	1.984	1.421	1.624	20	0.017	0.003	0.013
		100% load	1.992	1.452	1.658	19	0.014	0.003	0.011
Rajasekhar et al. (2014)	ABC-PID	20% load	0.926	1.141	1.423	18	0.093	0.006	0.052
		40% load	0.965	1.168	1.465	17	0.017	0.002	0.011
		60% load	1.023	1.241	1.482	17	0.014	0.002	0.010
		80% load	1.099	1.268	1.522	16	0.018	0.001	0.011
		100% load	1.124	1.232	1.562	15	0.015	0.001	00.10
Premkumar et al. (2016)	BA-PID	20% load	0.765	0.954	1.241	16	0.087	0.006	0.041
		40% load	0.865	0.886	1.342	14	0.017	0.003	0.012
		60% load	0.725	0.932	1.199	15	0.018	0.003	0.011
		80% load	0.897	0.945	1.241	14	0.017	0.002	0.011
		100% load	0.845	0.992	1.325	13	0.014	0.002	0.010
Chang et al. (2008)	MGA-FOPID	20% load	0.581	0.731	0.848	14	0.081	0.005	0.035
		40% load	0.615	0.735	0.862	13	0.018	0.002	0.012
		60% load	0.625	0.745	0.906	13	0.011	0.019	0.011
		80% load	0.661	0.770	0.943	12	0.018	0.001	0.012
		100% load	0.667	0.783	0.962	12	0.017	0.002	0.011
Vanchinathan et al. (2018)	ABC-FOPID	20% load	0.503	0.682	0.631	9	0.076	0.005	0.024
		40% load	0.569	0.692	0.691	9	0.018	0.001	0.011
		60% load	0.592	0.703	0.714	8	0.015	0.001	00.10
		80% load	0.602	0.713	0.780	8	0.014	0.001	0.009
		100% load	0.635	0.721	0.808	7	0.012	0.001	0.009
Vanchinathan et al. (2018)	BA-FOPID	20% load	0.036	0.065	0.088	4	0.072	0.004	0.064
		40% load	0.043	0.071	0.090	3	0.018	0.001	0.012
		60% load	0.053	0.075	0.091	3	0.017	0.002	0.011
		80% load	0.055	0.078	0.091	2	0.017	0.002	0.011
		100% load	0.060	0.081	0.096	2	0.014	0.002	0.010

$$ISE = \int e_2 \, dt = \int \left(N(t)_{set_i} - N(t)_{est_i} \right)_2 dt \tag{10.30}$$

Under varying load conditions, the speed control of sensorless BLDC motor with tuned PID/FOPID obtained using BA has better time-domain characteristics and performance indices in comparison with MGA and ABC algorithms, respectively.

10.5.4 Speed Response under Varying Set Speed Conditions

The time-domain characteristics like rise time, peak time, settling time and steady-state error are evaluated under varying speed conditions (Exp. 1, Exp. 2, Exp. 3, Exp. 4 and Exp. 5) of BLDC motor simulated with the tuned PID/FOPID controllers obtained using MGA, ABC and BA algorithms, which are shown in Table 10.8. From the results, it is

TABLE 10.8

Time-Domain Response for Varying Set Speed Conditions

Reference Paper & Year	Various Controllers	Varying Set Speed Conditions	Rise Time (s)	Peak Time (s)	Settling Time (s)	Steady-State Error (%)	IAE	ITAE	ISE
Bagis (2007)	MGA-PID	Exp. 1	1.012	1.195	1.421	8	0.025	0.005	0.016
		Exp. 2	1.035	1.186	1.487	8	0.017	0.021	0.011
		Exp. 3	0.994	1.184	1.496	11	0.015	0.012	0.010
		Exp. 4	1.087	1.654	1.465	10	0.013	0.027	0.008
		Exp. 5	1.065	1.624	1.452	12	0.009	0.004	0.006
Rajasekhar et al. (2014)	ABC-PID	Exp. 1	0.894	1.024	1.321	6	0.028	0.067	0.017
		Exp. 2	0.865	1.068	1.344	7	0.018	0.024	0.012
		Exp. 3	0.892	1.078	1.345	8	0.014	0.020	0.011
		Exp. 4	0.899	1.052	1.347	10	0.015	0.027	0.012
		Exp. 5	0.926	1.024	1.321	11	0.013	0.027	0.008
Premkumar et al. (2016)	BA-PID	Exp. 1	0.724	0.854	1.012	6	0.030	0.014	0.018
		Exp. 2	0.756	0.896	0.987	7	0.019	0.048	0.012
		Exp. 3	0.842	0.874	1.024	7	0.015	0.027	0.012
		Exp. 4	0.894	0.812	1.654	9	0.013	0.027	0.008
		Exp. 5	0.836	0.954	1.241	10	0.009	0.004	0.006
Wang et al. (2017)	MGA-FOPID	Exp. 1	0.551	0.641	0.840	5	0.025	0.027	0.015
		Exp. 2	0.551	0.641	0.840	5	0.022	0.021	0.012
		Exp. 3	0.682	0.726	0.882	9	0.021	0.020	0.010
		Exp. 4	0.642	0.749	0.865	9	0.015	0.027	0.012
		Exp. 5	0.682	0.726	0.892	9	0.013	0.027	0.008
Rajasekhar et al. (2014)	ABC-FOPID	Exp. 1	0.470	0.520	0.631	4	0.020	0.061	0.011
		Exp. 2	0.470	0.520	0.631	4	0.013	0.027	0.008
		Exp. 3	0.511	0.601	0.654	8	0.009	0.004	0.006
		Exp. 4	0.504	0.612	0.641	6	0.015	0.027	0.012
		Exp. 5	0.506	0.591	0.654	8	0.013	0.027	0.008
Vanchinathan et al. (2018)	BA-FOPID	Exp. 1	0.031	0.052	0.062	3	0.020	0.061	0.011
		Exp. 2	0.031	0.052	0.062	3	0.013	0.027	0.008
		Exp. 3	0.032	0.038	0.065	3	0.009	0.004	0.006
		Exp. 4	0.031	0.046	0.066	3	0.030	0.014	0.018
		Exp. 5	0.035	0.043	0.065	2	0.019	0.048	0.012

observed that BA-optimized FOPID controller is able to minimize the rise time, peak time, settling time, steady-state error and error indices compared with MGA and ABC-optimized PID/FOPID controllers, respectively.

10.5.5 Speed Response under Combined Operating Conditions

To validate the enhanced performance of the PID/FOPID control design, BLDC motor is subjected to both conditions (varying set speed and load) simultaneously. The following cases are selected for simulation:

a. **Case A**: combined Exp.1 and 20% load conditions
b. **Case B**: combined both Exp.1 and 60% load conditions
c. **Case C**: combined both Exp.1 and 100% load conditions
d. **Case D**: combined both Exp.3 and 20% load conditions
e. **Case E**: combined both Exp.3 and 60% load conditions
f. **Case F**: combined both Exp.3 and 100% load conditions
g. **Case G:** combined both Exp.5 and 20% load conditions
h. **Case H**: combined both Exp.5 and 60% load conditions
i. **Case I**: combined both Exp.5 and 100% load conditions

The corresponding time-domain characteristics and error indices are given in Table 10.9. From the results, it is observed that BA-optimized FOPID controller is able to minimize the rise time, peak time, settling time, steady-state error and error indices compared with other optimization techniques.

TABLE 10.9

Time-Domain Performance for the Combined Conditions

Operating Conditions	Time Response Algorithm	Rise Time t_r (s)	Peak Time t_p (s)	Settling Time t_s (s)	Steady-State Error e_{ss} (%)	Performance Indices		
						IAE	ITAE	ISE
Case A	MGA-PID	1.035	1.186	1.487	8	0.017	0.021	0.011
	ABC-PID	0.892	1.078	1.345	8	0.014	0.020	0.011
	BA-PID	0.842	0.874	1.024	7	0.015	0.027	0.012
	MGA-FOPID	0.716	0.913	0.999	5	0.016	0.018	0.013
	ABC-FOPID	0.611	0.806	0.920	4	0.012	0.010	0.008
	BA-FOPID	0.326	0.718	0.861	4	0.010	0.010	0.007
Case B	MGA-PID	1.035	1.186	1.487	8	0.017	0.021	0.011
	ABC-PID	0.926	1.024	1.321	11	0.013	0.027	0.008
	BA-PID	0.756	0.896	0.987	7	0.019	0.048	0.012
	MGA-FOPID	0.682	0.726	0.882	9	0.021	0.020	0.010
	ABC-FOPID	0.611	0.696	0.820	4	0.008	0.014	0.007
	BA-FOPID	0.430	0.640	0.792	3	0.007	0.013	0.005

(Continued)

TABLE 10.9 (*Continued*)

Time-Domain Performance for the Combined Conditions

Operating Conditions	Time Response Algorithm	Rise Time t_r (s)	Peak Time t_p (s)	Settling Time t_s (s)	Steady-State Error e_{ss} (%)	Performance Indices		
						IAE	ITAE	ISE
Case C	MGA-PID	1.148	1.384	1.635	21	0.017	0.003	0.013
	ABC-PID	1.023	1.241	1.482	17	0.014	0.002	0.010
	BA-PID	0.897	0.945	1.241	14	0.017	0.002	0.011
	MGA-FOPID	0.682	0.726	0.882	9	0.021	0.020	0.010
	ABC-FOPID	0.664	0.811	0.908	6	0.005	0.016	0.012
	BA-FOPID	0.564	0.809	0.894	5	0.005	0.013	0.010
Case D	MGA-PID	1.124	1.356	1.621	22	0.098	0.006	0.052
	ABC-PID	0.965	1.168	1.465	17	0.017	0.002	0.011
	BA-PID	0.724	0.854	1.012	6	0.030	0.014	0.018
	MGA-FOPID	0.716	0.913	0.999	8	0.016	0.018	0.013
	ABC-FOPID	0.611	0.806	0.920	6	0.012	0.010	0.008
	BA-FOPID	0.326	0.718	0.861	3	0.010	0.010	0.007
Case E	MGA-PID	0.716	0.833	0.994	9	0.014	0.016	0.008
	ABC-PID	0.892	1.078	1.345	8	0.014	0.020	0.011
	BA-PID	0.724	0.854	1.012	6	0.030	0.014	0.018
	MGA-FOPID	0.682	0.726	0.882	9	0.021	0.020	0.010
	ABC-FOPID	0.611	0.696	0.820	4	0.008	0.014	0.007
	BA-FOPID	0.430	0.640	0.792	3	0.007	0.013	0.005
Case F	MGA-PID	0.807	0.900	0.993	9	0.016	0.029	0.013
	ABC-PID	0.865	1.068	1.344	7	0.018	0.024	0.012
	BA-PID	0.842	0.874	1.024	7	0.015	0.027	0.012
	MGA-FOPID	0.682	0.726	0.892	9	0.013	0.027	0.008
	ABC-FOPID	0.664	0.811	0.908	6	0.005	0.016	0.012
	BA-FOPID	0.564	0.809	0.894	4	0.005	0.013	0.010
Case G	MGA-PID	1.992	1.452	1.658	19	0.014	0.003	0.011
	ABC-PID	0.756	0.896	0.987	7	0.019	0.048	0.012
	BA-PID	0.836	0.954	1.241	10	0.009	0.004	0.006
	MGA-FOPID	0.716	0.913	0.999	8	0.016	0.018	0.013
	ABC-FOPID	0.611	0.806	0.920	6	0.012	0.010	0.008
	BA-FOPID	0.326	0.718	0.861	3	0.010	0.010	0.007
Case H	MGA-PID	0.716	0.833	0.994	9	0.014	0.016	0.008
	ABC-PID	0.926	1.141	1.423	18	0.093	0.006	0.052
	BA-PID	0.724	0.854	1.012	6	0.030	0.014	0.018
	MGA-FOPID	0.667	0.783	0.962	12	0.017	0.002	0.011
	ABC-FOPID	0.611	0.696	0.820	5	0.008	0.014	0.007
	BA-FOPID	0.430	0.640	0.792	5	0.007	0.013	0.005
Case I	MGA-PID	0.807	0.900	0.993	9	0.016	0.029	0.013
	ABC-PID	0.899	1.052	1.347	10	0.015	0.027	0.012
	BA-PID	0.894	0.812	1.654	9	0.013	0.027	0.008
	MGA-FOPID	0.642	0.749	0.865	9	0.015	0.027	0.012
	ABC-FOPID	0.664	0.811	0.908	7	0.005	0.016	0.012
	BA-FOPID	0.564	0.809	0.894	7	0.005	0.013	0.010

10.5.6 Mean, Standard Deviation and Convergence

The mean and standard deviation for various operating conditions are presented in Table 10.10. From the results, it shows that mean and standard deviation values of BA-optimized PID/FOPID controller performance are better over those of MGA and ABC.

Figure 10.10 shows the convergence curve for various metaheuristic algorithms. From these results, the BA-optimized FOPID controller has a minimal output characteristic value, indicating superiority over MGA and ABC-optimized FOPID controller for the different operating conditions of BLDC motor. In BA-optimized FOPID controller for sensorless speed regulation of BLDC motor drives, the comparison of different optimization methods for fine-tuning FOPID controller under various optimization techniques observed that the time-domain characteristics, i.e. rise time, peak time, settling time, peak overshoot and steady-state error, are reduced (improved).

TABLE 10.10

Mean and Standard Deviation for Varying Set Speed Conditions

Various Controllers	MGA-PID	ABC-PID	BA-PID	MGA-FOPID	ABC-FOPID	BA-FOPID
Ref. Paper & Year	Bagis (2007).	Rajasekhar et al. (2014)	Premkumar et al. (2016)	Wang et al. (2017)	Vanchinathan et al. (2018)	Vanchinathan et al. (2018)
			Exp. 1			
Mean	0.6823541	0.6897562	0.525462	0.356215	0.354265	0.345684
Standard deviation	0.8256478	0.725486	0.526478	0.44325	0.42354	0.40213
			Exp. 2			
Mean	0.562482	0.542565	0.425874	0.363108	0.370482	0.362586
Standard deviation	0.635248	0.524725	0.425872	0.39525	0.40885	0.37985
			Exp. 3			
Mean	0.632654	0.5326541	0.425681	0.369254	0.361242	0.350124
Standard deviation	0.652412	0.598742	0.536214	0.44256	0.41356	0.38245
			Exp. 4			
Mean	0.465421	0.435142	0.365871	0.036254	0.23568	0.235421
Standard deviation	0.523641	0.568741	0.4.2568	0.365241	0..36589	0.236541
			Exp. 5			
Mean	0.568421	0.425861	0.456841	0.45621	0.36524	0.362581
Standard deviation	0.652487	0.562451	0.542582	0.42685	0.456981	0.545875
			20% Load			
Mean	0.532645	0.421456	0.42587	0.356582	0.350125	0.347352
Standard deviation	0.685245	0.625475	0.254825	0.45326	0.438276	0.374225
			40% Load			
Mean	0.6823541	0.6897562	0.525462	0.356215	0.354265	0.345684
Standard deviation	0.8256478	0.725486	0.526478	0.44325	0.42354	0.40213
			60% Load			
Mean	0.526478	0.4258467	0.365874	0.369322	0.360435	0.354153
Standard deviation	0.825645	0.6235478	0.425872	0.471516	0.40155	0.37153

(Continued)

TABLE 10.10 (*Continued*)

Mean and Standard Deviation for Varying Set Speed Conditions

Various Controllers	MGA-PID	ABC-PID	BA-PID	MGA-FOPID	ABC-FOPID	BA-FOPID
Ref. Paper & Year	Bagis (2007).	Rajasekhar et al. (2014)	Premkumar et al. (2016)	Wang et al. (2017)	Vanchinathan et al. (2018)	Vanchinathan et al. (2018)
80% Load						
Mean	0.632654	0.5326541\	0.425681	0.369254	0.361242	0.350124
Standard deviation	0.652412	0.598742	0.536214	0.44256	0.413563	0.382456
100% Load						
Mean	0.6524235	0.62547	0.542187	0.364258	0.371452	0.368452
Standard deviation	0.526254	0.42587	0.584752	0.42356	0.39564	0.372452
Case A						
Mean	0.562482	0.542565	0.425874	0.363108	0.370482	0.362586
Standard deviation	0.635248	0.524725	0.425872	0.39525	0.40885	0.379854
Case B						
Mean	0.626542	0.655412	0.452658	0.356582	0.350125	0.356248
Standard deviation	0.526412	0.562587	0.55421	0.45326	0.425871	0.425688
Case C						
Mean	0.465421	0.435142	0.365871	0.036254	0.23568	0.235421
Standard deviation	0.523641	0.568741	0.4.2568	0.365241	0..36589	0.236541
Case D						
Mean	0.452684	0.452651	0.425412	0.365241	0.36581	0.36254
Standard deviation	0.526478	0.526841	0.423561	0.326541	0.365879	0362541
Case E						
Mean	0.568421	0.425861	0.456841	0.45621	0.36524	0.362581
Standard deviation	0.652487	0.562451	0.54258	0.42685	0.456981	0.545875

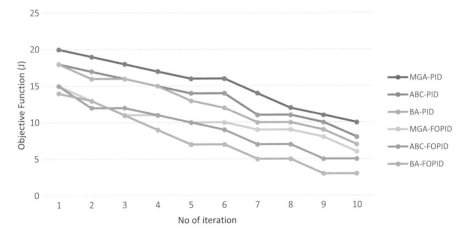

FIGURE 10.10

Convergence curve for various operating conditions.

10.6 Conclusion and Future Research Direction

A detailed industrial case study of the various metaheuristic optimization algorithm-based tuning of controllers was introduced to provide a clear perspective on the time-domain characteristics of the sensorless speed control of the BLDC motor drives. The transient and steady-state characteristics of PID/FOPID controllers were evaluated by applying the various operating conditions such as (i) constant load conditions, (ii) varying load conditions, (iii) changing set speed conditions and (iv) combined operating conditions. The FOPID controller was found to provide the quickest settling time, the fastest rise time, reduced overshoot and minimum steady-state error than the PID controller. From the results of the review and evaluation, it is inferred that the BA-optimized FOPID controller has improved efficiency and admirable speed controllability when compared with other optimized PID and FOPID controllers in terms of minimization of rising time, peak time, settling time and steady-state error for speed control of sensorless BLDC motor drives under the specified operating conditions.

The future research in BLDC motor drives is expected to focus on sensorless starting, reduction of motor and controller cost, comprehensive sensorless control, application-specific controller design, improved power quality and control effort controllers, and reduced cost controllers with power factor correction features. The economic viability and performance of BLDC motors in a wide range of applications are expected to grow in future. Controller stability analysis may be measured and recorded by different controllers. The suitability and effectiveness of the controllers under various vulnerable drive operating conditions also need further investigations.

References

Arpaci, H., & Ozguven, O. F. 2017. Design of adaptive fractional-order PID controller to enhance robustness by means of adaptive network Fuzzy inference system. *International Journal of Fuzzy Systems*, 19(4), 1118–1131.

Bagis, A. 2007. Determination of the PID controller parameters by modified genetic algorithm for improved performance. *Journal of Information Science and Engineering*, 23(5), 1469–1480.

Bouarroudj, N. 2015. A hybrid fuzzy fractional order PID sliding-mode controller design using PSO algorithm for interconnected nonlinear systems. *Journal of Control Engineering and Applied Informatics*, 17(1), 41–51.

Chang, F. K., & Lee, C. H. 2008. Design of fractional PID control via hybrid of electromagnetism-like and genetic algorithms. *In 2008 Eighth IEEE International Conference on Intelligent Systems Design and Applications*, 2, 525–530.

Damodharan, P., & Vasudevan, K. 2010. Sensorless brushless DC motor drive based on the zero-crossing detection of back electromotive force (EMF) from the line voltage difference. *IEEE Transactions on Energy Conversion*, 25(3), 661–668.

Dasari, M., Reddy, A. S., & Kumar, M. V. 2019. GA-ANFIS PID compensated model reference adaptive control for BLDC motor. *International Journal of Power Electronics and Drive Systems*, 10(1), 265.

De la Guerra, A., & Alvarez-Icaza, L. 2020. Robust control of the brushless DC motor with variable torque load for automotive applications. *Electric Power Components and Systems*, 48(2), 1–11.

Dokeroglu, T., Sevinc, E., Kucukyilmaz, T., & Cosar, A. 2019. A survey on new generation meta-heuristic algorithms. *Computers & Industrial Engineering*, 137, 106040.

El-Wakeel, A. S., Ellissy, E. K. M., & Abdel-hamed, A. M. 2015. A hybrid bacterial foraging-particle swarm optimization technique for optimal tuning of proportional-integral-derivative controller of a permanent magnet brushless DC motor. *Electric Power Components and Systems*, 43(3), 309–319.

Gan, M. G., Zhang, M., Zheng, C. Y., & Chen, J. 2018. An adaptive sliding mode observer over wide speed range for sensorless control of a brushless DC motor. *Control Engineering Practice*, 77, 52–62.

Gobinath, S., & Madheswaran, M. 2019. Deep perceptron neural network with fuzzy PID controller for speed control and stability analysis of BLDC motor. *Soft Computing*, 24, 10161–10180.

Hekimoğlu, B. 2019. Optimal tuning of fractional order PID controller for DC motor speed control via chaotic atom search optimization algorithm. *IEEE Access*, 7, 38100–38114.

Hemalatha, N., & Nageswari, S. 2020. A new approach of position sensorless control for brushless DC motor. *Current Signal Transduction Therapy*, 15(1), 65–76.

Ibrahim, M. A., Mahmood, A. K., & Sultan, N. S. 2019. Optimal PID controller of a brushless DC motor using genetic algorithm. *International Journal Power Electronics & Drives System*, 2088(8694), 8694.

Jia, Z., Zhang, Q., & Wang, D. 2020. A sensorless control algorithm for the circular winding brushless DC motor based on phase voltages and DC current detection. *IEEE Transactions on Industrial Electronics*.

Jigang, H., Hui, F., & Jie, W. 2019. A PI controller optimized with modified differential evolution algorithm for speed control of BLDC motor. *Automatika*, 60(2), 135–148.

Kesarkar, A. A., & Selvaganesan, N. 2015. Tuning of optimal fractional-order PID controller using an artificial bee colony algorithm. *Systems Science & Control Engineering*, 3(1), 99–105.

Kim, T., Lee, H. W., & Ehsani, M. 2007. Position sensorless brushless DC motor/generator drives: Review and future trends. *IET Electric Power Applications*, 1(4), 557–564.

Kiree, C., Kumpanya, D., Tunyasrirut, S., & Puangdownreong, D. 2016. PSO-based optimal fractional PI (D) controller design for brushless DC motor speed control with back EMF detection. *Journal of Electrical Engineering & Technology*, 11(3), 715–723.

Krishnan, R. 2017. *Permanent Magnet Synchronous and Brushless DC Motor Drives*. CRC Press, Boca Raton, FL.

Lee, A. C., Fan, C. J., & Chen, G. H. 2017. Current integral method for fine commutation tuning of sensorless brushless DC motor. *IEEE Transactions on Power Electronics*, 32(12), 9249–9266.

Lin, C. L., H. Y. Jan, and N.-C. Shieh, 2003. GA-optimized multi-objective PID control for a linear brushless DC motor. *IEEE Transactions on Mechatronics*, 8(1), 56–65.

Luo, H., Krueger, M., Koenings, T., Ding, S. X., Dominic, S., & Yang, X. 2017. Real-time optimization of automatic control systems with application to BLDC motor test rig. *IEEE Transactions on Industrial Electronics*, 64(5), 4306–4314.

Pillay, P., & Krishnan, R. 1988. Modeling of permanent magnet motor drives. *IEEE Transactions on Industrial Electronics*, 35(4), 537–541.

Premkumar, K., Manikandan, B. V. 2016. Bat algorithm optimized fuzzy PD based speed controller for brushless direct current motor. *Engineering Science Technology*, 19, 818–840.

Premkumar, K., & Manikandan, B. V. 2015. Fuzzy PID supervised online ANFIS based speed controller for brushless dc motor. *Neurocomputing*, 157, 76–90.

Rahideh, A., Korakianitis, T., Ruiz, P., Keeble, T., & Rothman, M. T. 2010. Optimal brushless DC motor design using genetic algorithms. *Journal of Magnetism and Magnetic Materials*, 322(22), 3680–3687.

Rajasekhar, A., Jatoth, R. K., & Abraham, A. 2014. Design of intelligent PID/PIλDμ speed controller for chopper fed DC motor drive using opposition based artificial bee colony algorithm. *Engineering Applications of Artificial Intelligence*, 29, 13–32.

Shah, P., & Agashe, S. 2016. Review of fractional PID controller. *Mechatronics* 38: 29–41.

Shanmugasundram, R., Zakariah, K. M., & Yadaiah, N. 2014. Implementation and performance analysis of digital controllers for brushless DC motor drives. *IEEE/ASME Transactions on Mechatronics,* 19(1), 213–224.

Valle, R. L., de Almeida, P. M., Ferreira, A. A., & Barbosa, P. G. 2017. Unipolar PWM predictive current-mode control of a variable-speed low inductance BLDC motor drive. *IET Electric Power Applications,* 11(5), 688–696.

Vanchinathan, K. & Valluvan, K. R., 2015. Improvement of time response for sensorless control of BLDC Motor drive using ant colony optimization technique, *International Journal of Applied Engineering Research,* 10(55), 3519–3524.

Vanchinathan, K., & Valluvan, K. R. 2018. A metaheuristic optimization approach for tuning of fractional-order PID controller for speed control of sensorless BLDC motor. *Journal of Circuits, Systems and Computers,* 27(08), 1850123.

Vanchinathan, K., & Valluvan, K. R. 2018. Tuning of fractional order proportional integral derivative controller for speed control of sensorless BLDC motor using artificial bee colony optimization technique. In Bhuvaneswari, M. C. & Saxena, J. (Eds.), *Intelligent and Efficient Electrical Systems,* pp. 117–127. Springer, Berlin, Germany.

Varshney, A., Gupta, D., & Dwivedi, B. 2017. Speed response of brushless DC motor using Fuzzy PID controller under varying load condition. *Journal of Electrical Systems and Information Technology,* 4(2), 310–321.

Verma, S. K., Yadav, S., Nagar, S. K. 2017. Optimization of fractional order PID controller using grey wolf optimizer. *Journal of Control Automation Electrical System* 28(3), 314–322.

Verma, V., Pal, N. S., & Kumar, B. 2019. Speed control of the sensorless BLDC motor drive through different controllers. In *Harmony Search and Nature Inspired Optimization Algorithms,* pp. 143–152. Springer, Singapore.

Wang, L., Wang, C., Yu, L., Liu, Y., & Sun, J. 2017. Design of fractional: Order PID controller based on genetic algorithm. *In 2017 29th Chinese Control and Decision Conference,* Chongqing, China, pp. 808–813.

Xia, C. L. 2012. *Permanent Magnet Brushless DC Motor Drives and Controls.* John Wiley & Sons, Hoboken, NJ.

Yao, X., Zhao, J., Lu, G., Lin, H., & Wang, J. 2019. Commutation error compensation strategy for sensorless brushless DC motors. *Energies,* 12(2), 203.

Zahir, A. A. M., Alhady, S. S. N., Wahab, A. A. A., & Ahmad, M. F. 2020. Objective functions modification of GA optimized PID controller for brushed DC motor. *International Journal of Electrical and Computer Engineering,* 10(3), 2426.

Zaky, M. S., Metwaly, M. K., Azazi, H. Z., & Deraz, S. A. 2018. A new adaptive SMO for speed estimation of sensorless induction motor drives at zero and very low frequencies. *IEEE Transactions on Industrial Electronics,* 65(9), 6901–6911.

Zheng, W., & Pi, Y. 2016. Study of the fractional-order proportional integral controller for the permanent magnet synchronous motor based on the differential evolution algorithm. *ISA Transaction;* 63: 387–393.

Zhou, X., Zhou, B., Guo, H., & Wei, J. 2016. Research on sensorless and advanced angle control strategies for doubly salient electro-magnetic motor. *IET Electric Power Applications,* 10(5), 375–383.

11

Predictive Analysis of Cellular Networks: A Survey

Nilakshee Rajule, Radhika Menon, and Anju Kulkarni

Dr. D. Y. Patil Institute of Technology

CONTENTS

DOI: 10.1201/9781003143505-11

List of Abbreviations

AP: Access Point
AARIMA: Adjusted ARIMA
AIC: Akaike Information Criterion
AUC: Area Under Curve

ACF:	Autocorrelation Functions
AE:	Autoencoders
AR:	Autoregressive
ACD:	Autoregressive Conditional Duration
ARCH:	Autoregressive Conditional Heteroscedasticity
ARIMA:	Autoregressive Integrated Moving Average
ARMA:	Autoregressive Moving Average
BS:	Base Station
BIC:	Bayesian Information Criterion
CDR:	Call Data Record
CADM:	CDR-Based Anomaly Detection Method
CTMC:	Continuous-Time Markov Chain
CDSA:	Control/Data Separation Architecture
CNN:	Convolutional Neural Networks
DBN:	Deep Belief Network
DL:	Deep Learning
DWT:	Discrete Wavelet Transform
DCCA:	Distance-Constrained Complementarity Aware
EE:	Energy Efficiency
eNB:	Evolved NodeB
EGARCH:	Exponential GARCH
EWMA:	Exponentially Weighted Moving Average
FARIMA:	Fractional ARIMA
GRU:	Gate Recurrent Unit
GARCH:	Generalized ARCH
GPS:	Global Positioning System
GSM:	Global System For Mobile Communications
HO:	Handover
HMM:	Hidden Markov Model
HEM:	History-Based Expectation-Maximization
HW:	Holt Winters
IP:	Internet Protocol
ISP:	Internet Service Providers
KPI:	Key Performance Indicator
LMSE:	Least-Mean Square Error
LB:	Load Balancing
LSAE:	Local Stacked Autoencoder
LBS:	Location-Based Service
LRD:	Long-Range Dependency
LSTM:	Long Short-Term Memory
LTE:	Long-Term Evolution
ML:	Machine Learning
MUBS-SAP:	Marginal Utility-Based BS-Sleeping Spatial Adaptive Play
MB-HUMM:	Markov-Based Hierarchical User Mobility Model
MLE:	Maximum-Likelihood Estimation
MAE:	Mean Absolute Error
MIA:	Mobile Intelligent Agent

MPBS:	Mobility Pattern-Based Scheme
MPE:	Mobility Prediction Entity
MLC:	Most Likely Cluster
MLP:	Multilayer Perceptron
NE:	Network Element
NARX:	Nonlinear Autoregressive With Exogenous
OC-NN:	One-Class Neural Network
PPM:	Prediction By Partial Matching
PRA:	Predictive Resource Allocation
PBS:	Profile-Based Scheme
QoS:	Quality of Service
RNN:	Recurrent Neural Network
ROI:	Region of Interests
RS:	Relay Stations
RA:	Resource Allocation
RBM:	Restricted Boltzmann Machines
SARIMA:	Seasonal ARIMA
SON:	Self-Organizing Networks
SRD:	Short-Range Dependency
SNA:	Social Network Analysis
STN:	Spatio-Temporal Neural Network
GSAE:	Stacked Autoencoder
SAM:	Stochastic Autoregressive Mean
SVM:	Support Vector Machine
VoIP:	Voice Over Internet Protocol

11.1 Introduction

In recent years, cellular networks are experiencing a sharp growth in data traffic owing to the rapid development of mobile user equipment, social networking applications and services, and so forth. A human life is highly influenced by the Internet-connected mobile devices as they are penetrating each and every aspect of individuals' life, work and entertainment. The increased number of smart phones and the emergence of evermore diverse applications trigger a surge in mobile data traffic. The annual worldwide IP traffic consumption is likely to reach 3.3 zeta bytes by 2021 as indicated by the latest industry forecasts, with smartphone traffic exceeding PC traffic by the same year [1].

Current mobile infrastructure faces great capacity demands as the system is getting changed from network-centric to user-centric. In response to this increased demand, early efforts propose to agilely provision resources and tackle mobility management in a distributive manner. In the long run, however, it is necessary for the Internet Service Providers (ISPs) to develop intelligent heterogeneous architectures and tools. Such architecture and tools can spawn the fifth generation of mobile systems (5G) and gradually can cope with the end-user application requirements.

To fulfil the increased user demands with limited resources, it is significant and necessary for the network operator to manage network resources properly and provide a good quality of service (QoS). Different behaviours of users will enable variations in usage of network resource changes with space and time. Analysis of the network traffic and prediction of the traffic load will help network operator to identify the base stations (BSs) with high traffic amount in advance. These predictions can help network planning and adjusting the resources to ensure good QoS in such regions [2].

Traffic prediction in communication networks plays an important role in intelligent network planning, operations and management. Accurate prediction helps to prevent network congestion and to maintain a reasonable utilization ratio of network resources. A network service provider needs to know the future trends of network parameters, routers and other devices information in order to proceed with the traffic variability. Various prediction approaches have been employed to manage the problem of predicting the future trends of network parameters, routers and other devices information in a real-time network. The predictability of network traffic is important in many areas, such as dynamic bandwidth allocation, network security, network planning, predictive congestion control and so on [3,4].

Regardless of emergent attention in the predictive analysis of cellular networks, current contributions are sprinkled across different research areas and a widespread survey is lacking. In this chapter, we present the comprehensive survey of predictive analysis and tried to narrow the differences between predictive analysis, deep learning (DL) and intelligence factor of cellular networks. In this chapter, we are going to present a thorough survey of research that resides at the intersection of these three fields. We wind up this chapter by discussing the role of intelligent networks in enhancing the performance of predictive analysis-enabled applications.

Motivation: Rapid development of user mobile equipment is resulting in an increased number of mobile users generating huge amount of data traffic. Due to the different behaviours of mobile users, this data traffic is multidimensional in nature. This multidimensional aspect and large amount of data is generating various data patterns. Study and analysis of these data patterns will help service providers to understand the expectations of their customers. Combining multiple parameter analysis will help service providers to offer the desired QoS to the users for the offered traffic load. The predictive analysis of these multiple parameters will provide multiobjective optimization of the cellular networks.

Survey Organization: This chapter is structured as shown in Figure 11.1. Our survey begins with the discussion of traffic and traffic characteristics. The description of various traffic characteristics is presented in Section 11.2. In Section 11.3, we present the overview of predictive analysis, and we introduce the parameters which are to be analysed. The traffic generated by mobile user possess spatio-temporal variations; techniques to analyse traffic are presented in Section 11.4; detailed call data record (CDR) analysis techniques and application areas are described in Section 11.5. Section 11.6 deals with user mobility and location analysis along with the various predicted outputs. In Section 11.7, we explain the various predicted outputs, which are obtained by the analysis of network data. In Section 11.8, we present the various applications of predictive analysis. In Section 11.9, we discuss the role of DL in predictive analysis along with brief introduction to different state-of-the-art techniques. We conclude this chapter with overview of emerging intelligent network in Section 11.10 and their relevance in predictive analysis. The diagrammatic representation of survey is given in Figure 11.1.

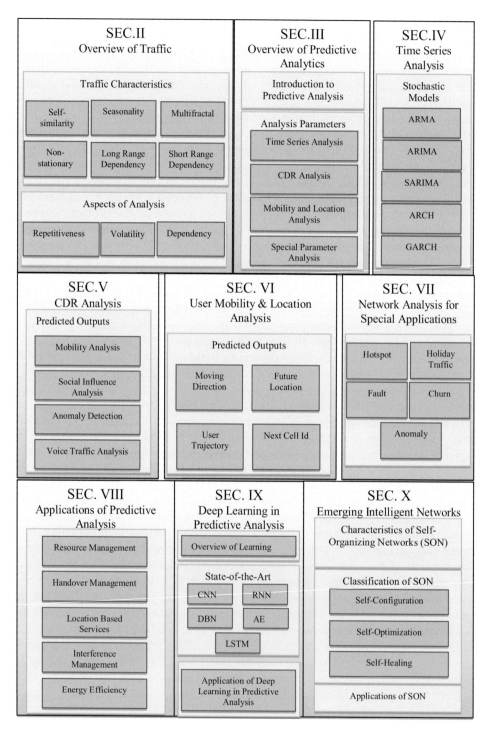

FIGURE 11.1
Survey organization.

11.2 Traffic Characteristics and Aspects of Analysis

In order to analyse the cellular networks, it is very important to study the various characteristics that are reflected by the network traffic and that are vital for accurate forecasting. In this section, we introduce six characteristics of network traffic from a network perspective. In this section, we have discussed the traffic characteristics, and along with various characteristics, we have also addressed the different aspects of traffic depending on which analysis can be done (Figure 11.2).

11.2.1 Traffic Characteristics

11.2.1.1 Self-Similarity

A number of studies have discussed self-similarity of network traffic [5,6]. In a very practical sense, self-similar means that a proportional segment of measuring tends to be observed in a different time scale. Measuring can be the number of packets or amount of data during certain period of time with a predefined granularity.

11.2.1.2 Seasonality

It is quite commonly observed that with a certain frequency in any domain of temporal measuring, traffic exhibits nearly the same patterns. Like typical weather conditions in each "season" of a year, network traffics also show quite similar patterns on weekends and weekdays, holidays or certain hours of a day, periodically. It is important to understand nature of the seasonality in data flows while analysing the nature of traffic, and also forecast the future traffic load possibly reflecting congruent patterns.

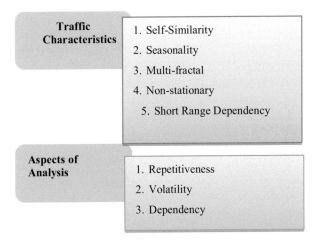

FIGURE 11.2
Network characteristics and analysis aspects.

11.2.1.3 Non-Stationarity

A very common assumption for time-series modelling and stochastic processes is mean, variance and correlation model that stays constant over time in a stationary process. However, changing statistical characteristic leading to a change in modelling may be shown by network traffic [7–9]. Therefore, in practice, a time-series model should be capable and sensitive to detect such changes that are observed frequently depending on various factors such as the number of users, connections and bandwidth utilization of related network elements (NEs) before forecasting.

11.2.1.4 Multifractal

In aggregated network traffics (i.e. consisting of multiple flows originated by multiple sources), it is possible to observe self-similar characteristics of individual network flows. This type of traffic flows is not only indicated as self-similar or fractal but also indicated as multifractal [10]. Detecting multifractal behaviours and fractal patterns of multiple flows simultaneously is naturally important yet challenging for forecasting.

11.2.1.5 Long-Range Dependency (LRD)

Various time-dependent systems or physical phenomena show a correlated behaviour during large time scales. For network traffic, especially the Internet, the LRD is observed in different significant studies [11,12]. Together with the self-similar properties, the LRD shifts the traffic modelling perspective from memory-less stochastic processes (e.g. Poisson) to long-memory time series.

11.2.1.6 Short-Range Dependency (SRD)

In comparison with the LRD, the correlation can be observed for shorter time scales in short-range-dependent processes. That is, the dependence among the observations quickly dissolves and it is related to quickly decaying correlations. Many traditional time-series modelling techniques examine the SRD; however, it is not enough to reflect today's network traffic pattern. SRD can be considered while forecasting the short-term traffic employing relatively low-complexity models.

11.2.2 Aspects of Analysis

Many studies follow similar statistical approaches but focus on different characteristics. The traffic characteristics can be divided into three aspects in order to handle such characteristics: repetitiveness, volatility and dependency [5]. The repetitiveness characterizes the cyclic and usual patterns, and it is directly associated with self-similarity and seasonality. However, non-stationary and multifractal are covered by the aspect volatility. Lastly, the time-dependent characteristics, which are LRD and SRD, are represented by the aspect dependency. The researchers can use one of the aspects or the combination of these aspects for analysis. The same is presented in Figure 11.3.

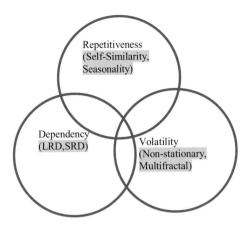

FIGURE 11.3
Aspects of analysis.

11.3 Overview of Predictive Analysis

In spite of the remarkable increase in volume of mobile cellular traffic data, the knowledge on its spatio-temporal patterns at the urban scale is very limited. The records of cellular traffic collected at cell towers can be widely adopted for daily network management and diagnosis. This large traffic data from cell towers are also advantageous to understand and predict urban cellular traffic, which can be extremely valuable for individual cell towers, cellular carriers and city authorities. It is very important to ensure the overall quality of service and network performance by reducing unnecessary operation cost by allocating energy and bandwidth tightly based on the future traffic demand. This can be facilitated by the accurate prediction of cellular traffic.

The term "predictive analytics" describes a variety of statistical and analytics techniques. Their main function is to analyse the current and historical facts and to predict the future based on the facts. Predictive analysis consists of various models that can predict the future events and behaviour of variables. The predictive analysis process consists of phases like data pre-processing, analysis, prediction and monitoring (Figure 11.4).

The first phase of predictive analysis consists of gathering and preparing the data. The data as per the requirement is collected, and then, the data is cleaned to construct the useful data. The second phase deals with analysis of the data. In this phase, the study of what happened, why that happened and what is happening is currently carried out.

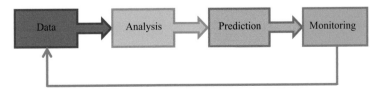

FIGURE 11.4
Phases of predictive analysis.

This study is further used as an input by the third phase, i.e. the prediction phase. In this phase, the models are developed that can predict the future events and behaviours of variables. Mostly, the predictive models return a score: a higher score indicates a higher likelihood of the behaviour given or event occurring, and vice versa. In the next section, we present various dimensions of predictive analysis.

11.3.1 Time-Series Analysis

It is extremely demanding that mobile cellular traffic should be predicted at both large scale and fine granularity. There can be a wide dynamic range of network traffic at an individual cell tower due to the diverse network demand of Internet-based applications (e.g. mobile videos, location-based games, VoIP) and user behaviours (e.g. at work, in transit, during sleep). The spatial dependencies are introduced due to user mobility into the cellular traffic among spatially distributed cell towers. There are various factors that can influence the geographical distribution of cellular traffic at the urban scale, such as population, land use, holidays and various social activities. The spatio-temporal dependencies are further complicated by these influential factors among cell tower traffic citywide [10].

Time-series modelling is a dynamic research area and has attracted attentions of researchers' community over the last few decades. The main objective of time-series modelling is to cautiously collect the data and rigorously study the past observations of a time series to develop an appropriate model. This model describes the inherent structure of the series and is then used to generate future values for the series, i.e., to make forecasts. Thus, time-series forecasting can be termed as the act of predicting the future by understanding the past.

Different approaches have been proposed for traffic forecasting till date, such as exponential smoothing, wavelets and hybrid methods, including multiple approaches. But stochastic models and artificial neural network models are two approaches, which are frequently used techniques for network traffic prediction in practice. They are the fundamental elements of traffic forecasting. The detailed explanation of time-series analysis is given in Section 11.4.

11.3.2 CDR Analysis

Mobile phones produce a lot of location data every day. Telecommunication industry deals with the extreme variety of data such as network data, CDR data, user's personal and billing data etc. CDR is basically a record that contains detailed information of a call, such as start and end time of a call, duration of a call, call parties, cell id, location etc. CDR is a most valuable source of a telecom industry, as it is used in many fundamental processes like billing, network efficiency, churn detection, fraud detection, value-added services etc.

11.3.3 Mobility and Location Analysis

The vision of future-generation wireless communication aims at anytime, anywhere communication. As the population of mobile users and user mobility is increasing, there is a huge demand of quality of service (QoS) and mobility management. The analysis of user mobility and location helps the service providers to facilitate users with desired QoS through various applications such as

- Resource management
- HO management
- Guaranteed QoS
- Data offloading.

Prediction of user mobility has been proven to be a necessary parameter for providing efficient resource management and services in wireless communication. Also, prior knowledge of users' mobility allows for more efficient HO management as it helps in reducing the amount of signalling and interruption time in order to improve performance and energy efficiency of the mobile system. In order to provide better QoS to mobile users, the prediction of individual mobile trajectory is very essential. Human mobility behaviour is far from random and influenced by their past behaviour, as individuals usually have a typical purpose and habitual route. Thus, the prediction of mobility trajectory is promising to gain high accuracy to guarantee efficient resource management and better service performance.

User mobility information can also be used to assist traffic routing in wireless networks to ease the bottleneck effect in overloaded BSs or access points (APs). Mobility analysis and prediction can also help mobile data traffic to move a portion of it to Wi-Fi networks, as the mobile data revenue significantly lags behind the exponential growth of data traffic.

11.4 Time-Series Analysis of Network Traffic

11.4.1 Stochastic Models

Autoregressive (AR) models are stochastic models that ingest the input values (of past) in a time sequence into a regression function to predict future values for related time series. The different techniques like autoregressive moving average (ARMA) [13], autoregressive integrated moving average (ARIMA) [14], fractional ARIMA (FARIMA) [15], seasonal ARIMA (SARIMA), autoregressive conditional heteroscedasticity (ARCH) [16], generalized ARCH (GARCH) [17], exponential GARCH (EGARCH), autoregressive conditional duration (ACD) [18], stochastic autoregressive mean (SAM) and nonlinear autoregressive with exogenous (NARX) are falling into this category. Indeed, there are statistical differences: for instance, while ARIMA models focus on conditional mean through temporal series, ARCH methods take conditional variance into consideration for modelling.

AR models comprise three phases: (i) statistical modelling with respect to some criteria, (ii) parameter estimation and (iii) forecasting (Figure 11.5).

The first phase is related to the detection of correlation in time series using autocorrelation functions (ACFs) and identification of the model based on widely used criteria such as Akaike Information Criterion (AIC) and Bayesian Information Criterion (BIC) [19]. Secondly, related coefficients of the identified model are estimated using the

FIGURE 11.5
Autoregressive process.

well-known estimation methods such as maximum-likelihood estimation (MLE) and least-mean square error (LMSE). After parameter estimation, future points of the time series are predicted and the accuracy is presented with respect to different metrics. AR techniques are further divided into SRD and LRD. The techniques like AR, ARMA and ARIMA fall into SRD, and FARIMA and multifractional wavelet model fall into LRD category. In the following section, we will discuss in brief the contribution of various researchers in time-series analysis using AR techniques.

11.4.2 Research Contribution

A combination of ARIMA/GARCH offers a new technique that addresses various network characteristics like LRD, SRD, multifractal and self-similarity. A one-step predictor is designed that uses ARIMA for linear traffic analysis and GARCH for changing variance [20]. Further, it is suggested that this predictor can be extended to make multistep predictions. A parameter estimation phase is incorporated in ARIMA/GARCH [21,22], which tunes the parameters of both ARIMA and GARCH [23] using MLE, which is based on BOX–COX method. The method performs better than FARIMA in terms of signal-to-error ratio and multistep predictions.

In Ref. [24], k-means clustering method is used to analyse user behaviour according to their hour call rate. The users are divided into three categories: low, medium and high rate of calling using clustering method. Each category is then separately modelled using SARIMA method. The traffic is modelled for daily and weekly patterns and used to make complete prediction of each user's behaviour. Further, it is concluded that the proposed method is more accurate for predictions than the prediction of aggregated traffic. The proposed prediction method shows 57% better results than the prediction of the aggregated traffic. The Internet traffic is modelled [25] at millisecond time scales using adjusted ARIMA (AARIMA) model. Traffic characteristics like self-similarity and LRD are addressed in Internet traffic. Results show that the AARIMA offers lower mean absolute error (MAE).

An attempt to eliminate seasonal patterns is made [22]. They have offered a learner ARIMA differencing process, which converts multiple stationarization operation (for trends and seasonal patterns both) to a single multiplicative process. The authors have analysed 12-month traffic data and presented single-step and multistep prediction models for each month.

TABLE 11.1

Summary of Autoregressive Techniques

Paper	Key Contribution	Technique	Traffic Characteristics
[20]	One-step predictor for linear traffic analysis	ARIMA + GARCH	LRD, SRD, multifractal and self-similarity
[21]	One-step predictor for nonlinear traffic analysis	GARCH	Self-similarity, LRD
[22]	Elimination of seasonal patterns	Learner ARIMA	LRD
[23]	Improved signal-to-error ratio and multistep predictions	ARIMA + GARCH	Self-similarity, LRD, multifractal
[24]	Analysis of user behaviour	SARIAM, K-means	Self-similarity
[25]	Offers lower MAE	AARIMA	Self-similarity, LRD

An enhanced (or generalized) ARCH model (GARCH) is introduced to develop a one-step predictor for nonlinear traffic models, e.g. Internet [21]. The authors point out those constant-variance models like ARIMA and its successors (e.g. FARIMA and SARIMA) cannot fit the bursty (and nonlinear) nature of the Internet traffic, whereas GARCH is taking conditional variance into account to react changing traffic patterns. To be able to determine GARCH parameters, MLE is deployed using the training data. The results show that the forecast error of GARCH is significantly less than that of the ARIMA-ARCH model for one-step prediction. However, its performance is open to validation in less aggregated traffics other than the Internet. The summary of time-series analysis is presented in Table 11.1.

11.5 CDR Analysis

11.5.1 Predicted Outputs

In this section, we are discussing the work done in CDR analysis for various application areas. The various applications of CDR analysis are shown in Figure 11.6.

11.5.1.1 Mobility Analysis

CDRs are also being used for user mobility analysis, travel distance analysis, user trajectory analysis, user's future location prediction etc. By collecting CDR data, it effectively avoids the traditional GPS data collection.

The CDR data is used to characterize the user mobility in Refs. [26,27]. Hadoop-based mobile big data processing platform is used along with systematic mobility analysis framework. A confidence-based method is proposed to reduce ping-pong effects that occurred while reconstructing user trajectories. However, the mobile usage analysis and user mobility analysis are combined together [28]. A method is proposed that uses palm calculus to analyse user movements. It is also found out that calls and user movements are correlated.

FIGURE 11.6
Applications of CDR analysis.

Travel distance between traffic zones is analysed using CDR data in Ref. [29]. CDR data is also used to divide the traffic analysis zones and to extract user activity data for trip generation. K-means clustering is used for dividing the traffic zones. The proposed method of pattern extraction helps in big data-driven transportation planning. The individual user behaviour is studied using CDR data to predict tourist's future location. Multiple classification algorithms are employed, including decision tree, neural network, Naïve Bayes, Random Forest, support vector machine (SVM) and long short-term memory (LSTM). Experimental results show that LSTM provides highest prediction accuracy (94.8%), while Random Forest/neural network gives the second best (85%).

11.5.1.2 Anomaly Detection

CDR data can also be used for detecting the unusual events of critical significance by analysing the subscriber's activities. CDR data analysis provides user's spatio-temporal information, which provides the exact location and time of a particular event as a massive increase occurs in CDR activities. Such an attempt is made in Ref. [31], where an idea of identifying anomalies in CDR data is presented. K-means clustering algorithm is used for verifying anomalies. Further, artificial neural network model is used to remove the anomalous data from anomaly-free data. The future traffic is also predicted using ARIMA model. In Ref. [79], the authors have proposed a CDR-based anomaly detection method (CADM). Anomalous behaviour of user movements in a cellular network was detected using CADM along with big data analytics tools.

11.5.1.3 Social Influence Analysis

In a telecommunication sector, it is very essential to understand the social influences among customers. Social network analysis (SNA) is usually employed to detect influencers and communities along with calling behaviours (profiles) by analysing the behaviours and relationships between customers.

The problem of social influencing is tackled in Ref. [32]; a communication influence model is proposed to model the influence probabilities. Influence measures are designed for estimating a pairwise influence strength using CDR along with web browsing histories. A model is proposed to analyse the nature of co-location events and frequent locations using social network contacts and CDR data [33]. In proposed work, locations are classified as frequent and non-frequent locations. By experimental results, it is concluded that non-frequent locations are associated with the social interaction. The location data is combined along with communication logs, i.e. CDR data to analyse the transport mode choices of mobile users [34]. The social influence on commute mode choice decision has been investigated. A method is proposed for data cleansing and for capturing influences in the network from CDR [35]. The proposed data cleansing process filters the anomaly numbers from CDR also with the help of proposed influence identification measure; it is possible to identify the key influencers.

11.5.1.4 Voice Traffic Analysis

CDR data is used for the analysis of international voice traffic in mobile networks. Along with analysing the traffic, the authors have further presented the traffic profile of users for different time periods as well as the factors affecting the traffic. The results show the long-term traffic stability and a daily/weekly traffic periodicity that reflects human activities [36].

11.5.2 Big Data Analysis of CDR

Many researchers have used big data analysis technique for CDR data analysis [37]. A parallel DBMS is used for database management, and MapReduce is used for data processing. In Ref. [38], big data analysis is used to analyse the user behaviour. To perform big data analysis and feature extraction, the call pattern analysis and network behaviour extraction approaches are designed. Several data mining techniques are presented to identify the hidden patterns from CDR data [39]. Further, they have used K-means clustering to categorize the usage statistics, and machine learning algorithms are used to predict the usage of telecom services. Various data mining algorithms are presented to characterize users' communication pattern and their influence on business-related activities [40]. A summary of work carried out in CDR analysis is presented in Table 11.2.

11.6 Mobility and Location Analysis

11.6.1 Predicted Outputs

Various outputs that can be predicted using mobility analysis [41] are shown in Figure 11.7.

11.6.1.1 Moving Direction

The first thing to come in our mind when it comes to user mobility is the moving direction. The user mobility is not purely random but direction-oriented, and it is easy to analyse in which direction the user is heading. The user cell transitions are predicted based on the estimation of user group [42]. In Ref. [43], the moving direction is predicted in order to find which region the user is most likely to go.

TABLE 11.2

Summary of CDR Analysis

Paper	Key Contribution	Predicted Output	Technique
[51]	Reduction of ping-pong effect occurred while reconstructing the user trajectories	Mobility	Hadoop, confidence-based model
[28]	Correlation between calls and user movements	Mobility	Palm calculus
[29]	Extraction of user activity data for trip generation	Distance between traffic zones	K-means, pattern extraction
[30]	Comparison of classification algorithms	Future location	Decision tree, neural networks, Random Forest
[31]	Removal of anomalous data from anomaly-free data	Anomaly	K-means
[32]	Estimation of pairwise influence strength	Social influence	Communication influence model
[33]	Classification of frequent and non-frequent locations	Frequent locations and social interaction	Analytical
[34]	Analysis of transport mode choice	Social influence	Analytical
[35]	Identification of key influence	Social influence	Data cleansing
[36]	Analysis of international voice traffic	Voice traffic	Analytical

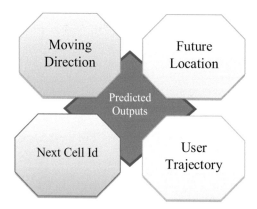

FIGURE 11.7
Mobility analysis-based predicted outputs.

11.6.1.2 Future Locations

One of the challenges nowadays is to provide highly accurate and reliable localization information anywhere and anytime as it directly determines measure of effective resource management. Therefore, information of location is the most important parameter in mobility analysis. Ariffin et al. [44] predicted the places where the users are going. The authors performed resource reservation prior to the actual HO in a long-term evolution (LTE) femto cell network.

11.6.1.3 User Trajectory

A trajectory is the path that a moving object follows through space as a function of time. By analysing the user trajectories, the network can anticipate the need of users on the move. Further, it is possible to reserve radio resource at cells along the path to the destination. A Markov-based model is proposed to forecast the large-distance trajectories of pedestrians. After analysing the trajectories, the authors have predicted the congested trajectories [45].

11.6.1.4 The Next Cell Id

When the users are on the move, they may access different cells, which cause the process of HO. The analysis of the next cell Id which will be visited by user can provide an input for resource management. It is possible to preconfigure bandwidth in the cell.

The mobility and location analysis can be carried out using various approaches [46] such as (i) Markov-based approach, (ii) probabilistic approach, (iii) analytical approach and (iv) pattern matching approach. In the following section, we have presented the summary of work carried out using these approaches [47–50].

11.6.2 Mobility Analysis

Markov-based approach has been adapted by many researchers to predict the user mobility in cellular communication. A user mobility model called as Markov-based hierarchical user mobility model (MB-HUMM) is presented, which adopts two hierarchical structures, i.e. GUM and LUM. These two hierarchies describe whether the mobile user is moving within cell or from a cell to another [51]. The GUM describes the macro mobility of mobile user, which is the mobility among cells, and LUM depicts the micro mobility of user, which is the mobility in cell. Further, the authors have defined the relation parameters that are used to express Gum and LUM. The authors concluded that MB-HUMM is superior to previous models like MC, MT, Markov model and hierarchical location model. The problem of efficient mobility management is focused in Ref. [52]. A model is developed to support the analysis and interference of a statistical pattern for the next service eNodeB prediction. In a given particular time period, the pattern captures the sequential relations between accessed eNodeB for a particular individual. A hidden Markov model (HMM) is proposed to predict the next service eNodeB. HMM is used to investigate the movement of about 2,800 users for more than a week, and accuracy of 53% is achieved with HMM.

A probabilistic approach for user mobility prediction is adapted in Ref. [53]. The authors have presented a probability distribution-format approach, which provides the mobility prediction. The main assistance of using PDF predictor is that it greatly increases services' control. The PDF predictor is evaluated using two predictors, namely HEM (history-based expectation-maximization) and G-Stat. HEM utilizes an expectation-maximization algorithm for prediction by taking account of past history of users' mobility. G-Stat helps HEM for providing better predictions by incorporating GPS trajectory-based predictions. The predictor is compared with Markov predictor, and the authors concluded that the proposed predictor can successfully predict with an average inaccuracy of 2.2%.

A predictive mobility support is used for providing the required QoS [54]. To support guaranteed timed-QoS in pico- and microcellular environment, a predictive and adaptive scheme is presented. The basic components of this framework are (i) a predictive service model, which supports timed-QoS guarantees; (ii) a mobility model to predict mobile's most likely cluster (MLC); and (iii) a call admission control model to verify the feasibility of supporting a call within the MLC. In the proposed scheme, continuity of the calls is ensured while mobile user is moving from one cell to another.

A new location-based scheme called the mobility pattern-based scheme (MPBS) is proposed, in which the time information is incorporated in user profile in order to determine the user's current location by his/her movement state [55]. The location update and paging traffic burden is minimized using the user mobility pattern. The system helps the users engaged in high-priority applications by assigning the resources in advance as the system can predict users' future location. The simulation results show that the MPBS performs better than profile-based scheme (PBS) as the paging delay of the MBPS is 55% less than that of PBS scheme.

A detailed evaluation of procedures that combines mobility prediction with prefetching is presented in Ref. [56] to enhance mobile data offloading. The authors have considered both delay-tolerant and delay-sensitive traffic in the paper. The evaluation is done in terms of offloaded traffic amount, the data transfer delay and the energy consumption. The authors have concluded that the performance of evaluation procedure depends on various factors like the data object size, mobile and Wi-Fi throughput and the number of Wi-Fi hotspots. Performance also depends on robustness of procedures to time and throughput estimation errors [57].

A mobility prediction model is developed for predictive HO management in control/data separation architecture (CDSA) networks [58]. The signalling cost is evaluated in predictive HO management and no-predictive HO management strategies. The mobile users' movement is tracked in terms of HO by the mobility prediction entity (MPE). To control the model's reaction to random and less frequent movements, a trajectory dependence parameter is proposed. Numerical results show that the signalling cost is reduced in predictive HO strategy. A novel analytical model for holistic HO cost evaluation is presented in Ref. [59], where the parameters like signalling overhead, call dropping, latency and radio resource wastage are taken into account. The proposed model is applicable to CDSA; further holistic HO cost is minimized by HO prediction algorithm. The authors proposed two learning-based mobility prediction models stacking LSTM and multilayer perceptron to evaluate the impact of prediction accuracy in order to estimate HO accuracy requirements. The authors concluded that the LSTM performs better than MLP and the developed data-driven and DL-based model can lead to reduced user dissatisfaction, signalling overhead, latency and resource wastage. The summary of the work done is presented in Table 11.3.

11.6.3 Location Analysis

Various LZ-based prediction algorithms are evaluated for location prediction [60]. These algorithms are divided into two independent phases: tree updating and probability calculation. Further, they have evaluated hit rate and resource consumption. The authors concluded that Active LeZi scheme has highest hit rate and PPM algorithm is a

TABLE 11.3

Summary of Mobility Analysis

Paper	Key Contribution	Predicted Output	Technique
[51]	Estimation of macro- and micro-mobility	User mobility	Markov-based hierarchical user mobility model
[52]	Analysis and interference of a statistical pattern for the next service eNodeB	Next service eNodeB	Hidden Markov model
[53]	GPS trajectory-based predictions	User mobility	Probabilistic approach
[54]	Ensuring continuity of calls	Mobile's most likely cluster	Predictive and adaptive scheme
[55]	Allocation of resources is done in advance for high-priority applications	User's current and future location	Mobility pattern-based scheme
[56]	Enhancement of mobile data offloading	Mobility	Analytical
[58]	Handover management	Signalling cost	Mobility prediction entity
[59]	Minimization of holistic HO cost by HO prediction algorithm	Handover	Stacked long short-term memory and multilayer perceptron

best probability calculation method. Bayesian learning-based neural networks are used for mobile user position prediction [61], and based on it, paging technique has been developed. The performance of the proposed model is compared with that of standard neural network techniques, and it is proven that the proposed model delivers better performance. The model is applicable to any arbitrary cell architecture. The authors concluded that the model enhances the mobility management and reduces the total location management cost. In Ref. [62], the authors have proposed neural network techniques to address the location management issue. The work carried out by the authors is purely analytical and based on past movement history of users. A multilayer perceptron is used for mobility management and to reduce the total cost. The past history of user movement is used to train the network for future predictions. The survey of recent advances in location prediction techniques is presented in Ref. [63]. The summary is presented in Table 11.4.

TABLE 11.4

Summary of Location Analysis

Paper	Key Contribution	Predicted Output	Technique
[60]	Evaluation of hit rate and resource consumption	User location	LZ-based algorithm
[61]	Enhancement of the mobility management and reduction of total location management cost	User location	Bayesian learning-based neural networks
[62]	Location management	User location	Neural networks
[63]	Survey of recent advances in location prediction techniques	User location	Analytical

11.7 Network Analysis for Special Parameters

The network traffic analysis also plays an important role in the prediction of different parameters such as

   i. Hotspot detection
   ii. Holiday traffic prediction
   iii. Customer churn prediction
   iv. Fault prediction
   v. Anomaly detection.

11.7.1 Hotspot Detection

As the number of mobile subscribers is increasing rapidly, massive numbers of devices are getting connected to Internet via BSs. These BSs have limited resources such as spectrum and signalling. A BS having a large number of subscribers and high traffic can be considered as hotspot. The identification of the hotspot is carried out by entropy weight-based method. This method helps to calculate the weight of features, and then using value of features, a score to identify hotspot is calculated. The ARMA and Holt-Winters models are used to predict the upload and download traffic amount to identify hotspots. From temporal analysis, the authors concluded that the number of hotspots changes with human activity and hotspots are identified from spatial analysis [64]. In Ref. [65], spatial dependencies and special heterogeneity are detected among various regions. The traffic is also estimated using CDR to balance the load. The authors have constructed a list of weighted neighbours of spatial objects and computed spatial dependencies by implementing different autocorrelation tests.

11.7.2 Holiday Traffic Prediction

Network traffic forecasting plays an important role for efficient congestion management and capacity planning. Solution to the problem of holiday traffic management is presented [66]. A decomposed model is provided which classifies the traffic data into trend, seasonality and holiday components. The holiday data is clustered using modified K-means algorithm. Further, the authors proposed the hybrid holiday traffic prediction algorithm for traffic prediction.

11.7.3 Customer Churn Prediction

Customer churn is the change of service provider, triggered by better services and benefits offered by a competitor. Customer relation is the main concern of telecom companies. By predicting the possibility of customer churn, the service providers can retain the customers by offering better services based on the predictions. The problem of churn detection and prediction is investigated in Ref. [67]; the performance of various state-of-the-art algorithms such as gradient boosting trees, Random Forest, long short-term memory and SVM is compared with that of churn prediction. They describe various types of features such as numerical, categorical and sequential to analyse the role of algorithms.

An advanced data mining methodology is proposed to predict customer churn in prepaid mobile industry [68]. The CDR data is used to reduce the dimensionality of the data, and they have used the principle component analysis algorithm. The authors concluded that the proposed algorithm provides about 99% accuracy for predicting churners as well as no-churners.

Various learning techniques have also been used by researchers to predict the customer churn. A framework is presented which incorporates hybrid learning in it. This hybrid learning method is a combination of genetic programming and tree induction system, which helps to derive the rule for classification based on customer behaviour [69]. The community effect of churn is investigated using game theory techniques. Further, the score is predicted which is nothing but the churn value of a mobile customer. The applications of deep belief networks, autoencoders (AEs) and multilayer feed forward networks are investigated for customer churn prediction [70]. The authors have reported results for predicting customer churn using the multilayer feed forward network. The efficient learning across multiple layers is enabled using the data representation architecture of detailed user behaviour representation. The experimental results show that the proposed models are effective for churn prediction as well as for capturing the complex dependency in the data. A model is proposed which uses machine learning techniques on big data platform [71]. Area under curve (AUC) standard measure is adapted to measure the performance of the model. The authors have used SNA features for making predictions and concluded that SNA helps to enhance the performance of the model.

11.7.4 Fault Prediction

Network faults refer to the partial or complete failure of NEs; these can be broadly classified as outages and malfunctions. Malfunction occurs when active NEs work with some errors, and outage occurs when active NE do not work at all or completely knocked out. Fault prediction delivers several benefits to cellular service providers such as helping in project planning and steering, helping in re-routing of network traffic in case of predicted problems in a route, taking corrective action before fault occurs, decision making, increasing quality of assurance, helping to reduce the operation cost as faults are identified earlier [72].

A Bayesian network is proposed to model the fault prediction. Probability associated with occurrences of faults is evaluated by the Bayesian model. Based on the probability, the predictions are made [73]. However in Ref. [74], proactive cellular network fault automation models are presented, which use mobile intelligent agent (MIA). Different artificial intelligent techniques are used to develop a robust agent that can report anomaly. The MIA delivers the success rate of about 78%.

A framework is presented which automatically detects and diagnoses root causes of faults that occurred in the network [75]. The proposed framework employs unsupervised learning on continuous signal and traffic Key Performance Indicators (KPIs); it is also designed to capture the knowledge on root cause analysis effectively. The results indicate that three specific KPIs, after analysis, can cover almost all faulty records in a GSM network. The proposed framework has proven as a decision support system for experts towards efficient resolution of cellular network faults. In Ref. [76], the authors have presented a proactive approach for failure prediction. For network failure, prediction techniques like SVM regression and multiple neural network variants were utilized. Continuous-time Markov chain (CTMC) analytical model is used along with prediction techniques to provide reliability analysis. The authors concluded that the pattern of these failures is most likely nonlinear.

Decision trees, rules and Bayesian classifiers are presented for the visualization of network faults [77]. Data mining techniques are used to classify optimization criteria based on the KPI metrics to identify network faults. The proposed method supports the most efficient optimization decisions. In Ref. [78], the authors have presented machine learning techniques for cell fault management. The authors have also discussed the different approaches and applications of ML and proposed a standard set of metrics that are used to evaluate faults.

11.7.5 Anomaly Detection

When a cellular network faces certain problem, the user would usually be the first one who realizes the service disruption and suffers the impact. Such an abnormal and disrupted service can be recognized by examining the CDR of the users in a specific area. In Ref. [79], the authors have proposed a CADM. Anomalous behaviour of user movements in a cellular network was detected using CADM along with big data analytics tools. A rule-based approach is used to detect location-based anomalies. The authors presented the advantages of the proposed anomaly detection method, which are as follows: (i) better ways to extract relevant rules without needing a training phase, (ii) fast lifecycle in order to reach target results and (iii) the ability to use modern cost-effective big data analytics technique. Similar work is carried out in Ref. [31], where CDR data is used for anomaly detection. K-means clustering algorithm is used for verifying anomalies. Further artificial neural network model is used to remove the anomalous data from anomaly-free data.

A mobility-based anomaly detection technique is proposed [80] that can effectively identify a group of especially harmful internal attackers. An exponentially weighted moving average (EWMA) is used to analyse user movement and update the user profile. Simulation results show that the proposed detection algorithm can achieve good performance in terms of false alarm rate and detection rate for users. Anomaly detection is combined with location prediction in Ref. [81]. The authors have exploited the location history traversed by a mobile user and anomaly detection using two schemes, namely the Lempel–Ziv, an LZ-based, scheme and Markov-based detection schemes. LZ-based scheme is used for location prediction, and Markov-based scheme is used for anomaly detection. Simulation results demonstrated that the LZ-based detection scheme can achieve more desirable performance in terms of false alarm rate and detection rate.

A data mining-based approach is proposed in Ref. [82] for detecting anomalies. The proposed data mining framework helps in detecting the sleeping cells, caused by RACH failure. The authors concluded that the proposed framework performs better than other methods. Chong Z. and Randy C. proposed a robust deep AE model for anomaly detection. Anomaly regularizing penalty is based upon either 1' or 2'; 1 norm is used by these methods. Resulting optimization problem was dealt with using ideas from proximal methods, backpropagation and the alternating direction of method of multipliers. The Authors demonstrated the effectiveness of proposed approaches and achieved a ~30% improvement over standard AEs.

A one-class neural network (OC-NN) model is proposed to detect anomalies in complex data sets [83]. The ability of deep networks to extract progressively rich representation of data is combined with the one-class objective of creating a tight envelope around normal data using OC-NN. One-class SVM is used to drive the training of the neural network. Experimental results demonstrate that OC-NN outperforms other state-of-the-art DL approaches for anomaly detection. A survey of DL techniques for anomaly detection is presented in Ref. [84]. Various research methods in DL-based anomaly detection along with its application across various domains are discussed in this chapter. The authors have also discussed the challenges in deep anomaly detection and presented several existing solutions to these challenges.

In the next section, we are going to highlight the various application areas of predictive analysis. We are presenting brief introduction to the various fields where predictive analysis can be effectively applicable.

11.8 Predictive Analysis-Enabled Applications

Analysis-based network control in cellular communication is gaining attention of researchers. To provide the QoS to the users, the service providers need to gain the knowledge of trends and patterns of the mobile traffic. Hence, analysing and forecasting the network traffic will help service providers to manage the resources properly and provide the desired QoS to the users. In this section, we have enlightened the various predictive analysis-enabled applications.

11.8.1 Resource Allocation

The aim of resource allocation (RA) in cellular networks is to allot BS bandwidth, and power varies with the density and usage patterns of mobile network subscribers. So for preparing network for sudden significant traffic fluctuations, it is necessary to predict where and when mobile users are using the network. The dynamic nature of traffic load comes from user behaviours such as routine activities and mobility, which has long been regarded as random in wireless system design. With the help of predictive resource allocation (PRA), one can take advantage of the prediction for the traffic load and mobility-related user behaviour.

The prediction of the traffic load for the next few hours will help the network to switch off lightly loaded BSs, thereby reducing the energy consumption. The user trajectories and traffic load can be predicted with various prediction algorithms. By predicting the next cells that a mobile user is likely to enter, QoS of the user can be improved with anticipatory HO [1]. The next eNodeB is predicted [59] using HMM and by analysing user trajectories. PRA framework is presented in Ref. [85] that uses the user rate predictions to improve network and user QoS. The probabilistic model is used [86] to predict the channel condition, and resource allocation vectors are configured accordingly. Predictive resource allocation is studied with end-to-end prediction [87]. A combination of clustering and packing algorithms has been proposed [88]. The proposed algorithm offers 34.8% better QoS than other algorithms. Thus, predictive analysis plays an important role in resource management.

11.8.2 Handover Management

HO assures the continuity of the service provided to the mobile user. But HO schemes in future networks are likely to face many challenges due to spatio-temporal variations of user traffic. The user mobility can cause frequent HOs, which results in high switching latency, thereby greatly degrading the performance of networks. The traditional HO approaches are passive as they lack the knowledge of user's mobility, which leads to extra signal overhead in the HO procedures, and there is a possibility of discontinuity of the service. Hence, predictive analysis provides the context information about the user and the user mobility. This knowledge can make the HO decisions more intelligent, and users can get more comfort in this way compared to traditional RSS-based HO scheme. Mobility prediction-based HO scheme needs to be proposed in future networks. The HO performance is enhanced by evaluating the impact of mobility prediction accuracy. Similarly, a mobility prediction model is developed for predictive HO management in CDSA networks [58].

11.8.3 Location-Based Services

The aim of location-based service (LBS) is to facilitate users with enhanced wireless services, and based on their geographical locations, some information is related to the specific location [46], such as sending target advertisements, local traffic information, instant communication with people nearby and merchant recommendation. To achieve this, we should have the knowledge of users' future location beforehand. Mobility prediction is able to provide future destinations of user, which is helpful in service preconfiguration [47]. Moreover, mobility and location prediction delivers the network with more intelligent and efficient way to manage travellers by traffic forecasting.

11.8.4 Interference Management

APs that use the same resource blocks can generate interference, and it is the main factor affecting the lifting of capacity. It has always been an essential building block to manage interference in the design and performance analysis of networks. Therefore, designing traffic prediction-based resource management scheme to coordinate the interference between different users also makes sense. A model is proposed which takes traffic prediction as an input and computes future interference [89]. The issue of co-channel interference management among small BSs in ultra-dense networks is identified and solution based on the heuristic approach for interference management is discussed in detail [90]. An interference management system for full duplex cellular communication system is proposed [91]. Various approaches like big data are used for interference analysis [92]. Interference aspect is merged with outage analysis in Ref. [93].

11.8.5 Energy Efficiency

One more area that is gaining attention of most of the researchers is green cellular networking, which mainly focuses on the reduction of energy consumption at BSs and utilization of energy in an effective manner. The reduction of energy consumption leads to the development of energy-efficient network architecture. The prediction of the future traffic load helps network/network operators to achieve energy efficiency with various techniques such as

- BS ON/OFF
- Cell breathing
- Cell zooming
- Reducing inter-user interference.

A marginal utility-based BS sleeping spatial adaptive play (MUBS-SAP) algorithm is proposed for BS sleeping mechanism. This algorithm considers user preference and BS load [94]. An adaptive cell zooming method is presented for energy efficiency and optimization purpose. This method analyses the varying traffic patterns and interferences, and reduces the number of active cells during the periods of low traffic [95]. A cell zooming algorithm is proposed, which is based on load balancing (LB) to improve EE [96]. This LB algorithm balances the traffic load of small and heavily loaded BSs by dispersing the traffic neighbouring BSs with light load. Shah et al. [97] have focused on the reduction of power consumption at BS. The energy consumption is reduced by reducing the size of data, which is to be transmitted. A power control algorithm is proposed with mobility prediction. The proposed algorithm helps to predict the user mobility, and remote radio head switching is carried out based on predictions [98].

11.9 Deep Learning in Predictive Analysis of Cellular Networks

In previous sections, we have discussed the analysis techniques for particular applications. In cellular networks, most of the times it is not feasible to apply a single well-defined model; also it is not possible always to apply a general purpose model even if the network characteristics are similar. In case of general purpose model, usually the size of training data, estimation and optimization steps become crucial and lead to uncountable different models with varying parameters for different applications. In few scenarios, a single well-defined model becomes incapable to draw accurate predictions; this leads to the requirement of hybrid models. Such hybrid models increase the complexity and required time for training. This issue leads to the incorporation of learning into the networks along with the application of network analysis techniques such as machine learning, DL etc. An area of machine learning that exhibits the nature of biological nervous system and performs learning through multilayer representations is DL. These multiple layers are composed of multiple linear and nonlinear transformations [1]. In the scenarios where the data size is increased and various parameters are involved, DL techniques have been given better results among all the existing machine learning techniques. Convolutional neural networks (CNNs), deep belief networks, recurrent neural networks (RNNs) and deep neural networks are some of the DL architectures and widely applied in the area of pattern recognition, computer vision and natural language processing [99–101].

11.9.1 Deep Learning State-of-the-Art

In this section, we have presented the brief introduction to DL state-of-the-art.

11.9.1.1 Convolutional Neural Networks

CNNs are the variants of multilayer perceptron and inspired by the biological process of visualization. Minimum amount of pre-processing is done in CNN. To capture the correlations between different data regions, CNNs employ a set of locally connected filters instead of full connections between layers. The representation of fully connected CNN is shown in Figure 11.8.

In CNN architecture, neurons are encompassed in three dimensions (width, height, and depth). In every layer, the 3D input volume is transformed to 3D output volume of neuron activations. CNN architecture consists of the following important features:

a. **Convolutional Layer:** This layer consists of learnable filters and is considered as the core building block of CNN. The network learns from these filters when the filters get activated. And the filters get activated when they encounter some specific type of features at some spatial input position.

b. **Local Connectivity**: In CNN, local connectivity indicates that the neuron of a layer is connected only to the neighbouring neurons of adjacent layers. This avoids the problem of connectivity of neurons when working with a high volume of input.

c. **Parameter Sharing**: Parameter sharing is followed in CNN at the convolutional layer to control the free parameters. For less parameter optimization and a faster convergence, weight vector and bias are shared among neurons.

d. **Pooling Layer**: Pooling is an important concept in CNN and is a form of nonlinear down-sampling. The input data is partitioned into a set of nonoverlapping slices and produces a maximum output for each set in this process. This helps in avoiding the problem of overfitting, which is an important issue with neural networks.

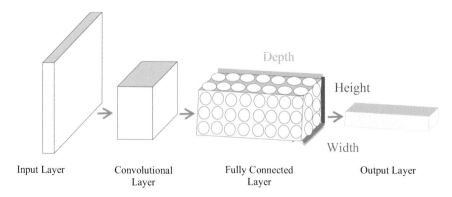

FIGURE 11.8
Convolutional neural networks.

11.9.1.2 Recurrent Neural Networks

RNN is the network that summons up the past, and the decisions are influenced by what it has learnt from the past. Like feed forward networks, RNNs learn similarly while training, but also they remember things learnt from prior input(s) when generating output(s). In RNNs, one or more input vectors are taken to produce one or more output vectors, and the outputs are influenced by weights applied on inputs like a regular NN, as well as by a "hidden"-state vector representing the context based on prior input(s)/output(s). So depending on previous inputs in the series, the same input could produce a different output (Figure 11.9).

11.9.1.3 Deep Belief Networks

The probabilistic generative models that are composed of multiple layers of stochastic latent variables are known as deep belief networks. These models are basically constructed from many layers of restricted Boltzmann machines (RBMs). An associative memory is formed by the top two layers, which have directed and symmetric connections between them. The upper layers provide connections to the lower layers. An efficient procedure is carried out for learning by degenerative weights that specify how the variables in one layer determine the probabilities of the variables in lower layer. Variables in every layer can be inferred by a single pass after learning multiple layers (Figure 11.10).

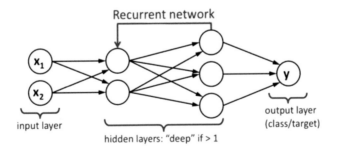

FIGURE 11.9
Representation of recurrent neural networks [99].

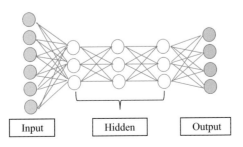

FIGURE 11.10
Representation of deep belief networks.

11.9.1.4 Autoencoders

AEs are a special type of feed forward neural networks designed for unsupervised learning. In AE, the input is the same as the output. The input is constructed into a lower-dimensional *code*, and then, the output is reconstructed from this representation. This lower-dimensional code is a compression of the input and called as the latent-space representation. AEs are usually used to learn compact representation of data for dimension reduction. An AE consists of three modules: encoder, code and decoder. The encoder compresses the input and generates the code; the decoder then reconstructs the input using the same code.

To build an AE, three things are necessary: an encoding method, decoding method and a loss function to compare the output with the target. The representation of AE is shown in Figure 11.11.

11.9.1.5 Long Short-Term Memory

LSTM is an exclusive type of RNN, which is capable of learning long-term dependencies and is useful for certain types of prediction required for the network to retain information over longer time periods. This is a task that traditional RNNs struggle with. Gated cells are used in LSTMs to store information outside the regular flow of the RNN. The network can manipulate the information in many ways, including storing information in the cells and reading from them with the help of these gates. The cells are individually proficient of making decisions regarding the information, and the decisions are executed by opening or closing the gates.

The architecture of LSTM permits it to contain information for longer time periods, and removes the issues of traditional RNNs. The three major parts of the LSTM include forget gate, input gate and output gate. The function of forget gate is to remove information which is no longer necessary for the task completion. This process is essential for optimizing the performance of network. The input gate is responsible for adding information to the cells. And lastly, the output gate is used to select and output the necessary information.

In the next subsection, we have tried to bridge the gap between predictive analysis and DL by discussing the work carried out in this area. We start our discussion with work carried out in time-series analysis using DL; further, we have highlighted the work carried out in CDR and mobility analysis using DL. At the end of the section, we have presented the summary of cross section of predictive analysis and DL in a tabular form.

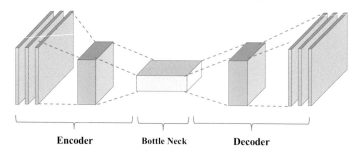

FIGURE 11.11
Representation of autoencoders.

11.9.2 Deep Learning-Based Time-Series Analysis

DL is an emerging technique for traffic forecasting. Nowadays, network traffic forecasting with NN and DL is becoming widespread as they provide different approach than traditional stochastic modelling. They are becoming useful for pattern detection and identification of structures in input data, learning through much iteration and with the help of this learning coming up with certain predictions. In the current scenario, NNs and DL techniques are widely used as the complex relationships in data are captured but in comparison with AR models, the data required for training are much higher. In this section, we have presented the different DL techniques of traffic forecasting.

In Ref. [102], the authors have identified spatial dependency among cell towers. To characterize the spatial dependency among cell towers, the cellular traffic is decomposed into in-tower and inter-tower traffic. In this work, the authors have jointly considered temporal and spatial dependency among cell towers and exploited graph neural network model for accurate cellular traffic prediction. They have compared the proposed model with NAIVE, ARIMA, LSTM and HW, and found that the proposed model performs better and offers lower prediction errors.

To achieve capacity optimization, a cluster of BSs with complementary traffic patterns is mapped to a data processing unit in C-RAN. This helps processing units to be utilized fully in different periods of time [102]. To learn the temporal dependency and spatial correlation among BS traffic patterns, a multivariate long short-term memory model is used, which helps to make accurate traffic forecast for future period of time. To optimize capacity utility and deployment cost, further authors have proposed a distance-constrained complementarity aware (DCCA) algorithm to achieve optimal BS clustering schemes. The authors used the proposed model for datasets of cities, i.e. Milan and Trentino, and found that capacity utility is increased to 83.4% and 76.7%, respectively.

A deep belief network architecture and compressive sensing method is presented in Ref. [103] to predict the traffic for wireless mesh backbone networks. A discrete wavelet transform (DWT) is used to divide the traffic into low-pass and high-pass components. The LRD of network traffic is expressed by low-pass component, and deep belief network is used to learn LRD. However, irregular fluctuations of traffic are expressed by high-pass component, and STCS method is used to analyse such traffic.

A hybrid DL model for spatio-temporal prediction is proposed in Ref. [104]. For spatial modelling, an AE-based model is used. The proposed model that consists of global stacked autoencoder (GSAE) and local stacked autoencoder (LSAE) allows users to capture common characteristics shared by different cells and specific location-dependent characteristics of a specific cell. Further authors have presented the LSTMs for the prediction of traffic. The proposed model is compared with ARIMA and SVR using MSE, MAE and log loss metrics.

A network-wide mobile traffic forecasting is carried out using spatio-temporal neural network (STN) architecture [105]. The model exploits the important correlations between user traffic patterns at different locations and times. ConvLSTM is used to model the long-term trends, and 3D-convNet structure is used to model the short-term variations of traffic. A multilayer perceptron is used for the prediction of future mobile traffic. The proposed model is compared with ARIMA and HW-Exps and found to be performing better than these two models.

DL was also employed in Refs. [106–109], where the authors employed CNNs, graph sequence and LSTMs to perform mobile traffic forecasting. Their proposals gain significantly higher accuracy than traditional approaches, such as ARIMA by effectively extracting spatio-temporal features.

11.9.3 Deep Learning-Based CDR Analysis

CDR analysis involves the extraction of knowledge from certain instances of telecommunication transactions such as phone number, session start/end time, cell ID, traffic consumption, etc. Using DL for CDR data analysis can serve a variety of functions. Liang et al. propose RNN-based approach to estimate metro density from streaming CDR data [110]. The trajectory of a mobile phone user is used as a sequence of locations; RNN models work well in handling such sequential data. Similarly, Felbo et al. use CDR data for studying demographics [111]. They use a CNN-based model to predict the age and gender of mobile users, and they demonstrate the superior accuracy of these structures over other machine learning tools. Chen et al. compare various models for predicting tourists' next locations of visit by analysing CDR data [112]. The experimental results suggest that RNN-based predictors significantly outperform traditional ML methods, including Naive Bayes, SVM, RF and MLP.

11.9.4 Deep Learning-Based Mobility & Location Analysis

Since DL is becoming a powerful tool for mobility analysis, it is able to capture spatial dependencies in sequential data. A DL approach for trajectory prediction is studied in Ref. [113]. The framework can perform multitask learning on both social networks and mobile trajectories modelling by sharing representations learned by RNN and gate recurrent unit (GRU). The DL is first used to reconstruct social network representations of users; subsequently to learn patterns of mobile trajectories with different time granularity, RNN and GRU models are employed. Importantly, the overall architecture is tightened and the efficient implementation is enabled as these two components jointly share the representations learned.

An online learning scheme is proposed to train a hierarchical CNN architecture, which allows model parallelization for data stream processing [114]. They proposed a framework called "DeepSpace" to predict individuals' trajectories with much higher accuracy as compared to naive CNNs by analysing usage records. In Ref. [115], to predict trajectories of individuals using mobile phone data, the authors design a neural Turing machine. The neural Turing machine holds two major components: a memory module and a controller. A memory module stores the historical trajectories, and a controller manages the "read" and "write" operations over the memory. Experimental results show that the proposed architecture attains superior generalization over stacked RNN and LSTM; also more accurate trajectory prediction is performed than the k-nearest neighbour and n-grams methods.

The mobility analysis is focused at a larger scale rather than focusing on individual trajectories [116]. LSTM networks are proposed for mobility analysis. These networks are used to model the city-wide movement patterns of a large group of people and vehicles jointly. This multitask architecture exhibits superior prediction accuracy over a standard LSTM. In Ref. [117], the authors have researched city-wide mobile patterns by architecting deep spatio-temporal residual networks. These networks forecast the movements of crowds. RNN-based models are used to capture the unique characteristics of spatio-temporal correlations associated with human mobility; three ResNets are constructed to extract nearby and distant spatial dependencies within a city. Temporal features are learned by the scheme and desired representations extracted by all models for the final prediction. The proposed scheme achieves the highest accuracy among all DL and non-DL methods studied. An RNN-based model is employed in Ref. [118], for short-term urban mobility forecasting on a huge dataset collected from a real-world deployment. The RNN model delivers superior accuracy over the Markovian approaches and n-gram.

Human movement chains are generated from cellular data, to support transportation planning [119]. Firstly, an input-output HMM is employed, which helps to label activity profiles for CDR data pre-processing. Further, an LSTM is designed for activity chain generation. Then using the generative model, urban mobility plans are synthesized and the simulation results reveal sound accuracy.

RNN-based model is presented for 24-h mobility prediction system [120]. For each hour, dynamic region of interests (ROIs) are employed to discover through divide-and-merge mining from raw trajectory database, leading to high prediction accuracy. To capture the complicated sequential transitions of human mobility, attention mechanisms are incorporated on RNN [121]. Heterogeneous transition regularity and multilevel periodicity are combined to provide up to 10% of accuracy improvement over the other state-of-the-art forecasting models.

The mobility of mobile devices in mobile ad hoc networks is predicted using MLP model, with previously observed knowledge of pause time, speed and movement direction [122]. Simulation results show that the proposed model achieves high prediction accuracy. An MLP model is also proposed in which high prediction accuracy is achieved by modelling the relationship between human mobility and personality [123]. Groups of similar trajectories are discovered to facilitate higher-level mobility-driven applications using RNNs [124]. A sequence-to-sequence AE is embraced to learn fixed-length representations of mobile users' trajectories. Experimental results show that the method can effectively capture spatio-temporal patterns in real dataset as well as synthetic datasets.

To understand the correlation between human mobility and traffic accidents, Chen et al. [125] combine GPS records and traffic accident data. The authors designed a stacked de-noising AE to learn a compact representation of the human mobility, and further, it is used to predict the traffic accident risk. Results show that the proposed method can deliver accurate, real-time prediction across large regions. Song et al. employed DBNs to predict mobility in natural disaster and simulate human emergency behaviour, based on learning from GPS records of 1.6 million users [126]. The proposed method can accurately predict different disaster scenarios such as tsunamis, earthquakes and nuclear accidents. GPS data is also utilized for mobility analysis in Ref. [127], where the potential of employing DL for urban traffic prediction using mobility data is studied.

11.10 Emerging Intelligent Networks

Incorporation of intelligence and autonomous adaptability into future cellular networks is the utmost need of today's era for the requirement to meet the needs of both users and operators in a cost-effective way. The networks that exhibit such properties are defined as self-organizing networks (SONs). Such networks continuously interact with the environment next to them [128]. Based on their continuous interaction with environments, SONs are able to decide when and how certain actions will be triggered and they are able to improve their performance by learning the previous actions taken by the system. SONs are mainly motivated by the following factors:

- Due to the varying nature of the wireless channel and mobility of users, systems face the issues of the under-utilization of resources resulting in low resource efficiency or over-utilization, which results in congestion and poor QoS, at varying times and locations.
- The system is facing longer recovery and restoration time as the increased complexity of systems is leading to greater human errors.
- SONs can eliminate the need for expensive skilled labour for configuration, commissioning, optimization, maintenance, troubleshooting and recovery of the system.

11.10.1 Characteristics of SON

11.10.1.1 Scalability

Scalability means that the system should remain operational even if a reasonable number of entities are added or removed from the system. It means when the scale or size of the system increases, the complexity of the system should not increase. Hence, to ensure the scalability networks should follow the same approach. In order to ensure that the system is adaptive and scalable, the two main factors that should be considered in the design process are minimal complexity and local cooperation.

11.10.1.2 Stability

Stability in the context of SO is an adaption mechanism or algorithm, which consistently passes through a finite number of states within acceptable finite time. This means that for a specific task, the algorithm should transition from its current state to any desired future state within a finite and feasible time frame.

11.10.1.3 Agility

Agility indicates how flexible or acutely responsive the algorithm is in its adaptation to the changes in its operational environment. In order to be self-organizing, algorithms along with carrying a capability to adapt and cope with its changing environment (stability) should also not be sluggish in its adaptation (agility). Subsequently, the algorithm should not be over-reactive to temporary changes in the system.

11.10.2 Classes of SON

The perception of SON can be divided into three classes: self-configuration, self-optimization and self-healing. The same is depicted in Figure 11.12.

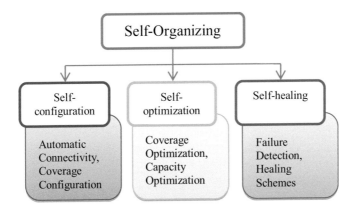

FIGURE 11.12
Classes of self-organizing networks.

11.10.2.1 Self-Configuration

The process of automatically configuring all parameters of network equipment can be defined as self-organization. During deployment/extension/upgrade of network terminals, configuration of BSs/eNodeBs (eNBs), relay stations (RS) and Femtocells is required. In future systems, the conventional configurations are also required when there is a failure of a node, a drop in network performance or a change is required in service type [129]. Also, it is necessary to replace the process of manual configuration with self-configuration. The fast installation and deployment of future evolved NodeBs (eNBs) is enabled by self-configuration capability, which reduces human involvement and deployment time. Thus, self-configuration is particularly useful at the preoperational stage of a wireless mobile network. The nodes in future cellular networks should likely be able to self-configure all of its initial parameters, including IP addresses, neighbour lists and radio access parameters [130].

11.10.2.2 Self-Optimization

To ensure efficient performance of the system and to maintain all its optimization objectives, it is necessary to continuously optimize system parameters after the initial configuration phase. The self-optimization concept basically enlightens three application areas: LB, capacity and coverage and interference control.

11.10.2.2.1 Self-Optimization for Load Balancing

To mitigate the effects of natural spatio-temporally varying user distributions and traffic, there is a severe need for LB mechanisms in cellular networks. The LB concept is further divided into three prominent areas: resource adaption, coverage adaption and traffic shaping.

Resource Adaption is adapting the amount of resource allocated to a cell in order to match it to the offered traffic load to that cell. In case cell is overloaded, it can borrow channels from other cells which are less loaded or from free channels.

Coverage Adaption is a mechanism of changing the effective coverage area of the cell to match the traffic offered with the available resources. This coverage adaption can be achieved either through power adaptation, through antenna adaptation or by hybrid techniques.

Traffic Shaping is a mechanism that matches the offered traffic load with available resources either through pre-emptive and strategic admission control or through forced HO of the ongoing calls.

11.10.2.2.2 Self-Optimization for Capacity and Coverage

The capacity and coverage optimization is achieved through relaying mechanism. Relaying is a means of delivering cost-effective throughput enhancement and coverage optimization. When the RS is deployed, it is desired to meet the two main objectives, i.e. capacity optimization and coverage extension.

11.10.2.2.3 Self-Optimization for Interference Control

Diverse ways are provided by self-optimization to control interference, thereby improving capacity. The simplest approach would be to switch off cells or to put them into idle mode when not in use. Interference troll is further divided into long-term intercell interference coordination (ICIC) and short-term ICIC. Long-term ICIC is achieved using integer frequency reuse and fractional frequency reuse. Short-term ICIC is achieved using dynamic ICIC.

The summary of self-optimization schemes is shown in Figure 11.13.

11.10.2.3 Self-Healing

In wireless cellular system, the network component malfunction or natural disasters can cause faults and failures in the system. These failures can be hardware-oriented or

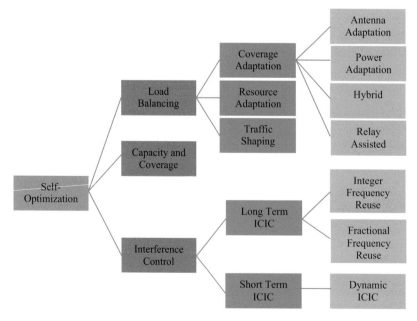

FIGURE 11.13
Summary chart of self-optimization [128].

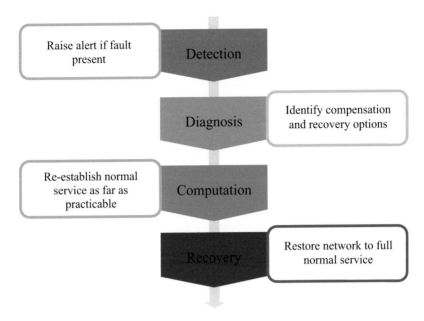

FIGURE 11.14
Self-healing life cycle.

software-related. The process of detecting and resolving these faults or failure manually is time-consuming. In some cases, some faults and failure remain undetected and cannot be addressed until the customer lodges a formal complaint. In SONs, this process is expected to be enhanced by incorporating self-healing function. Self-healing process involves the remote detection and diagnosis and providing compensation or recovery actions to diminish the effect of faults in the networks equipment [131–133].

Self-healing process can be logically divided into three phases: monitoring, diagnosis and compensation. The self-healing system continuously monitors the network for any anomaly or if any faults in order to satisfy the self-healing conditions. In case such conditions are detected, the analysis of data is carried out using intelligent or expert system to specify the type of fault. The optimum compensation action to be taken and neighbours that are required to compensate are recommended by the self-healing system. The self-healing life cycle is represented in Figure 11.14.

11.10.3 Applications of SON

11.10.3.1 Coverage and Capacity Optimization

In 3GPP, coverage and capacity optimization techniques are currently under study and aim to provide continuous coverage and optimal capacity of the network. The performance of the network can be obtained via key measurement data, and adjustments can then be made to improve the network performance. For example, call drop rates will give an initial indication of the areas within the network that have insufficient coverage and traffic counters; this can be used for the identification of capacity problems. The network can optimize the performance by trading off capacity and coverage based on these measurements.

11.10.3.2 Mobility Robustness Optimization

The primary aim of mobility robustness is to reduce the number of handover-related radio link failures by optimally setting the handover parameters. Secondarily, it aims at avoiding the ping-pong effect or prolonged connection to a non-optimal cell.

11.10.3.3 Mobility Load Balancing

LB is related to mobility robustness, which targets to optimize the reselection of cell and HO parameters to deal with uneven traffic loads. The goal of the study is to achieve this while minimizing the number of HOs and redirections needed to achieve the LB.

11.10.3.4 RACH Optimization

In order to improve the access to the system, RACH optimization such as RACH load and the uplink interference have been proposed to optimize the system parameters based upon monitoring the network conditions. The objective is to minimize the RACH load and the access delays for all the UEs in the system.

11.11 Conclusion

Predictive analysis plays an increasingly important role in the mobile and wireless networking domain. In this chapter, we provide a wide-ranging survey of recent work carried out for the analysis of various parameters and traffic forecasting, which shows that the time-series analysis is carried out to forecast the future traffic. This helps to identify the BSs with low and high amount of traffic. CDR data is useful for predicting the various parameters such as user mobility and location, anomaly, social influence, voice traffic etc. The user mobility analysis helps to predict moving direction of a user, possible future location of a user, user trajectories etc.

We summarize both basic models and advanced DL-based models for network analysis. Further, we review the different application areas of predictive analysis, i.e. dynamic resource allocation, HO management, LBS, interference management and energy efficiency, which in turn will help to fulfil the desired quality of experience of a user. We discuss the role of DL models in the predictive analysis of cellular networks. Further, we present the emerging intelligent networks and their application areas. We conclude our paper by pinpointing the relevance of emerging intelligent networks in predictive analysis, which may lead to valuable future research results. Our aim is that this chapter will become a definite guide to researchers and practitioners interested in working in the area of predictive analysis of cellular networks.

References

1. C. Zhang, P. Patras, and H. Haddadi, Deep learning in mobile and wireless networking: A survey, *IEEE Communications Surveys & Tutorials*, vol. 21, no. 3, pp. 2224–2287, Third Quarter, 2019. doi: 10.1109/COMST.2019.2904897.

2. Z. Fadlullah, F. Tang, B. Mao, N. Kato, O. Akashi, T. Inoue, and K. Mizutani, State-of-the-art deep learning: Evolving machine intelligence toward tomorrow's intelligent network traffic control systems. *IEEE Communications Surveys & Tutorials*, 19(4): 2432–2455, 2017.
3. J. S. Marcus, The economic impact of internet traffic growth on network operators, 2014, WIK-Consult Google Inc., Bad Honnef, Germany, October 2014. [Online] Available: http://www.wik.org/ uploads/media/Google_Two-Sided_Mkts.pdf.
4. M. Barabas, G. Boanea, and V. Dobrota, Multipath routing management using neural networks-based traffic prediction, *In Proceedings of the 3rd International Conference on Emerging Networks and Systems Intelligence*, Lisbon, Portugal, November 2011, pp. 118–124.
5. A. Erramilli, M. Roughan, D. Veitch, and W. Willinger, Self-similar traffic and network dynamics, *Proceedings of the IEEE*, vol. 90, no. 5, pp. 800–819, 2002.
6. K. Park and W. Willinger, *Self-Similar Network Traffic and Performance Evaluation*, 1st ed. New York: John Wiley & Sons, Inc., 2000.
7. T. Karagiannis, M. Molle, M. Faloutsos, and A. Broido, A nonstationary poisson view of internet traffic, *Proceedings of the IEEE INFOCOM 2004*, vol. 3, pp. 1558–1569, 2004.
8. J. Cao, W. S. Cleveland, D. Lin, and D. X. Sun, On the nonstationarity of Internet traffic, *In Proceedings of the 2001 ACM SIGMETRICS international conference on Measurement and modeling of computer systems (SIGMETRICS '01)*. Association for Computing Machinery, New York, pp. 102–112, 2001. doi: 10.1145/378420.378440.
9. R. Morris and D. Lin, Variance of aggregated web traffic, *In Proceedings of the IEEE INFOCOM 2000. Conference on Computer Communications. Nineteenth Annual Joint Conference of the IEEE Computer and Communications Societies*, Tel Aviv, Israel, vol. 1, pp. 360–366, March 2000.
10. M. S. Taqqu, V. Teverovsky, and W. Willinger, Is network traffic self-similar or multifractal? *Fractals*, vol. 5, no. 1, pp. 63–73, 1997. doi: 10.1142/S0218348X97000073.
11. M. Grossglauser and J. Bolot, On the relevance of long-range dependence in network traffic, *IEEE/ACM Transactions on Networking*, vol. 7, no. 5, pp. 629–640, 1999.
12. T. Karagiannis, M. Molle, and M. Faloutsos, Long-range dependence ten years of internet traffic modeling, *IEEE Internet Computing*, vol. 8, no. 5, pp. 57–64, 2004.
13. J. A. Cadzow, ARMA time series modeling: An effective method, *IEEE Transactions on Aerospace and Electronic Systems*, vol. AES-19, no. 1, pp. 49–58, 1983.
14. G. E. P. Box, *Time Series Analysis: Forecasting and Control*. New York: John Wiley & Sons, 2015.
15. O. W. W. Yang, Traffic prediction using FARIMA models, *1999 IEEE International Conference on Communications*, Vancouver, BC, Canada, pp. 3–7, 1999.
16. H. Chen, J. Wu, and S. Gao, A study of autoregressive conditional heteroscedasticity model in load forecasting, *In Proceedings of the 2006 International Conference on Power System Technology*, Chongqing, China, October 2006, pp. 1–8.
17. R. F. Engle and K. F. Kroner, Multivariate simultaneous generalized ARCH, *Econometric Theory*, vol. 11, no. 1, pp. 122–150, 1995. [Online] Available: http://www.jstor.org/stable/3532933.
18. R. F. Engle and J. R. Russell, Autoregressive conditional duration: A new model for irregularly spaced transaction data, *Econometrica*, vol. 66, no. 5, pp. 1127–1162, 1998. [Online]. Available: https://ideas.repec.org/a/ecm/emetrp/v66y1998i5p1127-1162.html.
19. J. Ding, V. Tarokh, and Y. Yang, Bridging AIC and BIC: A new criterion for autoregression, *IEEE Transactions on Information Theory*, vol. 64, no. 6, pp. 4024–4043, 2018.
20. B. Zhou, D. He, Z. Sun, and W. Ng, Network traffic modeling and prediction with ARIMA/GARCH, *In Proceedings of theHET-NETs' 06 Conference*, 2006.
21. N. C. Anand, C. Scoglio, and B. Natarajan, GARCH - Non-linear time series model for traffic modeling and prediction, *In Proceedings of the IEEE/IFIP Network Operations and Management Symposium: Pervasive Management for Ubiquitous Networks and Services, NOMS 2008*, Salvador, Brazil, 2008, pp. 694–697.
22. Y. Yu, J. Wang, M. Song, and J. Song, Network traffic prediction and result analysis based on seasonal ARIMA and correlation coefficient, *in Proceedings of the 2010 International Conference on Intelligent System Design and Engineering Application, ISDEA 2010*, Changsha, China, vol. 1, no. 1, IEEE, 2011, pp. 980–983.

23. G. E. Box and D. R. Cox, An analysis of transformations, *Journal of the Royal Statistical Society: Series B (Methodological)*, vol. 26, no. 2, pp. 211–243, 1964.
24. B. Vujicic, and L. Trajkovic, Prediction of traffic in a public safety network, *in IEEE International Symposium on Circuits and Systems*, Kos, Greece, May 2006, p. 4.
25. H. M. El Hag and S. M. Sharif, An adjusted ARIMA model for internet traffic, 2007.
26. X. Wang, Z. Zhou, F. Xiao, K. Xing, Z. Yang, Y. Liu, and C. Peng. Spatio-temporal analysis and prediction of cellular traffic in metropolis. *IEEE 25th International Conference on Network Protocols (ICNP)*, 2017, pp. 1–10, doi: 10.1109/ICNP.2017.8117559.
27. W. Sun, D. Miao, X. Qin, and G. Wei, Characterizing user mobility from the view of 4G cellular network, *2016 17th IEEE International Conference on Mobile Data Management (MDM)*, Porto, 2016, pp. 34–39. doi: 10.1109/MDM.2016.19.
28. Y. Leo, A. Busson, C. Sarraute, and Y. E. Fleury, Call detail records to characterize usages and mobility events of phone users. *Computer Communications*, vol. 95, pp. 43–53, 2016.
29. X. Wang, H. Dong, Y. Zhou, K. Liu, L. Jia, and Y. Qin, Travel distance characteristics analysis using call detail record data, *2017 29th Chinese Control And Decision Conference (CCDC)*, Chongqing, 2017, pp. 3485–3489. doi: 10.1109/CCDC.2017.7979109.
30. N. C. Chen, W. Xie, R. E. Welsch, K. Larson, and J. Xie. Comprehensive predictions of tourists' next visit location based on call detail records using machine learning and deep learning methods. *In Proceedings of IEEE International Congress on Big Data (BigData Congress)*, Honolulu, HI, USA, pp. 1–6, 2017.
31. K. Sultan, H. Ali, and Z. Zhang, Call detail records driven anomaly detection and traffic prediction in mobile cellular networks, *IEEE Access*, vol. 6, pp. 41728–41737, 2018. doi: 10.1109/ACCESS.2018.2859756.
32. J. Li, M. Yeh, M. Chen, and J. Lin, Modeling social influences from call records and mobile web browsing histories, *2015 IEEE International Conference on Big Data (Big Data)*, Santa Clara, CA, 2015, pp. 1357–1361. doi: 10.1109/BigData.2015.7363895.
33. M. Picornell, T. Ruiz, M. Lenormand, J. J. Ramasco, T. Dubernet, and E. Frias-Martinez. Exploring the potential of phone call data to characterize the relationship between social network and travel behavior. *Transportation*, 42, 2015. doi: 10.1007/s11116-015-9594-1.
34. S. Phithakkitnukoon, T. Sukhvibul, M. Demissie, et al. Inferring social influence in transport mode choice using mobile phone data. *EPJ Data Science*, vol. 6, p. 11, 2017. doi: 10.1140/epjds/s13688-017-0108-6.
35. N. Werayawarangura, T. Pungchaichan, and P. Vateekul, Social network analysis of calling data records for identifying influencers and communities, *2016 13th International Joint Conference on Computer Science and Software Engineering (JCSSE)*, Khon Kaen, 2016, pp. 1–6. doi: 10.1109/JCSSE.2016.7748864.
36. Z. Aziz and R. Bestak, Analysis of call detail records of international voice traffic in mobile networks, *2018 Tenth International Conference on Ubiquitous and Future Networks (ICUFN)*, Prague, 2018, pp. 475–480. doi: 10.1109/ICUFN.2018.8436669.
37. S. B. Elagib, A. A. Hashim, and R. F. Olanrewaju, CDR analysis using big data technology, *2015 International Conference on Computing, Control, Networking, Electronics and Embedded Systems Engineering (ICCNEEE)*, Khartoum, 2015, pp. 467–471. doi: 10.1109/ICCNEEE.2015.7381414.
38. D. Jiang, Y. Wang, Z. Lv, S. Qi, and S. Singh, Big data analysis based network behavior insight of cellular networks for industry 4.0 applications, *IEEE Transactions on Industrial Informatics*, vol. 16, no. 2, pp. 1310–1320, 2020. doi: 10.1109/TII.2019.2930226.
39. M. Ved and B. Rizwanahmed, Big data analytics in telecommunication using state-of-the-art big data framework in a distributed computing environment: A case study, *2019 IEEE 43rd Annual Computer Software and Applications Conference (COMPSAC)*, Milwaukee, WI, 2019, pp. 411–416. doi: 10.1109/COMPSAC.2019.00066.
40. A. Bascacov, C. Cernazanu, and M. Marcu, Using data mining for mobile communication clustering and characterization, *2013 IEEE 8th International Symposium on Applied Computational Intelligence and Informatics (SACI)*, Timisoara, 2013, pp. 41–46. doi: 10.1109/SACI.2013.6609004.

41. H. Zhang and L. Dai, Mobility prediction: A survey on state-of-the-art schemes and future applications, *IEEE Access*, vol. 7, pp. 802–822, 2019. doi: 10.1109/ACCESS.2018.2885821.

42. N. P. Kuruvatti, A. Klein, J. Schneider, and H. D. Schotten, Exploiting diurnal user mobility for predicting cell transitions, *In Proceedings of IEEE Globecom Workshops (GC Wkshps)*, Atlanta, GA, USA, December 2013, pp. 293–297.

43. B. Li, H. Zhang, and H. Lu, User mobility prediction based on Lagrange's interpolation in ultra-dense networks, *In Proceedings of IEEE 27th Annual International Symposium on Personal, Indoor and Mobile Radio Communications (PIMRC)*, September 2016, pp. 1–6.

44. N. A. Amirrudin, S. H. S. Ariffin, N. N. N. A. Malik, and N. E. Ghazali, Mobility prediction via Markov model in LTE femtocell, *International Journal of Computer Applications*, vol. 65, no. 18, pp. 40–44, 2013.

45. M. Karimzadeh, F. Gerber, Z. Zhao, and T. Braun, Pedestrians trajectory prediction in Urban environments, *2019 International Conference on Networked Systems (NetSys)*, Munich, Germany, 2019, pp. 1–8. doi: 10.1109/NetSys.2019.8854506.

46. H. Si, Y. Wang, J. Yuan, and X. Shan, Mobility prediction in cellular network using hidden Markov model, *2010 7th IEEE Consumer Communications and Networking Conference*, Las Vegas, NV, 2010, pp. 1–5. doi: 10.1109/CCNC.2010.5421684.

47. Y. Zhang, User mobility from the view of cellular data networks, *IEEE INFOCOM 2014-IEEE Conference on Computer Communications*, Toronto, ON, 2014, pp. 1348–1356. doi: 10.1109/INFOCOM.2014.6848068.

48. H. Tabassum, M. Salehi, and E. Hossain, Mobility-aware analysis of 5G and B5G cellular networks: A tutorial. arXiv:1805.02719, 2018.

49. Y. Ye, M. Xiao, Z. Zhang, and Z. Ma, Performance analysis of mobility prediction based proactive wireless caching, *2018 IEEE Wireless Communications and Networking Conference (WCNC)*, Barcelona, 2018, pp. 1–6. doi: 10.1109/WCNC.2018.8377018.

50. S. Danafar, M. Piorkowski, and K. Krysczcuk, Bayesian framework for mobility pattern discovery using mobile network events, *2017 25th European Signal Processing Conference (EUSIPCO)*, Kos, 2017, pp. 1070–1074. doi: 10.23919/EUSIPCO.2017.8081372.

51. W. Wang, Y. Cao, D. Li, and Z. Qin, Markov-based hierarchical user mobility model, *2007 Third International Conference on Wireless and Mobile Communications (ICWMC'07)*, Guadeloupe, 2007, pp. 47–47. doi: 10.1109/ICWMC.2007.52.

52. Q. Lv, Z. Mei, Y. Qiao, Y. Zhong, and Z. Lei, Hidden Markov model based user mobility analysis in LTE network, *2014 International Symposium on Wireless Personal Multimedia Communications (WPMC)*, Sydney, NSW, 2014, pp. 379–384. doi: 10.1109/WPMC.2014.7014848.

53. D. Stynes, K. N. Brown, and C. J. Sreenan, A probabilistic approach to user mobility prediction for wireless services, *2016 International Wireless Communications and Mobile Computing Conference (IWCMC)*, Paphos, 2016, pp. 120–125. doi: 10.1109/IWCMC.2016.7577044.

54. A. Aljadhai and T. F. Znati, Predictive mobility support for QoS provisioning in mobile wireless environments, *IEEE Journal on Selected Areas in Communications*, vol. 19, no. 10, pp. 1915–1930, 2001. doi: 10.1109/49.957307.

55. W. Ma, Y. Fang, and P. Lin, Mobility management strategy based on user mobility patterns in wireless networks, *IEEE Transactions on Vehicular Technology*, vol. 56, no. 1, pp. 322–330, 2007. doi: 10.1109/TVT.2006.883743.

56. V. A. Siris and M. Anagnostopoulou, Performance and energy efficiency of mobile data offloading with mobility prediction and prefetching, *2013 IEEE 14th International Symposium on A World of Wireless, Mobile and Multimedia Networks (WoWMoM)*, Madrid, 2013, pp. 1–6. doi: 10.1109/WoWMoM.2013.6583450.

57. S. Tian, X. Li, H. Ji, and H. Zhang, Mobility prediction scheme for optimized load balance in heterogeneous networks, *2018 IEEE Globecom Workshops (GC Wkshps)*, Abu Dhabi, United Arab Emirates, 2018, pp. 1–6. doi: 10.1109/GLOCOMW.2018.8644289.

58. A. Mohamed, O. Onireti, S. A. Hoseinitabatabaei, M. Imran, A. Imran, and R. Tafazolli, Mobility prediction for handover management in cellular networks with control/data separation, *2015 IEEE International Conference on Communications (ICC)*, London, 2015, pp. 3939–3944. doi: 10.1109/ICC.2015.7248939.

59. M. Ozturk, M. Gogate, O. Onireti, A. Adeel, A. Hussain, M. Imran, A novel deep learning driven low-cost mobility prediction approach for 5G cellular networks: The case of the Control/Data Separation Architecture (CDSA). *Neurocomputing*, 2019. doi: 10.1016/j.neucom.2019.01.031.

60. A. Rodriguez-Carrion, C. Garcia-Rubio, and C. Campo, Performance evaluation of LZ-based location prediction algorithms in cellular networks, *IEEE Communications Letters*, vol. 14, no. 8, pp. 707–709, 2010. doi: 10.1109/LCOMM.2010.08.092033.

61. S. Akoush and A. Sameh, Bayesian learning of neural networks for mobile user position prediction, *2007 16th International Conference on Computer Communications and Networks*, Honolulu, HI, 2007, pp. 1234–1239. doi: 10.1109/ICCCN.2007.4317989.

62. S. Parija, R. K. Ranjan, and P. K. Sahu, Location prediction of mobility management using neural network techniques in cellular network, *2013 International Conference on Emerging Trends in VLSI, Embedded System, Nano Electronics and Telecommunication System (ICEVENT)*, Tiruvannamalai, 2013, pp. 1–4. doi: 10.1109/ICEVENT.2013.6496540.

63. C. L. Leca, L. Tută, I. Nicolaescu, and C. I. Rincu, Recent advances in location prediction methods for cellular communication networks, *2015 23rd Telecommunications Forum Telfor (TELFOR)*, Belgrade, 2015, pp. 898–901. doi: 10.1109/TELFOR.2015.7377610.

64. A. Cao, Y. Qiao, K. Sun, H. Zhang, and J. Yang, Network traffic analysis and prediction of Hotspot in cellular network. 2018 International Conference on Network Infrastructure and Digital Content (IC-NIDC), Guiyang, China, 2018, pp. 452–456. doi: 10.1109/ICNIDC.2018.8525553.

65. M. Ghahramani, M. Zhou, and C. T. Hon, Mobile phone data analysis: A spatial exploration toward hotspot detection, *IEEE Transactions on Automation Science and Engineering*, vol. 16, no. 1, pp. 351–362, 2019. doi: 10.1109/TASE.2018.2795241.

66. M. Xu, Q. Wang, and Q. Lin, Hybrid holiday traffic predictions in cellular networks, *NOMS 2018-2018 IEEE/IFIP Network Operations and Management Symposium*, Taipei, 2018, pp. 1–6. doi: 10.1109/NOMS.2018.8406291.

67. F. Khan and S. S. Kozat. Sequential churn prediction and analysis of cellular network users: A multi-class, multi-label perspective. *2017 25th Signal Processing and Communications Applications Conference (SIU)*, Antalya, Turkey, 2017, pp. 1–4.

68. I. Brânduşoiu, G. Toderean, and H. Beleiu, Methods for churn prediction in the pre-paid mobile telecommunications industry, *2016 International Conference on Communications (COMM)*, Bucharest, 2016, pp. 97–100. doi: 10.1109/ICComm.2016.7528311.

69. V. Yeshwanth, V. Raj, and S. Mohan, Evolutionary churn prediction in mobile networks using hybrid learning. *Proceedings of the 24th International Florida Artificial Intelligence Research Society, FLAIRS-24*, Florida, USA, 2011.

70. F. Castanedo, Using deep learning to predict customer churn in a mobile telecommunication network, 2014.

71. A. K. Ahmad, A. Jafar, and K. Aljoumaa, Customer churn prediction in telecom using machine learning in big data platform. *Journal of Big Data*, vol. 6, p. 28, 2019. doi: 10.1186/s40537-019-0191-6.

72. P. Casas et al., Predicting QoE in cellular networks using machine learning and in-smartphone measurements, *2017 Ninth International Conference on Quality of Multimedia Experience (QoMEX)*, Erfurt, 2017, pp. 1–6. doi: 10.1109/QoMEX.2017.7965687.

73. O. P. Kogeda and J. I. Agbinya. Prediction of faults in cellular networks using Bayesian network model, 2007.

74. O. P. Kogeda and S. Nyika, Automating cellular network faults prediction using mobile intelligent agents, *2009 Fifth International Joint Conference on INC, IMS and IDC*, Seoul, 2009, pp. 1805–1810. doi: 10.1109/NCM.2009.257.

75. S. Rezaei, H. Radmanesh, P. Alavizadeh, H. Nikoofar, and F. Lahouti, Automatic fault detection and diagnosis in cellular networks using operations support systems data, *NOMS 2016-2016 IEEE/IFIP Network Operations and Management Symposium*, Istanbul, 2016, pp. 468–473. doi: 10.1109/NOMS.2016.7502845.

76. Y. Kumar, H. Farooq, and A. Imran, Fault prediction and reliability analysis in a real cellular network, *2017 13th International Wireless Communications and Mobile Computing Conference (IWCMC)*, Valencia, 2017, pp. 1090–1095. doi: 10.1109/IWCMC.2017.7986437.

77. E. Rozaki, Network fault diagnosis using data mining classifiers. *Computer Science & Information Technology*, pp. 29–40, 2015. doi: 10.5121/csit.2015.50703.

78. D. Mulvey, C. H. Foh, M. A. Imran, and R. Tafazolli, Cell fault management using machine learning techniques, *IEEE Access*, vol. 7, pp. 124514–124539, 2019. doi: 10.1109/ACCESS.2019.2938410.

79. I. A. Karatepe and E. Zeydan, Anomaly detection in cellular network data using big data analytics, *Proceedings of 20th European Wireless Conference, European Wireless 2014*, Barcelona, Spain, 2014, pp. 1–5.

80. B. Sun, and F. Yu, Mobility-based anomaly detection in cellular mobile networks. *International Conference on WiSe 2004*, Philadelphia, PA, pp. 61–69, 2004.

81. B. Sun, F. Yu, K. Wu, Y. Xiao, and V. C. M. Leung, Enhancing security using mobility-based anomaly detection in cellular mobile networks, *IEEE Transactions on Vehicular Technology*, vol. 55, no. 4, pp. 1385–1396, 2006. doi: 10.1109/TVT.2006.874579.

82. S. Chernov, M. Cochez, and T. Ristaniemi, Anomaly detection algorithms for the sleeping cell detection in LTE networks. *IEEE Vehicular Technology Conference*, 2015. doi: 10.1109/VTCSpring.2015.7145707.

83. J. Hall, M. Barbeau, E. Kranakis, Using mobility profiles for anomaly-based intrusion detection in mobile networks. *In Proceedings of the Wireless and Mobile Computing, Networking and Communications*, Montreal, Canada, pp. 22–24, August 2005, Preliminary version in NDSS 2005 Preconference Workshop on Wireless and Mobile Security.

84. R. Chalapathy, and S. Chawla, Deep learning for anomaly detection: A survey. 2019.

85. H. Abou-zeid, H. S. Hassanein, and S. Valentin, Optimal predictive resource allocation: Exploiting mobility patterns and radio maps, *2013 IEEE Global Communications Conference (GLOBECOM)*, Atlanta, GA, 2013, pp. 4877–4882. doi: 10.1109/GLOCOMW.2013.6855723.

86. S. Gayathri, and S. Rajagopal, Resource allocation in downlink of LTE using bandwidth prediction through statistical information. *Indonesian Journal of Electrical Engineering and Computer Science*, vol. 10, pp. 680–686, 2018. doi: 10.11591/ijeecs.v10.i2.

87. J. Guo, C. Yang, and I. Chih-Lin, Exploiting future radio resources with end-to-end prediction by deep learning, *IEEE Access*, vol. 6, pp. 75729–75747, 2018. doi: 10.1109/ACCESS.2018.2882815.

88. U. Karneyenka, K. Mohta, and M. Moh, Location and mobility aware resource management for 5G cloud radio access networks, *2017 International Conference on High Performance Computing & Simulation (HPCS)*, Genoa, 2017, pp. 168–175. doi: 10.1109/HPCS.2017.35.

89. J. F. Schmidt, et al. Interference prediction in wireless networks: Stochastic geometry meets recursive filtering. ArXiv abs/1903.10899, 2019.

90. J. Liu, M. Sheng, L. Liu, and J. Li, Interference management in ultra-dense networks: challenges and approaches. *IEEE Network*, pp. 1–8, 2017. doi: 10.1109/MNET.2017.1700052.

91. Y. Feng, X. Shen, R. Zhang, and P. Zhou, Interference-area-based resource allocation for full-duplex communications, *2016 IEEE International Conference on Communication Systems (ICCS)*, Shenzhen, 2016, pp. 1–5. doi: 10.1109/ICCS.2016.7833655.

92. J. Gao, X. Cheng, L. Xu, and H. Ye, An interference management algorithm using big data analytics in LTE cellular networks, *2016 16th International Symposium on Communications and Information Technologies (ISCIT),* Qingdao, 2016, pp. 246–251. doi: 10.1109/ISCIT.2016.7751630.

93. S. Kusaladharma, Z. Zhang, and C. Tellambura, Interference and outage analysis of random D2D networks underlaying millimeter-wave cellular networks, *IEEE Transactions on Communications,* vol. 67, no. 1, pp. 778–790, 2019. doi: 10.1109/TCOMM.2018.2870378.

94. Y. Xu and J. Chen, An optimal load-aware base station sleeping strategy in small cell networks based on marginal utility, *2017 3rd International Conference on Big Data Computing and Communications (BIGCOM),* Chengdu, 2017, pp. 291–296. doi: 10.1109/BIGCOM.2017.65.

95. Y.-H. Choi, J. Lee, J. Back, S. Park, Y. Chung, and H. Lee, Energy efficient operation of cellular network using on/off base stations. *International Journal of Distributed Sensor Networks,* 2015, 1–7, 2015. doi: 10.1155/2015/108210.

96. Z. Zhang, F. Liu, and Z. Zeng, The cell zooming algorithm for energy efficiency optimization in heterogeneous cellular network, *2017 9th International Conference on Wireless Communications and Signal Processing (WCSP),* Nanjing, 2017, pp. 1–5. doi: 10.1109/WCSP.2017.8171171.

97. S. W. H. Shah, A. T. Riaz, and Z. Fatima. Energy-efficient mechanism for smart communication in cellular networks. *Networking and Internet Architecture,* 2019, pp. 1–5.

98. H. Park, and Y. Lim. Energy-effective power control algorithm with mobility prediction for 5G heterogeneous cloud radio access network. *Sensors (Basel, Switzerland)* vol. 18, no. 9, p. 2904, 2018. doi:10.3390/s18092904.

99. V. Kumar and M. L. Garg, Deep learning in predictive analytics: A survey, *2017 International Conference on Emerging Trends in Computing and Communication Technologies (ICETCCT),* Dehradun, 2017, pp. 1–6. doi: 10.1109/ICETCCT.2017.8280331.

100. M. Usama et al., Unsupervised machine learning for networking: Techniques, applications and research challenges, *IEEE Access,* vol. 7, pp. 65579–65615, 2019. doi: 10.1109/ACCESS.2019.2916648.

101. U. Challita, H. Ryden, and H. Tullberg, When machine learning meets wireless cellular networks: Deployment, challenges, and applications, *IEEE Communications Magazine,* vol. 58, no. 6, pp. 12–18, 2020. doi: 10.1109/MCOM.001.1900664.

102. L. Chen, D. Yang, D. Zhang, C. Wang, J. Li, et al. Deep mobile traffic forecast and complementary base station clustering for C-RAN optimization. *Journal of Network and Computer Applications,* 121, 59–69, 2018.

103. L. Nie, D. Jiang, S. Yu, and H. Song. Network traffic prediction based on deep belief network in wireless mesh backbone networks. *In Proceedings of IEEE Wireless Communications and Networking Conference (WCNC),* San Francisco, CA, USA, pp. 1–5, 2017.

104. J. Wang, J. Tang, Z. Xu, Y. Wang, G. Xue, X. Zhang, and D. Yang. Spatiotemporal modeling and prediction in cellular networks: A big data enabled deep learning approach. *In Proceedings of 36th Annual IEEE International Conference on Computer Communications (INFOCOM),* Atlanta, GA, USA, 2017.

105. C. Zhang and P. Patras. Long-term mobile traffic forecasting using deep spatio-temporal neural networks. *In Proceedings of Eighteenth ACM International Symposium on Mobile Ad Hoc Networking and Computing,* pp. 231–240, 2018.

106. C.-W. Huang, C.-T. Chiang, and Q. Li. A study of deep learning networks on mobile traffic forecasting. *In Proceedings of 28th IEEE Annual International Symposium on Personal, Indoor, and Mobile Radio Communications (PIMRC),* Montreal, QC, Canada, pp. 1–6, 2017.

107. C. Zhang, H. Zhang, D. Yuan, and M. Zhang. Citywide cellular traffic prediction based on densely connected convolutional neural networks. *IEEE Communications Letters,* vol. 22, no. 8, pp. 1656–1659, Aug. 2018, doi: 10.1109/LCOMM.2018.2841832.

108. Z. Wang. The applications of deep learning on traffic identification. BlackHat, USA, 2015.

109. L. Fang, X. Cheng, H. Wang, and L. Yang. Mobile demand forecasting via deep graph-sequence spatiotemporal modeling in cellular networks. *IEEE Internet of Things Journal*, 2018.

110. V. C Liang, R. T. B. Ma, W. S. Ng, L. Wang, M. Winslett, H. Wu, S. Ying, and Z. Zhang. Mercury: Metro density prediction with recurrent neural network on streaming CDR data. *In Proceedings of IEEE 32nd International Conference on Data Engineering (ICDE)*, pp. 1374–1377, 2016.

111. B. Felbo, P. Sundsøy, A. S. Pentland, S. L. and Y.-A. de Montjoye. Using deep learning to predict demographics from mobile phone metadata. *In Proceedings of workshop track of the International Conference on Learning Representations (ICLR)*, 2016.

112. N. C. Chen, W. Xie, R. E. Welsch, K. Larson, and J. Xie. Comprehensive predictions of tourists' next visit location based on call detail records using machine learning and deep learning methods. *In Proceedings of IEEE International Congress on Big Data (BigData Congress)*, Honolulu, HI, USA, pp. 1–6, 2017.

113. C. Yang, M. Sun, W. X. Zhao, Z. Liu, and E. Y. Chang. A neural network approach to jointly modeling social networks and mobile trajectories. *ACM Transactions on Information Systems (TOIS)*, vol. 35, no. (4), p. 36, 2017.

114. X. Ouyang, C. Zhang, P. Zhou, and H. Jiang. DeepSpace: An online deep learning framework for mobile big data to understand human mobility patterns. arXiv preprint arXiv:1610.07009, 2016.

115. J. Tkacík and P. Kordík. Neural turing machine for sequential learning of human mobility patterns. *In Processing's of IEEE International Joint Conference on Neural Networks (IJCNN)*, Vancouver, BC, Canada, pp. 2790–2797, 2016.

116. X. Song, H. Kanasugi, and R. Shibasaki. DeepTransport: Prediction and simulation of human mobility and transportation mode at a citywide level. *In Proceedings of International Joint Conference on Artificial Intelligence*, pp. 2618–2624, 2016.

117. J. Zhang, Y. Zheng, and D. Qi. Deep spatio-temporal residual networks for citywide crowd flows prediction. *In Proceedings of National Conference on Artificial Intelligence (AAAI)*, 2017.

118. R. Jiang, X. Song, Z. Fan, T. Xia, Q. Chen, S. Miyazawa, and R. Shibasaki. Deep Urban momentum: An online deep-learning system for short-term urban mobility prediction. *In Proceedings of National Conference on Artificial Intelligence (AAAI)*, 2018.

119. Z. Lin, M. Yin, S. Feygin, M. Sheehan, J.-F. Paiement, and A. P. Deep generative models of urban mobility. *IEEE Transactions on Intelligent Transportation Systems*, 2017.

120. R. Jiang, X. Song, Z. Fan, T. Xia, Q. Chen, Q. Chen, and R. Shibasaki. Deep ROI-based modeling for urban human mobility prediction. *Proceedings of ACM on Interactive, Mobile, Wearable and Ubiquitous Technologies (IMWUT)*, vol. 2, no. 1, p. 14, 2018.

121. J. Feng, Y. Li, C. Zhang, F. Sun, F. Meng, A. Guo, and D. Jin. DeepMove: Predicting human mobility with attentional recurrent networks. *In Proceedings of World Wide Web Conference on World Wide Web*, pp. 1459–1468, 2018.

122. Y. Yayeh, H.-P. Lin, G. Berie, A. B. Adege, L. Yen, and S.-S. Jeng. Mobility prediction in mobile Ad-hoc network using deep learning. *In Proceedings of IEEE International Conference on Applied System Invention (ICASI)*, Chiba, Japan, pp. 1203–1206, 2018.

123. D. Y. Kim and H. Y. Song. Method of predicting human mobility patterns using deep learning. *Neurocomputing*, vol. 280, pp. 56–64, 2018.

124. D. Yao, C. Zhang, Z. Zhu, J. Huang, and Jingping Bi. Trajectory clustering via deep representation learning. *In Proceedings of IEEE International Joint Conference on Neural Networks (IJCNN)*, Anchorage, AK, USA, pp. 3880–3887, 2017.

125. Q. Chen, X. Song, H. Yamada, and R. Shibasaki. Learning deep representation from big and heterogeneous data for traffic accident inference. *In Proceedings of National Conference on Artificial Intelligence (AAAI)*, pp. 338–344, 2016.

126. X. Song, R. Shibasaki, N. J. Yuan, X. Xie, T. Li, and R. Adachi, DeepMob: Learning deep knowledge of human emergency behavior and mobility from big and heterogeneous data. *ACM Transactions on Information Systems (TOIS)*, vol. 35, no. 4, p. 41, 2017.

127. Z. Liu, Z. Li, K. Wu, and M. Li. Urban traffic prediction from mobility data using deep learning. *IEEE Network*, vol. 32, no. 4, pp. 40–46, 2018.

128. O. G. Aliu, A. Imran, M. A. Imran, and B. Evans, A survey of self organisation in future cellular networks, *IEEE Communications Surveys & Tutorials*, vol. 15, no. 1, pp. 336–361, First Quarter, 2013. doi: 10.1109/SURV.2012.021312.00116.

129. A. Roy, N. Saxena, B. Sahu, & S. Singh, BiSON: A Bioinspired Self-Organizing Network for Dynamic Auto-Configuration in 5G Wireless. *Wireless Communications and Mobile Computing*, 2018, 1–13, 2018. doi: 10.1155/2018/2632754.

130. M. Peng, D. Liang, Y. Wei, J. Li, and H. Chen, Self-configuration and self-optimization in LTE-advanced heterogeneous networks, *IEEE Communications Magazine*, vol. 51, no. 5, pp. 36–45, 2013. doi: 10.1109/MCOM.2013.6515045.

131. T. Zhang, K. Zhu, and E. Hossain, Data-driven machine learning techniques for self-healing in cellular wireless networks: Challenges and solutions, 2019.

132. M. Selim, A. Kamal, K. Elsayed, H. Abd-El-Atty, and M. Alnuem, A novel approach for backhaul self healing in 4G/5G HetNets, *2015 IEEE International Conference on Communications (ICC)*, London, 2015, pp. 3927–3932. doi: 10.1109/ICC.2015.7248937.

133. A. Asghar, H. Farooq, and A. Imran, Self-healing in emerging cellular networks: Review, challenges, and research directions, *IEEE Communications Surveys & Tutorials*, vol. 20, no. 3, pp. 1682–1709, Third Quarter 2018. doi: 10.1109/COMST.2018.2825786.

12

Optimization Techniques and Algorithms for Dental Implants – A Comprehensive Review

Niharika Karnik and Pankaj Dhatrak

Dr. Vishwanath Karad MIT-WPU

CONTENTS

DOI: 10.1201/9781003143505-12

12.1 Introduction

Dental implants are a key mode of treatment for dental restoration in patients. They are inserted in place of a missing tooth and mainly consist of the screw, abutment and crown. The materials used for these tiny screws are generally titanium and titanium alloys. The implant is surgically screwed into its place in the jaw by the dentist. Implant design, mechanical loading, patient and surgical technique and implant environment are some of the major factors that can cause early failure of the dental implant. Once the implant is inserted in the jaw, it needs to fuse with the surrounding bone. This process is called osseointegration and was first termed by Bränemark. Generally, 95% of the dental implants function well for around 15 years; however, there is still some failure involved, and thus, it becomes necessary to optimize the dental implants. Optimization of the dental implant can mainly fall into three categories.

12.1.1 Structural Optimization

Structural optimization deals with maximizing or minimizing the structural parameters of the dental implant to achieve the best possible implantation results. Three branches are clearly distinguished in structural optimization, which are discussed in the following section.

a. **Sizing of Implant Parameters:** In sizing, optimization is carried out in order to find effective component dimensions like the implant's length or diameter [1]. Some authors defined eight geometric parameters with regard to sizing as real variables for optimization: implant head diameter (P_1), taper surface angle of implant head (P_2), implant head height (P_3), implant diameter (P_4), hexagonal slot height (P_5), surface opening angle (P_6), thread shapes of dental implant (right) (P_7) and pitch of thread (P_8) as shown in Figure 12.1 [2,3]. With reference to sizing, this chapter carried out the minimization of the diameter of a dental implant using the genetic algorithm (GA) and Hooke-Jeeves algorithm [4].

b. **Shape Optimization:** In shape optimization, the shape of the structure is obtained by changing the shape of the used components with other components of different shapes (cylindrical and/or conical shape), in order to improve a desired variable within the system [1,5]. The purpose of a study was to employ topology optimization technology to derive alternate shapes for dental implants, with the aim of optimizing the stress distribution along the bone-implant interface [6]. Another study used the optimization theory to find the best shape and size of a dental implant in order to minimize cortical bone strains in the peri-implant crest [7].

c. **Topology Optimization:** This category optimizes the material distribution within a specified domain by varying contours and diameter of the implant [1]. Arbitrary holes and different connections may also develop within the structure during optimization. An article took advantage of the topology optimization method to look for redundant material distribution on a dental implant and

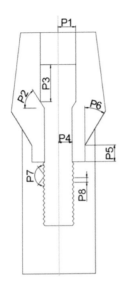

P$_1$: Implant head diameter
P$_2$: Taper surface angle of implant head
P$_3$: Implant head height
P$_4$: Implant diameter
P$_5$: Hexagonal slot height
P$_6$: Surface opening angle
P$_7$: Thread shapes of dental implant
P$_8$: Pitch of thread

FIGURE 12.1
Parameters selected for sizing of dental implant [2].

restructured a new implant macro-geometry with the assessment of its biomechanical functions [8]. Another article took advantage of this method to reduce material distribution on a threaded premolar implant [5].

12.1.2 Surface Morphology Optimization

This optimization category deals with the surface characteristics of the dental implants. The dental implants are generally made of titanium alloys, and their surface properties play an important role in osseointegration and antibacterial property of the implant. A study was carried out to describe the surface composition of titanium and titanium-zirconium dental implant substrates subjected to surface treatments developed by Murphy et al. [9]. An article aimed to carry out a multiobjective optimization framework for the coating of the surface of porous implants [10]. Another article reviewed the current status of surface optimization techniques for titanium implants and placed a strong emphasis on the improved biological responses to titanium implants, which occur due to the surface finishing process [11].

12.1.3 Material Properties Optimization

Despite dental implantation being a great success, one of its main problems is the disparity of mechanical properties between engineered biomaterials and the surrounding tissue, which makes osseointegration and bone remodelling difficult [12]. Therefore, the modulus of elasticity of the implant should be very close to that of the native tissue. An article has been examining patient-specific design to modify implant design and material composition according to the patient's anatomy [13]. In some literature, the goal was to create an optimized elastic modulus to reduce the magnitude of stress concentration for the dental implant models [14].

12.2 FEA Aspect of Optimization Techniques

One of the most commonly used numerical techniques in the study of dental implants is the finite element analysis (FEA). This method mainly focuses on studying the various mechanical features that influence the behaviour of the implant. Most of the studies use the finite element (FE) method with static loads to simulate biomechanical behaviour but the process of mastication includes dynamic loads and consequently induced stresses that are higher than those for static loads [15,16]. Hence, considering dynamic loads in the FE modelling would yield more accurate results. Previous research makes use of linear isotropic material models during FEA for ease of the modelling and analysing process [17]. However, FE results show that the orthotropic model is more suitable than the isotropic one to find the stress along the bone-implant interface [18]. The first step of modelling is to use CAD to define the wanted bone and implant geometry. This procedure is followed by defining the material behaviour according to the Young's modulus and Poisson's ratio for bone components and the implant. After applying the load and boundary conditions, the different parameters and their effect on the stress profile can be estimated [14,19]. Additionally, the outcomes obtained from the FEA of implants are used as objective functions for the optimization algorithms.

In a certain study, 3-D models of implants with different lengths, diameters and bone quality were created using the finite element method (FEM) in ANSYS, and the principal stress and micromotion were evaluated by applying a load of 150 N [20]. Consequently, topology optimization was applied to remove redundant materials from the implants, and the biomechanical functions were further evaluated with the FEA technique [21]. Another study carried out optimization of the neck design of implants by performing numerical simulations using the FEM models of bone-level dental implants and by applying a load of 100 N [22]. The results of FEM analysis were compared to the results from in vivo studies with New Zealand rabbits for four different neck designs: screwed, four-ring, three-ring and even surfaced design. Few other papers discuss the use of FEA to evaluate the best thread shape for tapered cylindrical implants between triangular and square or triangular and trapezoidal shapes [23–25]. These studies used FEM to compare the stress induced in the cortical and cancellous bone due to each thread shape and concluded that stresses in the bone are found to be reduced as the thread taper angle increases. Three-dimensional FEA of a distraction implant in the posterior mandible was carried out to optimize the diameter of the distraction screw of the implant in another study [26,27]. The diameter varied from 1.0 to 3.0 mm, and the stresses and displacements generated were analysed.

The fact that the intricate FEAs as the objective function make the optimization computationally costly led to substituting it through a surrogate model [28]. The design of experiments (DOE) is an alternative method in which a statistical model can be made to formulate responses of a given design [29,30]. The main aim of DOE is that the resultant statistical model should have low uncertainty in its approximation and a high accuracy in prediction. This is done by spreading out the samples under consideration. Response surface methodology (RSM) is an important technique that makes use of both mathematical and statistical techniques to understand the effect of experimental variables on the response variable with the smallest number of samples [31].

This is one of the main DOE techniques used for the optimization of dental implants. The major application of this method in design optimization is to reduce the cost of analysis. A study carried out optimization of the dental implant using response surface optimization by using the screening method, which supports multiple objectives and constraints [32].

12.3 Optimization Techniques and Algorithms

Optimization is a technique of maximizing or minimizing the various parameters of a model in order to obtain the best performance from the model [33]. A dental implant is used for restoring the function of chewing and biting, and therefore eating, which is the most essential activity of human beings [12]. Optimization methods can thus be a useful tool to provide comprehensive scientific strategies for the design of dental implants as the optimization can help preserve osseointegration and lengthen the survival of dental implants [6]. Many studies carried out the optimization of the implant using built-in algorithms in analysis software like ANSYS and ABAQUS [34,35]. Some other studies have carried out the optimization of the implant through optimization algorithms. Therefore, in this section, the various optimization algorithms used in the field of dental implants have been highlighted.

12.3.1 Genetic Algorithm

One of the most popular optimization methods today is the genetic algorithm (GA), which is a stochastic search approach. Genetic algorithm, also known as GA, is the most widely used evolutionary optimization technique, which can deliver the global optimal solution for multifaceted problems linking a huge number of variables and constraints [13]. GAs have been stimulated by evolutionary biology and integrate techniques such as inheritance, mutation, selection and crossover to find a better solution [3]. These algorithms focus on problem solving over analytical relations and can thus prove very advantageous. A study used the Kriging interpolation method and a genetic algorithm (KIGA) to optimize the geometric features of an implant, in order to yield an optimized one-piece dental implant model [36]. Another study employed a GA for optimum designing of patient-specific dental implants with varying dimensions and porosity [28]. In this paper, the data that was generated by FEA of the dental implant was converted into an artificial neural network (ANN) model. After converting the output of the neural network model into a desirability function, they were used as an objective function for GA. Another research group carried out an optimization study using the GA to explore the optimum setting of processing parameters necessary for SLM (selective laser melting) of a Ti_6Al_4V dental implant [37]. One more study followed an incorporation procedure, including uniform design method, KIGA to optimize both exogenous factors for implant shape (like thread pitch, thread depth, diameter of implant neck and body size) and endogenous features (like bone density, cortical bone thickness and non-osseointegration) [15]. Figure 12.2 shows the basic flowcharts used for implementing a GA.

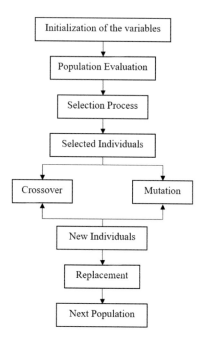

FIGURE 12.2
Flowchart for genetic algorithm [3].

The basic steps for the implementation of the GA are as follows:

1. Generation of a set of initial solutions which are called the initial population.
2. Every population undergoes an evaluation process where it is assigned a fitness value that corresponds to an objective function and FEA results.
3. Individuals with the highest fitness values proceed to form the next generation.
4. These individuals undergo crossover and mutation in order to create new design parameters and to maintain diversity of the population.
5. GA terminates when the population consists of individuals with the highest fitness values or a maximum number of generations is reached.

Mathematically, the string representing the chromosome can be characterized as:

$$C = c_1 c_2 c_3, \ldots, c_i, \ldots, c_l, \tag{12.1}$$

In Equation (12.1), c_i shows a specific bit at position i and l denotes the length of the string. A schema is introduced to gather chromosomes with the same features as:

$$H = h_1 h_2 h_3, \ldots, h_i, \ldots, h_l \tag{12.2}$$

In Equation (12.2), h_i represents a particular element at position i and l denotes the length of the schema. Now, by representing the fitness value for the jth individual with respect to the schema H as f_j^H and their number after t^{th} iteration by $m(H,t)$, the average fitness for the schema is given as:

$$f(H,t) = \frac{\Sigma f_j^H}{m(H,t)} \tag{12.3}$$

The average fitness value $\bar{f}$ for the entire population is given as:

$$\bar{f} = \frac{\Sigma f_j}{n} \tag{12.4}$$

In Equation (12.4), n stands for the population size and f_j stands for the jth representation of individuals. Using these equations, we can formulate the GA used for the optimization of dental implants.

12.3.2 Topology Optimization Algorithm

It is a technique used in structural optimization that proposes the best distribution of material through continuous iterative calculations which are built on set load and boundary conditions in a design region to achieve optimal performance of the structure [38]. Structural topology and shape optimization have been an important research area where FE methods are the main tools used [6]. Topology optimization algorithm makes use of design variable, an objective function and a constraint condition. A study used the topology optimization technique to look for redundant material distribution on a dental threaded implant, and a new implant was designed with the assessment of its biomechanical functions [5]. Another paper presented the modification and optimization of three-dimensional (3D) dental implants with the root shapes of natural teeth [14]. The topology optimization algorithm flowchart is shown in Figure 12.3. There are mainly two algorithms, namely soft kill option (SKO) and solid topology optimization, and they are described briefly in the following section.

12.3.2.1 SKO (Soft Kill Option)

This method was first suggested by Baumgartner in a paper to find an optimum structural topology by simulating adaptive bone mineralization [39]. This is one of the simplest and most accepted methods used as a topological shape optimization technique. SKO impersonates the biological mineralization process in living bone via FE modelling and the non-load-bearing material is factored out in SKO conferring to the load distribution, thus giving a preoptimized lightweight design [6]. The element's stiffness is connected to the system's stress response, and the elements with higher stresses are strengthened and the elements with lower stresses are weakened [1]. In order to decide a reference, a specific stress value is defined. Elements with a stress value above the reference are strengthened, and those below are weakened. The algorithm for the SKO is as given in Figure 12.4.

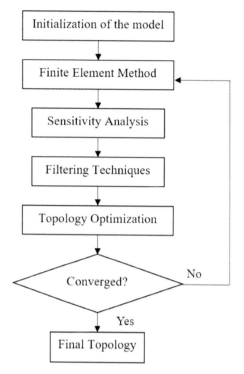

FIGURE 12.3
Topology optimization algorithm flowchart [38].

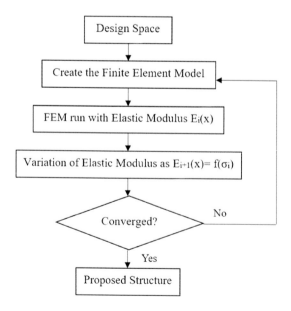

FIGURE 12.4
SKO flowchart [6].

12.3.2.2 Solid Isotropic Material with Penalization (SIMP)

This topology optimization technique was first suggested by M.P. Bendsoe in a paper that recommended the introduction of a density function that is a continuous design variable for optimization [40]. In this technique, the material distribution is varied and a structural response is obtained in order to achieve an optimized topology. The element's stiffness and material properties are related to the relative material density, which is used as the optimization variable [1]. This technique was utilized for the optimization of a dental implant with a factor of penalization (p), and the process presumed that the material for the dental implant is isotropic and the Poisson's ratio is constant [38]. The relative Young's modulus of each element was parameterized as a function of the relative density of each element, which is the penalization factor that dictates the optimization process.

12.3.3 Particle-Swarm Optimization

The particle swarm optimization (PSO) is a fairly simple technique that can be used for optimizing complex functions. This technique was first introduced by Kennedy and Eberhart in their paper in 1995 where they stated that the PSO is a population-based optimization algorithm based on the premise that social sharing of information among species provides an evolutionary advantage [41]. The population is known as a swarm in the PSO, and each individual in the population is called a particle. The movement of the particles is stochastic; but it is also dictated by the particle's own memories as well as the memories of its peers, and each particle tracks its coordinates in the problem space [42]. The algorithm keeps a track of the best solution achieved so far by ever particle and also the best solution achieved for all the particles. A study performed on dental implants used a PSO to recognize optimal designs of zirconia-porcelain multilayered dental restorations where the thickness of each interlayer and compositional distribution are the design variables [43]. In another study, the authors have used a multiobjective particle swarm optimization (MOPSO) algorithm to formulate the ideal configuration of surface coating for dental implants that uses the particle size and the volume fraction of the beads or particles as design variables [44]. This optimization technique is an extended version of the PSO, which brings together the ability to include multiple objectives without losing the speedy convergence of the PSO.

12.3.4 Multiobjective Optimization Algorithm

Real-world problems generally do not have just one optimization objective but are multiobjective in nature. Thus, multiobjective optimization algorithms have gained popularity among scientists and researchers. Multiobjective optimization provides a chance to find a solution for which the design variables and constraints satisfy the criteria for design in the best way in accordance with all anticipated aspects [45]. A study was carried out using MATLAB software to run the multiobjective genetic algorithm (MOGA) and the Suppapitnarm and Parks multiobjective simulated annealing (SMOSA) method in design optimization of a FGM dental implant [12]. The aim of another paper was to formulate the design procedure of a multi objective function by recognizing best weights of the sum function with respect to shape and design target using GA. Further, this multiobjective function was used in topographic optimization to deliver a shape-optimized design of the dental implant [45]. In another study, a multiobjective optimization framework is used to maximize the average apparent density of the implant material [10]. Thus, multiobjective algorithms use the traditional optimization algorithms like the genetic or topology optimization algorithm to achieve multiple objectives of optimization. For example, the algorithm can be applied to two-objective functions: a displacement function and a cortical density function for the implants as done in a certain study [12].

12.3.5 Approximate Optimization

In this optimization technique, FEA, numerical optimization algorithm and Monte Carlo simulation are used together to decrease the cost of in vivo tests and to design a dependable dental implant [2]. Following this approach, the formulation of the geometric design of the implant is done with a numerical optimization algorithm. After carrying out FEA of the implant, the objective and restrictive functions that govern the optimization problem are changed appropriately and the optimal design is obtained at the end. Figure 12.5 gives the approximate optimization algorithm flowchart.

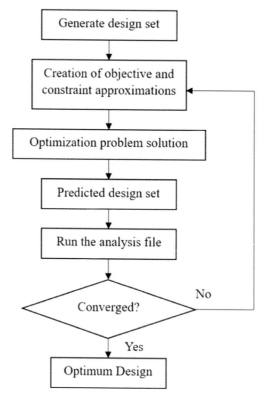

FIGURE 12.5
Flowchart of the approximate optimization process [2].

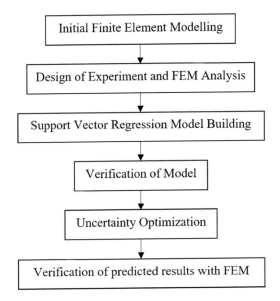

FIGURE 12.6
Flowchart for the uncertainty optimization algorithm used in [46].

12.3.6 Uncertainty Optimization Algorithm

Uncertainty exists in the design variables of dental implants because of the machining and manufacturing errors. Hence, the uncertainty optimization of dental implants is essential, and a prevalent uncertainty optimization method built on the stochastic design variables (mean value and standard deviation), constraints and optimization objectives is the "k-sigma-method" [46]. Another uncertainty optimization technique is the interval method, which only needs the upper and lower bounds of an uncertain parameter and thus has strong potential in engineering practices [47]. In a study, an uncertainty optimization approach to decrease the stress at the implant-bone interface is proposed, and it uses the above-mentioned methods [46]. The results obtained are compared with the results obtained from a GA optimization, and the flowchart for the same is as given in Figure 12.6.

12.3.7 Memetic Search Optimization

Memetic algorithm is one of the most useful and flexible metaheuristic methods for solving difficult optimization problems [48]. This is done by encouraging the use of multiple heuristics which act together and use all available sources of information for a problem. While making metaheuristics, broadening of the search space and strengthening of the optimal solution are the two contradictory criteria. The memetic algorithm, which is a mix of the local and global search algorithm, can be used to maintain a balance between these two criteria and can help improve their performance [49]. In a certain study, this optimization technique along with genetic-scale recurrent neural network was used to identify the successful measure of the dental implant treatment process [50]. The flowchart used for this optimization is as shown in Figure 12.7.

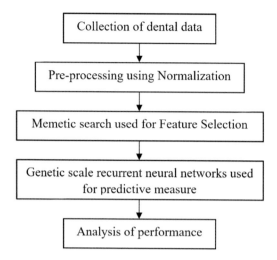

FIGURE 12.7
Memetic search optimization algorithm flowchart [50].

12.4 Parameters for Optimization

The surface quality, peri-implantitis, micro-gap, surgical trauma, occlusal overload, biologic width and implant crest module work towards the osseointegration and long-term clinical success of dental implants [51,52]. It thus becomes very important to optimize the dental implant design or design features to yield lower stresses in the surrounding bone. Some of the parameters that play a major role in successful implantation and are thus mainly used as optimization parameters in optimization techniques and algorithms are given below.

12.4.1 Structural Parameters

Screw design (i.e., diameter, length, thread pitch, thread shape, cutting flutes, and so forth) and host bone features (quantity and quality of jawbone) are the two important factors that affect the primary stability of the screw [53]. The mini-screw has a major disadvantage because it sometimes loses its primary stability during the first few weeks of treatment. A study was carried out that aimed to assess the simultaneous variation of implant diameter and length, to find their optimal ranges in the posterior mandible under biomechanical consideration [54]. Another study concluded that the maximum equivalent elastic strain (MES) was influenced by the diameter of the implant more than the length of the dental implant [31]. Thus, a large amount of research is targeted at obtaining an optimal diameter for the dental implant. The thread configuration is another important objective in biomechanical optimization of dental implant as threads are used to increase initial contact, improve initial stability, broaden implant surface area and cause dissipation of interfacial stress [55]. A study considered two different dental implant systems: IS-II implant system and IS-III for obtaining an optimal design whose thread characteristics vary as shown in Figure 12.8 [56]. Another study observed the effect of variation of thread depth, implant body shapes (cylindrical and tapered),

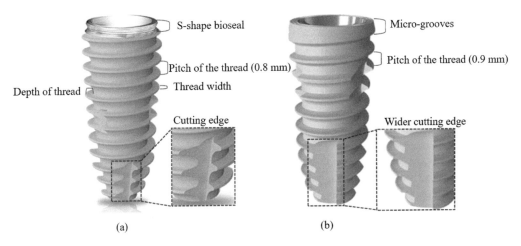

FIGURE 12.8
Implant design features with varying thread types: (a) IS-II implant; (b) IS-III implant [57].

in two different regions (interior and posterior) of the mandible bone for obtaining an optimal implant [57]. Thus, based on all these studies, it is important to note that the variation of the structural parameters of the implant during optimization is one of the leading areas of research.

12.4.2 Material Properties and Surface Morphology

Surface properties such as roughness greatly influence the implant's osseointegration as they control the adhesion and proliferation of the osteoblasts on the implant's surface [58]. A certain study used design optimization and the rule of mixture to evaluate the Young's modulus and the porosity of the coating of implant that causes optimal stress shielding of the peri-implant bone [35]. Another study examines the factors that influence the tissue-implant interface with respect to implant factors like materials, design, surface texture, surface topography and surface biochemistry [59]. As Carl Misch mentions, another key factor for clinical success is the analysis of the bone density in a possible implant site, as the strength of bone and the elastic modulus are directly proportional to bone density [60,61]. Hence, an important aspect for dental implants is the optimization of the materials used for making the implant so that the modulus of elasticity of the implant material can be as close as possible to that of the bone surrounding it. Table 12.1 gives the values of the modulus of elasticity of the cortical and cancellous bone along with that of the most commonly used titanium alloy for dental implants.

TABLE 12.1

Bone and Material Properties for Dental Implants [63]

Material	Young's Modulus (MPa)	Poisson's Ratio
Cortical bone	13,700	0.3
Cancellous bone	1,370	0.3
Titanium alloy (Ti-6Al-4V)	110,000	0.35

A study aimed at optimizing the micro-computerized tomography technique for the measurement of bone mineral density around titanium dental implants [62]. Hence, surface morphology optimization and material properties optimization can go hand in hand in many research areas for the optimization of dental implants.

12.4.3 Osseointegration, Implant Design, Surgical Technique and Excessive Loading

Success of implants depends greatly on long-term osseointegration and initial stability, and thus, various studies have tried to find the optimum thread type in implants that can prevent poor osseointegration by enhancing initial stability, increasing area of surface contact and reducing micro-movements [64,65]. Fatigue failure and micro-gap formation that can be caused by wrong implant design can cause early failure of the implant [66]. A study evaluated these parameters for various implant diameters, connection types and bone densities using the FEM, and concluded that two-piece implants, which have diameters lower than 3.5 mm, should be avoided in the posterior mandibular area [67]. A review tried to study the effect of faulty surgical technique and thus compared the effects of high, regular and low insertion torque in dental implant placement, and concluded that a high insertion torque does not affect the rate of marginal bone loss or implant stability [68]. Excessive loading on dental implants before enough osseointegration has taken place can cause early failure [69,70]. The main purpose of another study was to examine the radiological outcome and clinical success of delayed or immediate loading of dental implants in premolar and anterior sites [71].

12.5 Complementary Techniques Used with Optimization Algorithms

Optimization algorithms require a predefined data set in order to carry out the various optimization tasks assigned to them. These data sets can be built through FEA of the implant model. However, most of these algorithms require a large number of iterations to find the appropriate solutions. Hence, it is necessary to find a new method to replace FEM to ease the computational cost [46]. In this section, methods used to replace FEM for optimization have been discussed.

12.5.1 Surrogate Models for Optimization Algorithms

These are fast-running approximations of time-consuming computer simulations, which can provide an alternate to the mathematical formulation of the connection between the input parameters and the output response [72]. They reduce computational cost greatly. It has been observed that the computational cost and time of FEA for continuous optimization are very high, and thus, surrogate models have been used to replace it. Many different surrogate models have been developed, of which some of the most commonly used ones are given below.

12.5.1.1 Artificial Neural Networks (ANN)

ANNs have established themselves as one of the most useful tools to obtain satisfactory relationships between the structure and properties of various systems. ANNs are a family of models that are derived from biological neural networks and are used for the approximation of complex functions that depend on various input factors and are generally unfamiliar [28]. It consists of a collection of simple processing units which are interconnected by sending signals to each other over weighted connections. A neural network has to be organized such that a set of inputs produces the desired set of outputs. Various methods to strengthen the connections exist where one way is to define the weights clearly using prior knowledge and another way is to train the neural network with the help of teaching patterns and allow it to change its weights with respect to a learning rule [73]. In one of the papers, an approach for designing the shape and geometry of a dental implant with the use of ANNs has been developed, where the aim was to reduce the micro-strain [28]. Figure 12.9 shows the basic flowchart used to design and optimize the implant.

12.5.1.2 Kriging Interpolation

Kriging interpolation is a geo-statistical interpolation technique as it can evaluate the desirable values in an unknown area with respect to both the distance and the degree of variation between known data points [74]. It can identify the association of primary and secondary parameters and hence has been used as a surrogate model for optimization quite often. In a certain study, the KIGA are used to optimize the geometric features to produce an optimized dental implant model as shown in Figure 12.10. [36]. This study

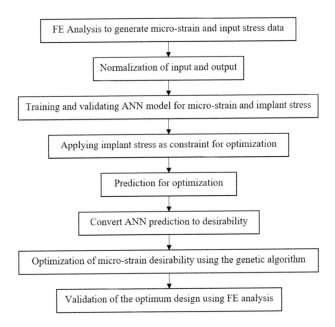

FIGURE 12.9
Use of artificial neural networks as a surrogate model for optimization [28].

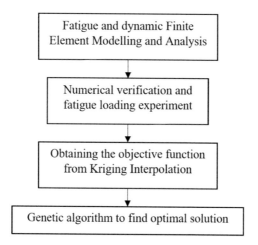

FIGURE 12.10
Use of Kriging interpolation with genetic algorithm [36].

used a Kriging interpolation function because it allows the consideration of both exogenous and endogenous factors. Another study was conducted where FEM was used to calculate the stress at the implant-bone interface, and two surrogate models, namely support vector regression (SVR) and Kriging interpolation (KRG), were built to replace FEM for the multidisciplinary optimization of the implant [72].

12.5.1.3 Use of Support Vector Regression (SVR)

SVR is based on spatial geometry, and it constructs a mixture built on the support vectors (the points that are known on the boundaries) and hence is very useful in problems where it needs to provide restricted sampling points [72]. To build SVR, some sampling points need to be produced first to explore the design space, and then, these sampling points and their responses which are calculated by FEM are used to make the SVR [48]. SVR has been derived from support vector machines (SVMs) where the main difference between the two is that SVR maps the data into a higher-dimensional feature space, but then does regression in this space, whereas for SVMs, after mapping into the higher-dimensional space, they create a hyperplane which maps the nonlinear problems in the lower-dimensional feature space to the linear problems in the higher-dimensional feature space. In one of the studies, an uncertainty optimization approach for dental implant is planned to decrease the stress at the implant-bone interface where the FEM is applied to analyse the stress and SVR is used to substitute the FEM to ease the computational cost [46].

12.6 Conclusion

Optimization algorithms for implants can provide a solution that will prevent poor osseointegration and excessive loading, and thus inhibit early failure. It can help make implants that have reduced micromotion and increased stability. This chapter discusses various optimization algorithms, complementary techniques like surrogate models that increase the speed of optimization and parameters to be considered. Hence, it can provide researchers with a base to understand and apply these algorithms and techniques for making optimized implants. The major optimization algorithms used for dental implants are the GA, topology optimization and the PSO technique. The GA is used mainly when the thread parameters, structural parameters of the implant and the material models for the implant are considered as optimization parameters. This algorithm is generally used in conjunction with the surrogate models like the neural networks or the SVR models for ease of computation and for saving time. The topology optimization technique is used for shape optimization and is directly used with the FEA results. The PSO technique can be used for material model optimization and for shape and structural parameters. Various algorithms like approximate optimization, uncertainty optimization and memetic search have also been used for optimizing the dental implants. Generally, as the optimization of dental implants has more than one objective function, multiobjective optimization algorithms have been developed with the help of traditional genetic or PSO techniques. These algorithms can at once optimize various features of the implant, leading to better results and higher rates of implantation success. Thus, by optimizing various features of the implant simultaneously, the success rate of implantation can increase as failures due to improper optimized implant parameters can be reduced. The various surrogate models can be used to reduce computational cost and time while giving more accurate results of optimization.

Acknowledgement

The authors received no financial support for the research, authorship and/or publication of this chapter.

References

1. K. F. Seitz, J. Grabe, T. Köhne, (2018), A three-dimensional topology optimization model for tooth-root morphology, *Computer Methods in Biomechanics and Biomedical Engineering*, 21(2), 177–185. doi: 10.1080/10255842.2018.1431778.
2. O. Kayabasi, (2020), Design methodology for dental implant using approximate solution techniques, *Journal of Stomatology, Oral and Maxillofacial Surgery*, 1–12. doi: 10.1016/j.jormas.2020.01.003.
3. T. Łodygowski, K. Szajek, M. Wierszycki, (2009), Optimization of dental implant using genetic algorithm. *Journal of Theoretical and Applied Mechanics*, 47, 573–598.
4. K. Szajek, M. Wierszycki, T. Łodygowski, (2011), Reduction of the tooth-implant components dimensions by optimization procedure, *Computer Methods in Mechanics*. 19th International Conference on Computer Methods in Mechanics, CMM-2011, May 9–12, 2011, Warsaw, Poland.

5. A. Suresh Babu, M. Jaivignesh, R. Subramanian, M. P. Dilipkumar, R. Loganathan, S. Vijayakumar, (2000), Topological optimization of direct metal laser sintered Ti_{64} dental implants, 4–7.
6. L. Shi, H. Li, A. S. L. Fok, C. Ucer, H. Devlin, K. Horner, (2008), Shape optimization of dental restorations, *Dental Biomaterials: Imaging, Testing and Modelling*, 226–238. doi: 10.1533/9781845694241.226.
7. C. S. Petrie, J. L. Williams, (2002), Shape optimization of dental implant designs under oblique loading using the p-version finite element method, 339–345, https://www.world scientific.com/doi/abs/10.1142/S0219519402000435.
8. C.-L. Chang, C.-S. Chen, C.-H. Huang, M.-L. Hsu, (2012), Finite element analysis of the dental implant using a topology optimization method, *Medical Engineering and Physics*, 34(7), 999–1008, doi: 10.1016/j.medengphy.2012.06.004.
9. M. Murphy, M.S. Walczak, A.G. Thomas, N. Silikas, S. Berner, R. Lindsay, (2017), Toward optimizing dental implant performance: Surface characterization of Ti and TiZr implant materials, *Dental Materials*, 33(1), 43–53, doi: 10.1016/j.dental.2016.10.001.
10. C. Rungsiyakull, Q. Li, G. Sun, W. Li, M.V. Swain, (2010), Surface morphology optimization for osseointegration of coated implants, *Biomaterials*, 31(27), 7196–7204, doi: 10.1016/j.biomaterials.2010.05.077.
11. Y. Shibata, Y. Tanimoto, (2015), A review of improved fixation methods for dental implants. Part I: Surface optimization for rapid osseointegration, *Journal of Prosthodontic Research*, 59(1), 20–33, doi: 10.1016/j.jpor.2014.11.007.
12. A. Sadollah, A. Bahreininejad, (2011), Optimum gradient material for a functionally graded dental implant using metaheuristic algorithms, *Journal of the Mechanical Behavior of Biomedical Materials*, 4(7), 1384–1395, doi: 10.1016/j.jmbbm.2011.05.009.
13 M.C.H. Chua, C.-K. Chui, (2016), Optimization of patient-specific design of medical implants for manufacturing, *Procedia CIRP*, 40, 402–406, doi: 10.1016/j.procir.2016.01.078.
14. Y. Yoon, X. Sun, J.-K. Huang, G. Hou, K. Rechowicz, F.D. McKenzie, (2013), Designing natural-tooth-shaped dental implants based on soft-kill option optimization, *Computer-Aided Design and Applications*, 10(1), 59–72, doi: 10.3722/cadaps.2013.59-72.
15. Y.-C. Cheng, D.-H. Lin, C.-P. Jiang, Y.-M. Lin, (2017), Dental implant customization using numerical optimization design and 3-dimensional printing fabrication of zirconia ceramic, *International Journal for Numerical Methods in Biomedical Engineering*, 33(5), doi: 10.1002/cnm.2820.
16. P. Dhatrak, U. Shirsat, S. Sumanth, V. Deshmukh, (2018), Finite element analysis and experimental investigations on stress distribution of dental implants around implant-bone interface. *Materials Today: Proceedings*, 5(2), 5641–5648, doi: 10.1016/j.matpr.2017.12.157.
17. M. Pirmoradian, H.A. Naeeni, M. Firouzbakht, D. Toghraie, M.K. Khabaz, R. Darabi, (2020), Finite element analysis and experimental evaluation on stress distribution and sensitivity of dental implants to assess optimum length and thread pitch, *Computer Methods and Programs in Biomedicine*, 187, 105258, doi: 10.1016/j.cmpb.2019.105258.
18. P. Dhatrak, V. Girme, U. Shirsat, S. Sumanth, V. Deshmukh, (2019), Significance of orthotropic material models to predict stress around bone-implant interface using numerical simulation, *BioNanoScience*, 9(3), 652–659, doi: 10.1007/s12668-019-00649-5.
19 T. Li, K. Hu, L. Cheng, Y. Ding, Y. Ding, J. Shao, L. Kong, (2011), Optimum selection of the dental implant diameter and length in the posterior mandible with poor bone quality: A 3D finite element analysis, *Applied Mathematical Modelling*, 35(1), 446–456, doi: 10.1016/j.apm.2010.07.008.
20. Y. Gupta et al., (2020), Design of dental implant using design of experiment and topology optimization: A finite element analysis study, *Proceedings of the Institution of Mechanical Engineers, Part H: Journal of Engineering in Medicine*, doi: 10.1177/0954411920967146.

21. A. Arra et al., (2020), A review on techniques employed for topology optimization in implant dentistry, *AIP Conference Proceedings*, Cleveland, OH.

22. E. Dávila, M. Ortiz-Hernández, R.A. Perez, M. Herrero-Climent, M. Cerrolaza, F.J. Gil, (2019), Crestal module design optimization of dental implants: finite element analysis and in vivo studies, *Journal of Materials Science: Materials in Medicine*, 30(8), 1–10, doi: 10.1007/s10856-019-6291-1.

23. R. Patil, S. Mahajan, (2015) Optimizing dental implant model by comparison of three dimensional finite element, *International Journal of Innovative Science, Engineering and Technology*, 4(12), 26–33, doi: 10.15680/IJIRSET.2015.0411067.

24. L. Jadhav et al., (2020) Design of Experiments (DoE) based optimization of dental implants: A review, *AIP Conference Proceedings*, Medchal (M), Hyderabad, India.

25. R. Chowdhary, A. Halldin, R. Jimbo, A. Wennerberg, (2013), Evaluation of stress pattern generated through various thread designs of dental implants loaded in a condition of immediately after placement and on osseointegration: An FEA study, *Implant Dentistry*, 22(1), 91–96, doi: 10.1097/ID.0b013e31827daf55.

26. S. Lu et al., (2013), Biomechanical optimization of the diameter of distraction screw in distraction implant by three-dimensional finite element analysis, *Computers in Biology and Medicine*, 43(11), 1949–1954, doi: 10.1016/j.compbiomed.2013.08.019.

27. R. Eazhil, S. Swaminathan, M. Gunaseelan, G. Kannan, C. Alagesan, (2016), Impact of implant diameter and length on stress distribution in osseointegrated implants: A 3D FEA study, *Journal of International Society of Preventive and Community*, 6(6), 590–596, doi: 10.4103/2231-0762.195518.

28. S. Roy, S. Dey, N. Khutia, A.R. Chowdhury, S. Datta, (2018), Design of patient specific dental implant using FE analysis and computational intelligence technique, *Applied Soft Computing Journal*, 65, 272–279, doi: 10.1016/j.asoc.2018.01.025.

29. M. Geramizadeh, H. Katoozian, R. Amid, M. Kadkhodazadeh, (2018), Three-dimensional optimization and sensitivity analysis of dental implant thread parameters using finite element analysis, *Journal of the Korean Association of Oral and Maxillofacial Surgeons*, 44(2), 59–65, doi: 10.5125/jkaoms.2018.44.2.59.

30. P. Dhatrak, U. Shirsat, S. Sumanth, V. Deshmukh, (2020), Numerical investigation on stress intensity around bone-implant interface by 3-dimensional FEA and experimental verification by optical technique, *Materials Today: Proceedings*, doi: 10.1016/j.matpr.2020.06.097.

31. N. Ueda, Y. Takayama, A. Yokoyama, (2017), Minimization of dental implant diameter and length according to bone quality determined by finite element analysis and optimized calculation, *Journal of Prosthodontic Research*, 61(3), 324–332, doi: 10.1016/j.jpor.2016.12.004.

32. M.R. Niroomand, M. Arabbeiki, (2020), Effect of the dimensions of implant body and thread on bone resorption and stability in trapezoidal threaded dental implants: a sensitivity analysis and optimization, *Computer Methods in Biomechanics and Biomedical Engineering*, 1–9, doi: 10.1080/10255842.2020.1782390.

33 S.P. Gosavi, P.N. Dhatrak, K.M. Narkar, (2015), Optimisation of dental implant, *International Engineering Research Journal*, 2, 4319–4323, www.ierjournal.org.

34. T. Łodygowski, M. Wierszycki, K. Szajek, W. Hędzelek, R. Zagalak, (2010), Tooth-implant life cycle design, In: Kuczma M., Wilmanski K., (Eds), *Computer Methods in Mechanics, Advanced Structured Materials*, vol 1. Springer, Berlin, Heidelberg, doi: 10.1007/978-3-642-05241-5_21.

35. S. Mahajan, R. Patil, (2016), Application of finite element analysis to optimizing dental implant, *International Research Journal of Engineering and Technology (IRJET)*, 3(2), 850–856.

36. Y.-C. Cheng, C.-P. Jiang, D.-H. Lin, (2019), Finite element-based optimization design for a one-piece zirconia ceramic dental implant under dynamic loading and fatigue life validation, *Structural and Multidisciplinary Optimization*, 59(3), 835–849, doi: 10.1007/s00158-018-2104-2.

37. M. Elsayed, M. Ghazy, Y. Youssef, K. Essa, (2018), Optimization of SLM process parameters for Ti_6Al_4V medical implants, *Rapid Prototyping Journal*, doi: 10.1108/RPJ-05-2018-0112.
38. N. Dai, J.-F. Zhu, M. Zhang, L.-Y. Meng, X.-L. Yu, Y.-H. Zhang, B.-Y. Liu, S.-L. Zhang, (2018), Design of a maxillofacial prosthesis based on topology optimization, *Journal of mechanics in medicine and biology*, 18(3), 1–14, doi: 10.1142/s0219519418500240.
39. A. Baumgartner, L. Harzheim, C. Mattheck, (1992), SKO (soft kill option): The biological way to find an optimum structure topology, *International Journal of Fatigue*, 14(6), 387–393, doi: 10.1016/0142-1123(92)90226-3.
40. M.P. Bendsøe, (1989), Optimal shape design as a material distribution problem, *Structural Optimization*, 1(4), 193–202, doi: 10.1007/BF01650949.
41. J. Kennedy, R. Eberhart, (1995), Particle swarm optimization, *Tsinghua Science and Technology*, 21(2), 221–230, doi: 10.1109/TST.2016.7442504.
42. M. Cai, X. Zhang, G. Tian, J. Liu, (2007), Particle swarm optimization system algorithm, In: Huang D.S., Heutte L., Loog M. (Eds). *Advanced Intelligent Computing Theories and Applications, with Aspects of Contemporary Intelligent Computing Techniques, ICIC 2007. Communications in Computer and Information Science*, vol. 2. Springer, Berlin, Heidelberg, doi: 10.1007/978-3-540-74282-1_44.
43. E. Askari, P. Flores, F. Silva, (2018), A particle swarm-based algorithm for optimization of multi-layered and graded dental ceramics, *Journal of the Mechanical Behavior of Biomedical Materials*, 77(October 2017), 461–469, doi: 10.1016/j.jmbbm.2017.10.005.
44. W. Li, J. Chen, C. Rungsiyakull, M.V. Swain, Q. Li, (2017), Multiscale remodelling and topographical optimisation for porous implant surface morphology design, doi: 10.1007/978-3-662-53574-5.
45. M. Al Ali, A.Y. Sahib, M. Al Ali, (2018), Teeth implant design using weighted sum multi-objective function for topology optimization and real coding genetic algorithm, 181–188, doi: 10.12792/iciae2018.037.
46. H. Li, M. Shi, X. Li, Y. Shi, (2019), Uncertainty optimization of dental implant based on finite element method, global sensitivity analysis and support vector regression, *Proceedings of the Institution of Mechanical Engineers, Part H: Journal of Engineering in Medicine*, 233(2), 232–243, doi: 10.1177/0954411918819116.
47. L. Wang, Z. Chen, G. Yang, Q. Sun, J. Ge, (2020), An interval uncertain optimization method using back-propagation neural network differentiation, *Computer Methods in Applied Mechanics and Engineering*, 366, 113065, doi: 10.1016/j.cma.2020.113065.
48. C. Cotta, L. Mathieson, P. Moscato, (2016), *Handbook of Heuristics*, doi: 10.1007/978-3-319-07153-4.
49. S.K. Baliarsingh, W. Ding, S. Vipsita, S. Bakshi, (2019), A memetic algorithm using emperor penguin and social engineering optimization for medical data classification, *Applied Soft Computing Journal*, 85, 105773, doi: 10.1016/j.asoc.2019.105773.
50. A. Alarifi, A.A. AlZubi, (2018), Memetic search optimization along with genetic scale recurrent neural network for predictive rate of implant treatment, *Journal of Medical Systems*, 42(11), doi: 10.1007/s10916-018-1051-1.
51. N. Tatarakis, J. Bashutski, H.L. Wang, T.J. Oh, (2012), Early implant bone loss: Preventable or inevitable? *Implant Dentistry*, 21(5), 379–386, doi: 10.1097/ID.0b013e3182665d0c.
52. P. Dhatrak, U. Shirsat, V. Deshmukh, (2015), Fatigue life prediction of commercial dental implants based on biomechanical parameters: A review, *Journal of Materials Science & Surface Engineering*, 3(2), 221–226.
53. F.M. Dastenaei, A. Hajarian, O. Zargar, M.M. Zand, S. Noorollahian, (2019), Effects of thread shape on strength and stability of dental miniscrews against orthodontic forces, *Procedia Manufacturing*, 35, 1032–1038, doi: 10.1016/j.promfg.2019.06.053.
54. T. Li, K. Gulati, N. Wang, Z. Zhang, S. Ivanovski, (2018), Bridging the gap: Optimized fabrication of robust titania nanostructures on complex implant geometries towards clinical translation, *Journal of Colloid and Interface Science*, 529, 452–463, doi: 10.1016/j.jcis.2018.06.004.

55. L. Kong, Y. Zhao, K. Hu, D. Li, H. Zhou, Z. Wu, B. Liu, (2009), Selection of the implant thread pitch for optimal biomechanical properties: A three-dimensional finite element analysis, *Advances in Engineering Software*, 40(7), 474–478, doi: 10.1016/j.advengsoft.2008.08.003.

56. W.H. Kim, J.C. Lee, D. Lim, Y.K. Heo, E.S. Song, Y.J. Lim, B. Kim, (2019), Optimized dental implant fixture design for the desirable stress distribution in the surrounding bone region: A biomechanical analysis, *Materials*, 12(7), doi: 10.3390/ma12172749.

57. F.A. Hussein, K.N. Salloomi, B.Y. Abdulrahman, A.R. Al-Zahawi, L.A. Sabri (2019), Effect of thread depth and implant shape on stress distribution in anterior and posterior regions of mandible bone: A finite element analysis, *Dental Research Journal*, 16(3), 200–207, doi: 10.4103/1735-3327.255745.

58. T.A. Dantas, P. Pinto, P.C.S. Vaz, F.S. Silva, (2020), Design and optimization of zirconia functional surfaces for dental implants applications, *Ceramics International*, 46(10), 16328–16336, doi: 10.1016/j.ceramint.2020.03.190.

59. W.L Chai, M. Razali, K. Moharamzadeh, M.S. Zafar, (2020), The hard and soft tissue interfaces with dental implants, In *Dental Implants*, Elsevier Ltd., Amsterdam, Netherlands. doi: 10.1016/b978-0-12-819586-4.00010-x.

60. C.E. Misch, (2004), Dental implant prosthetics, doi: 10.1016/B978-0-323-07845-0.00038-5.

61. D.J. Thakur, T.Z. Quazi, P.N. Dhatrak, (2016), Impact of implant pitch and length variation on stress intensity around dental implant-bone interface-By 3D finite element analysis, *International Journal of Engineering Sciences and Research Technology*, 5(3), 261–269, doi: 10.5281/zenodo.47038.

62. C. Park, M. Swain, W. Duncan, (2010), Micro-computerised tomography optimisation for the measurement of bone mineral density around titanium dental implants, *Journal of Biomechanical Science and Engineering*, 5(1), 2–10, doi: 10.1299/jbse.5.2.

63. C.H. Li, C.H. Wu, C.L. Lin, (2020). Design of a patient-specific mandible reconstruction implant with dental prosthesis for metal 3D printing using integrated weighted topology optimization and finite element analysis. *Journal of the Mechanical Behavior of Biomedical Materials*, 105(October 2019), 103700, doi: 10.1016/j.jmbbm.2020.103700.

64. Z. Arsalanloo, R. Telchi, K.G. Osgouie, (2014), Selection of optimum thread type in implants to achieve optimal biomechanical properties by using 3D finite element method, *International Journal of Bioscience, Biochemistry and Bioinformatics*, 4(3), 185–190, doi: 10.7763/ijbbb.2014.v4.336.

65. U. Narendrakumar, A.T. Mathew, N. Iyer, F. Rahman, I. Manjubala, (2018), A 3D finite element analysis of dental implants with varying thread angles, *Materials Today: Proceedings*, 5(5), 11900–11905, doi: 10.1016/j.matpr.2018.02.163.

66. N. Lioubavina-Hack, N.P. Lang, T. Karring, (2006), Significance of primary stability for osseointegration of dental implants, *Clinical Oral Implants Research*, 17, 3, 244–250, doi: 10.1111/j.1600-0501.2005.01201.x.

67. H. Lee, M. Jo, G. Noh, (2020), Biomechanical effects of dental implant diameter, connection type, and bone density on microgap formation and fatigue failure: A finite element analysis, *Computer Methods and Programs in Biomedicine*, 105863, doi: 10.1016/j.cmpb.2020.105863.

68. C.A.A. Lemos et al., (2020), Clinical effect of the high insertion torque on dental implants: A systematic review and meta-analysis, *Journal of Prosthetic Dentistry*, 1–7, doi: 10.1016/j.prosdent.2020.06.012.

69. O.E. Ogle, (2015), Implant surface material, design, and osseointegration, *Dental Clinics of North America*, 59(2), 505–520, doi: 10.1016/j.cden.2014.12.003.

70. M. Ghadiri, N. Shafiei, S.H. Salekdeh, P. Mottaghi, T. Mirzaie, (2016), Investigation of the dental implant geometry effect on stress distribution at dental implant–bone interface, *The Journal of the Brazilian Society of Mechanical Sciences and Engineering*, 38(2), 335–343, doi: 10.1007/s40430-015-0472-8.

71. A. Henningsen et al., (2017), Immediate loading of subcrestally placed dental implants in anterior and premolar sites, *Journal of Cranio-Maxillofacial Surgery*, 45(11), 1898–1905, doi: 10.1016/j.jcms.2017.08.017.

72. M. Shi, H. Li, X. Liu, (2017), Multidisciplinary design optimization of dental implant based on finite element method and surrogate models, *Journal of Mechanical Science and Technology*, 31(10), 5067–5073, doi: 10.1007/s12206-017-0955-x.

73. P. van der Smagt, B. Kröse (1996), Introduction to neural networks, *International Journal for the Joining of Materials*, 6(1), 4–6.

74. W. Yun, Z. Lu, X. Jiang, (2018), An efficient reliability analysis method combining adaptive Kriging and modified importance sampling for small failure probability, *Structural and Multidisciplinary Optimization*, 58(4), 1383–1393, doi: 10.1007/s00158-018-1975-6.

Index

Note: **Bold** page numbers refer to tables and *italic* page numbers refer to figures.